The Monte Carlo Methods - Recent Advances, New Perspectives and Applications

Edited by Abdo Abou Jaoudé

Published in London, United Kingdom

IntechOpen

Supporting open minds since 2005

The Monte Carlo Methods - Recent Advances, New Perspectives and Applications
http://dx.doi.org/10.5772/intechopen.96413
Edited by Abdo Abou Jaoudé

Contributors
Abdo Abou Jaoudé, Mirko Salomón Alva-Sánchez, Ana Quevedo, Iury Mergen Knoll, Maher Abdelghani, Chad L. Pope, Alison Wells, Gabriela Hoff, Raquel S. Thomaz, Leandro I. Gutierres, Sven Müller, Viviana Fanti, Elaine E. Streck, Ricardo M Papaleo, Omaima Essaad Belhaj, Hamid Boukhal, El Mahjoub Chakir, Tokunbo Ogunfunmi, Manas Deb

First published in London, United Kingdom, 2022 by IntechOpen
IntechOpen is the global imprint of INTECHOPEN LIMITED, registered in England and Wales, registration number: 11086078, 5 Princes Gate Court, London, SW7 2QJ, United Kingdom
Printed in Croatia

British Library Cataloguing-in-Publication Data
A catalogue record for this book is available from the British Library

Additional hard and PDF copies can be obtained from orders@intechopen.com

The Monte Carlo Methods - Recent Advances, New Perspectives and Applications
Edited by Abdo Abou Jaoudé
p. cm.
Print ISBN 978-1-83968-759-4
Online ISBN 978-1-83968-760-0
eBook (PDF) ISBN 978-1-83968-761-7

For EU product safety concerns:
IN TECH d.o.o., Prolaz Marije Krucifikse Kozulić 3, 51000 Rijeka, Croatia, info@intechopen.com or visit our website at intechopen.com.

$$Chf_{Z_U^*} = 2iP_r^*|_{Z_U^*}P_m^*|_{Z_U^*} = 2i\left(1-\frac{1}{N}\right)i\left(\frac{1}{N}\right) = \frac{-2(N-1)}{N^2}$$

$$\Leftrightarrow Chf_{z_j} = Chf_{z_j^*} = Chf_{Z_U} = Chf_{Z_U^*}$$

The Magnitude of the Chaotic Factor or $MChf_{z_j}$ of z_j is:

$$MChf_{z_j} = \left|Chf_{z_j}\right| = \left|\frac{-2(N-1)}{N^2}\right| = \frac{2(N-1)}{N^2}, \forall j: 1 \le j \le N$$

The Magnitude of the Chaotic Factor or $MChf_{z_j^*}$ of z_j^* is:

$$MChf_{z_j^*} = \left|Chf_{z_j^*}\right| = \left|\frac{-2(N-1)}{N^2}\right| = \frac{2(N-1)}{N^2}, \forall j: 1 \le j \le N$$

The Magnitude of the Chaotic Factor or $MChf_{Z_U}$ of $\frac{Z_U}{N}$ is:

$$MChf_{Z_U} = \left|Chf_{Z_U}\right| = \left|\frac{-2(N-1)}{N^2}\right| = \frac{2(N-1)}{N^2}$$

The Magnitude of the Chaotic Factor or $MChf_{Z_U^*}$ of $\frac{Z_U^*}{N}$ is:

$$MChf_{Z_U^*} = \left|Chf_{Z_U^*}\right| = \left|\frac{-2(N-1)}{N^2}\right| = \frac{2(N-1)}{N^2}$$

$$\Leftrightarrow MChf_{z_j} = MChf_{z_j^*} = MChf_{Z_U} = MChf_{Z_U^*}$$

The probability Pc_{z_j} in $\mathcal{C} = \mathcal{R} + \mathcal{M}$ of z_j is:

$$Pc_{z_j}^2 = \left(P_{rj} + P_{mj}/i\right)^2 = \left[\frac{1}{N} + \frac{\left(1-\frac{1}{N}\right)i}{i}\right]^2 = \left[\frac{1}{N} + 1 - \frac{1}{N}\right]^2 = 1^2 = 1 = Pc_{z_j}, \forall j: 1 \le j \le N$$

The probability $Pc_{z_j^*}^*$ in $\mathcal{C} = \mathcal{R} + \mathcal{M}$ of z_j^* is:

$$Pc_{z_j^*}^*\Big|^2 = \left(P_{rj}^* + P_{mj}^*/i\right)^2 = \left[\left(1-\frac{1}{N}\right) + \frac{\frac{i}{N}}{i}\right]^2 = \left[1 - \frac{1}{N} + \frac{1}{N}\right]^2 = 1^2 = 1 = Pc_{z_j^*}^*, \forall j: 1 \le j \le N$$

The probability $Pc|_{Z_U}$ in $\mathcal{C} = \mathcal{R} + \mathcal{M}$ of $\frac{Z_U}{N}$ is:

$$Pc^2|_{Z_U} = \left(P_r|_{Z_U} + \frac{P_m|_{Z_U}}{i}\right)^2 = \left[\frac{1}{N} + \frac{\left(1-\frac{1}{N}\right)i}{i}\right]^2 = \left[\frac{1}{N} + 1 - \frac{1}{N}\right]^2 = 1^2 = 1 = Pc|_{Z_U}$$

The probability $Pc^*|_{Z_U^*}$ in $\mathcal{C} = \mathcal{R} + \mathcal{M}$ of $\frac{Z_U^*}{N}$ is:

$$Pc^*|_{Z_U^*}^2 = \left(P_r^*|_{Z_U^*} + \frac{P_m^*|_{Z_U^*}}{i}\right)^2 = \left[\left(1-\frac{1}{N}\right) + \frac{\frac{i}{N}}{i}\right]^2 = \left[1 - \frac{1}{N} + \frac{1}{N}\right]^2 = 1^2 = 1$$
$$= Pc^*|_{Z_U^*}$$

$$\Leftrightarrow Pc_{z_j} = Pc_{z_j^*}^* = Pc|_{Z_U} = Pc^*|_{Z_U^*} = 1$$

Meet the editor

Abdo Abou Jaoudé has been teaching for many years and has a passion for researching and teaching mathematics. He is currently an Associate Professor of Mathematics and Statistics at Notre Dame University-Louaizé (NDU), Lebanon. He holds a BSc and an MSc in Computer Science from NDU, and three Ph.D.'s in Applied Mathematics, Computer Science, and Applied Statistics and Probability, all from Bircham International University through a distance learning program. He also holds two PhDs in Mathematics and Prognostics from the Lebanese University, Lebanon, and Aix-Marseille University, France. Dr. Jaoudé's broad research interests are in the field of applied mathematics. He has published twenty-three international journal articles and six contributions to conference proceedings, in addition to seven books on prognostics, applied mathematics, and computer science.

Contents

Preface

The Monte Carlo Methods - Recent Advances, New Perspectives and Applications illustrates the famous Monte Carlo methods and the computer simulation of random experiments in different areas of science. As such, the book will be of interest to all scholars, researchers, and undergraduate and graduate students in mathematics and science in general.

In applied mathematics, the name Monte Carlo is given to the method of solving problems by means of experiments with random numbers. This name, after the casino at Monaco, was first applied around 1944 to the method of solving deterministic problems by reformulating them in terms of a problem with random elements, which could then be solved by large-scale sampling. But, by extension, the term has come to mean any simulation that uses random numbers.

The development and proliferation of computers has led to the widespread use of Monte Carlo methods in virtually all branches of science, ranging from nuclear physics (where computer-aided Monte Carlo was first applied) to astrophysics, biology, engineering, medicine, operations research, and the social sciences.

The Monte Carlo method of solving problems by using random numbers in a computer, either by direct simulation of physical or statistical problems or by reformulating deterministic problems in terms of one incorporating randomness, has become one of the most important tools of applied mathematics and computer science. A significant proportion of articles in technical journals in such fields as physics, chemistry, and statistics contain articles reporting results of Monte Carlo simulations or suggestions on how they might be applied. Some journals are devoted almost entirely to Monte Carlo problems in their fields. Studies in the formation of the universe or of stars and their planetary systems use Monte Carlo techniques. Studies in genetics, the biochemistry of DNA, and the random configuration and knotting of biological molecules are studied by Monte Carlo methods. In number theory, Monte Carlo methods play an important role in determining primality or factoring of very large integers far beyond the range of deterministic methods. Several important new statistical techniques such as "bootstrapping" and "jackknifing" are based on Monte Carlo methods.

Hence, the role of Monte Carlo methods and simulation in all the sciences has increased in importance during the past several years. These methods play a central role in the rapidly developing subdisciplines of the computational physical sciences, computational life sciences, and other computational sciences. Therefore, the growing power of computers and evolving simulation methodology has led to the recognition of computation as a third approach for advancing the natural sciences, together with theory and traditional experimentation. Knowing that at the kernel of Monte Carlo simulation is random number generation.

Moreover, the book develops methods for simulating simple or complicated processes or phenomena. If the computer can be made to imitate an experiment or a process, then by repeating the computer simulation with different data, we can draw statistical conclusions. Thus, a simulation of a spectrum of mathematical processes on computers was conducted. The result and accuracy of all the algorithms are truly amazing and delightful; hence, this confirms two complementary accomplishments: first, the triumphs of the theoretical calculations already established using different theorems and second, the power and success of modern computers to verify them.

Additionally, each time I work in the field of mathematical probability and Monte Carlo methods I find pleasure in tackling the knowledge, theorems, proofs, and applications of the theory. In fact, each problem is like a riddle to be solved, a conquest to be won, and I become relieved and extremely happy when I reach the end of the solution. This proves two important facts: first, the power of mathematics and its models to deal with such kinds of problems and second, the power of the human mind to understand such class of problems and to tame such wild concepts that are randomness, probability, stochasticity, uncertainty, chaos, chance, and nondeterminism.

I am truly astonished by the power of probability and these random techniques to deal with random data and phenomena, and this feeling and impression never left me from the first time I was introduced to this branch of science and mathematics. I hope that in this book I am able to convey and share this feeling with readers. I hope also that readers will discover and learn about the concepts and applications of the probabilistic and Monte Carlo paradigm.

Abdo Abou Jaoudé, Ph.D.
Notre Dame University-Louaizé,
Zouk Mosbeh, Lebanon

Chapter 1

The Paradigm of Complex Probability and Thomas Bayes' Theorem

Abdo Abou Jaoudé

"Simple solutions seldom are. It takes a very unusual mind to undertake analysis of the obvious."

Alfred North Whitehead.

"Nothing in nature is by chance… Something appears to be chance only because of our lack of knowledge."

Baruch Spinoza.

"Fundamental progress has to do with the reinterpretation of basic ideas."

Alfred North Whitehead.

"Mathematics, rightly viewed, possesses not only truth but supreme beauty… "

Bertrand Russell.

Abstract

The mathematical probability concept was set forth by Andrey Nikolaevich Kolmogorov in 1933 by laying down a five-axioms system. This scheme can be improved to embody the set of imaginary numbers after adding three new axioms. Accordingly, any stochastic phenomenon can be performed in the set $\mathcal{C}$ of complex probabilities which is the summation of the set $\mathcal{R}$ of real probabilities and the set $\mathcal{M}$ of imaginary probabilities. Our objective now is to encompass complementary imaginary dimensions to the stochastic phenomenon taking place in the "real" laboratory in $\mathcal{R}$ and as a consequence to gauge in the sets $\mathcal{R}$, $\mathcal{M}$, and $\mathcal{C}$ all the corresponding probabilities. Hence, the probability in the entire set $\mathcal{C} = \mathcal{R} + \mathcal{M}$ is incessantly equal to one independently of all the probabilities of the input stochastic variable distribution in $\mathcal{R}$, and subsequently the output of the random phenomenon in $\mathcal{R}$ can be evaluated totally in $\mathcal{C}$. This is due to the fact that the probability in $\mathcal{C}$ is calculated after the elimination and subtraction of the chaotic factor from the degree of our knowledge of the nondeterministic phenomenon. We will apply this novel paradigm to the classical Bayes' theorem in probability theory.

Keywords: Chaotic factor, degree of our knowledge, complex random vector, imaginary probability, probability norm, complex probability set

1. Introduction

The crucial job of the theory of classical probability is to compute and to assess probabilities. A deterministic expression of probability theory can be attained by adding supplementary dimensions to nondeterministic and stochastic experiments. This original and novel idea is at the foundations of my new paradigm of complex probability. In its core, probability theory is a nondeterministic system of axioms that means that the phenomena and experiments outputs are the products of chance and randomness. In fact, a deterministic expression of the stochastic experiment will be realized and achieved by the addition of imaginary new dimensions to the stochastic phenomenon taking place in the real probability set $\mathcal{R}$ and hence this will lead to a certain output in the set $\mathcal{C}$ of complex probabilities. Accordingly, we will be totally capable to foretell the random events outputs that occur in all probabilistic processes in the real world. This is possible because the chaotic phenomenon becomes completely predictable. Thus, the job that has been successfully completed here was to extend the set of real and random probabilities which is the set $\mathcal{R}$ to the complex and deterministic set of probabilities which is $\mathcal{C} = \mathcal{R} + \mathcal{M}$. This is achieved by taking into account the contributions of the imaginary and complementary set of probabilities to the set $\mathcal{R}$ and that we have called accordingly the set $\mathcal{M}$. This extension proved that it was effective and consequently we were successful to create an original paradigm dealing with prognostic and stochastic sciences in which we were able to express deterministically in $\mathcal{C}$ all the nondeterministic processes happening in the 'real' world $\mathcal{R}$. This innovative paradigm was coined by the term "The Complex Probability Paradigm" and was started and established in my seventeen earlier publications and research works [1–17].

At the end, and to conclude, this research work is organized as follows: After the introduction in section 1, the purpose and the advantages of the present work are presented in section 2. Afterward, in section 3, the extended Kolmogorov's axioms and hence the complex probability paradigm with their original parameters and interpretation will be explained and summarized. Moreover, in section 4, the complex probability paradigm axioms are applied to Bayes' theorem for a discrete binary random variable and for a general discrete uniform random variable and which will be hence extended to the imaginary and complex sets. Additionally, in section 5, the flowchart of the new paradigm will be shown. Furthermore, the simulations of the novel model for a discrete random distribution and for a continuous stochastic distribution are illustrated in section 6. Finally, we conclude the work by doing a comprehensive summary in section 7, and then present the list of references cited in the current research work.

2. The purpose and the advantages of the current publication

The advantages and the purpose of this current work are to:

1. Extend the theory of classical probability to encompass the complex numbers set, hence to bond the theory of probability to the field of complex variables and analysis in mathematics. This mission was elaborated and initiated in my earlier seventeen papers [1–17].

2. Apply the novel probability axioms and paradigm to the classical Bayes' theorem.

3. Show that all nondeterministic phenomena can be expressed deterministically in the complex probabilities set which is $\mathcal{C}$.

4. Compute and quantify both the degree of our knowledge and the chaotic factor of all the probabilities in the sets $\mathcal{R}$, $\mathcal{M}$, and $\mathcal{C}$.

5. Represent and show the graphs of the functions and parameters of the innovative paradigm related to Bayes' theorem.

6. Demonstrate that the classical concept of probability is permanently equal to one in the set of complex probabilities; hence, no randomness, no chaos, no ignorance, no uncertainty, no nondeterminism, no unpredictability, and no disorder exist in:

$$\mathcal{C}\text{ (complex set)} = \mathcal{R}\text{ (real set)} + \mathcal{M}\text{ (imaginary set)}.$$

7. Prepare to implement this creative model to other topics in prognostics and to the field of stochastic processes. These will be the job to be accomplished in my future research publications.

Concerning some applications of the novel founded paradigm and as a future work, it can be applied to any nondeterministic phenomenon using Bayes' theorem whether in the continuous or in the discrete cases. Moreover, compared with existing literature, the major contribution of the current research work is to apply the innovative paradigm of complex probability to Bayes' theorem. The next figure displays the major purposes and goals of the Complex Probability Paradigm (*CPP*) (**Figure 1**).

3. The complex probability paradigm

3.1 The original Andrey Nikolaevich Kolmogorov system of axioms

The simplicity of Kolmogorov's system of axioms may be surprising. Let E be a collection of elements $\{E_1, E_2, \ldots\}$ called elementary events and let F be a set of subsets of E called random events [18–22]. The five axioms for a finite set E are:

Axiom 1: F is a field of sets.

Axiom 2: F contains the set E.

Axiom 3: A non-negative real number $P_{rob}(A)$, called the probability of A, is assigned to each set A in F. We have always $0 \le P_{rob}(A) \le 1$.

Axiom 4: $P_{rob}(E)$ equals 1.

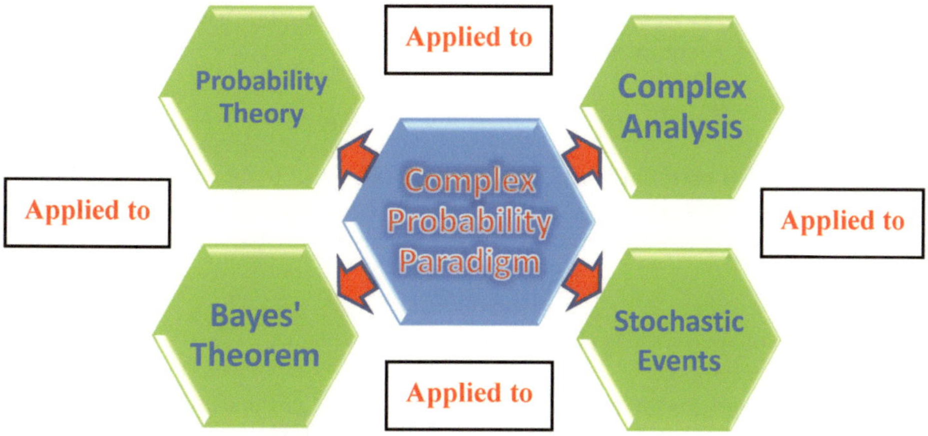

Figure 1.
The diagram of the Complex Probability Paradigm major goals.

Axiom 5: If A and B have no elements in common, the number assigned to their union is:

$$P_{rob}(A \cup B) = P_{rob}(A) + P_{rob}(B)$$

hence, we say that A and B are disjoint; otherwise, we have:

$$P_{rob}(A \cup B) = P_{rob}(A) + P_{rob}(B) - P_{rob}(A \cap B)$$

And we say also that: $P_{rob}(A \cap B) = P_{rob}(A) \times P_{rob}(B/A) = P_{rob}(B) \times P_{rob}(A/B)$ which is the conditional probability. If both A and B are independent then: $P_{rob}(A \cap B) = P_{rob}(A) \times P_{rob}(B)$.

Moreover, we can generalize and say that for N disjoint (mutually exclusive) events $A_1, A_2, \ldots, A_j, \ldots, A_N$ (for $1 \leq j \leq N$), we have the following additivity rule:

$$P_{rob}\left(\bigcup_{j=1}^{N} A_j\right) = \sum_{j=1}^{N} P_{rob}(A_j)$$

And we say also that for N independent events $A_1, A_2, \ldots, A_j, \ldots, A_N$ (for $1 \leq j \leq N$), we have the following product rule:

$$P_{rob}\left(\bigcap_{j=1}^{N} A_j\right) = \prod_{j=1}^{N} P_{rob}(A_j)$$

3.2 Adding the Imaginary Part $\mathcal{M}$

Now, we can add to this system of axioms an imaginary part such that:

Axiom 6: Let $P_m = i \times (1 - P_r)$ be the probability of an associated complementary event in $\mathcal{M}$ (the imaginary part) to the event A in $\mathcal{R}$ (the real part). It follows that $P_r + P_m/i = 1$ where i is the imaginary number with $i = \sqrt{-1}$ or $i^2 = -1$.

Axiom 7: We construct the complex number or vector $z = P_r + P_m = P_r + i(1 - P_r)$ having a norm $|z|$ such that:

$$|z|^2 = P_r^2 + (P_m/i)^2.$$

Axiom 8: Let Pc denote the probability of an event in the complex probability universe $\mathcal{C}$ where $\mathcal{C} = \mathcal{R} + \mathcal{M}$. We say that Pc is the probability of an event A in $\mathcal{R}$ with its associated event in $\mathcal{M}$ such that:

$$Pc^2 = (P_r + P_m/i)^2 = |z|^2 - 2iP_rP_m \text{ and is always equal to } 1.$$

We can see that by taking into consideration the set of imaginary probabilities we added three new and original axioms and consequently the system of axioms defined by Kolmogorov was hence expanded to encompass the set of imaginary numbers [1–17].

3.2.1 A concise interpretation of the original paradigm

As a summary of the new paradigm, we declare that in the universe $\mathcal{R}$ of real probabilities we have the degree of our certain knowledge is unfortunately incomplete and therefore insufficient and unsatisfactory, hence we encompass in our analysis the set $\mathcal{C}$ of complex numbers which integrates the contributions of both

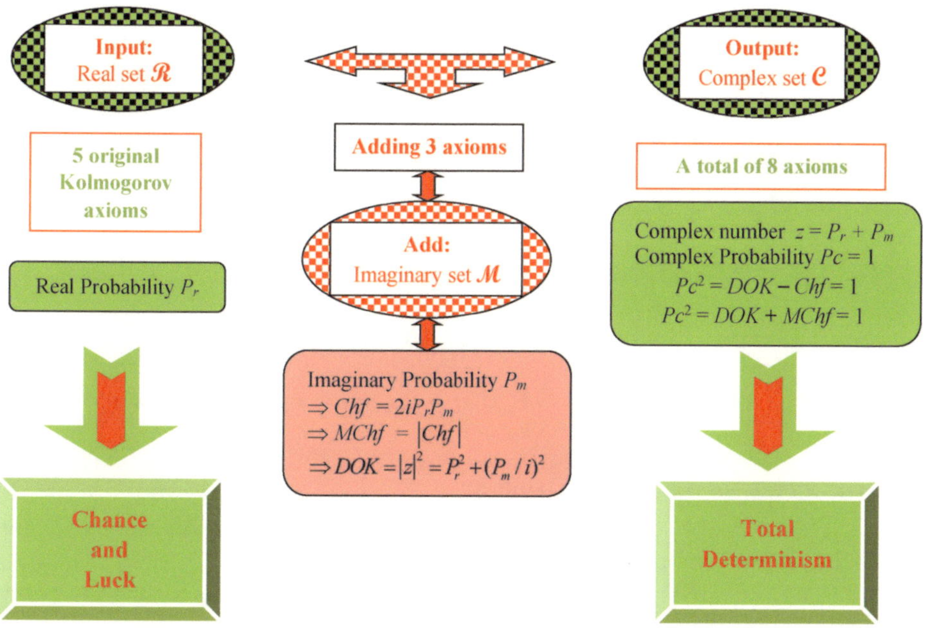

Figure 2.
The EKA *or the* CPP *diagram.*

the real set $\mathcal{R}$ of probabilities and its complementary imaginary probabilities set that we have called accordingly $\mathcal{M}$ [1–17]. Subsequently, a perfect and an absolute degree of our knowledge is obtained and achieved in the universe of probabilities $\mathcal{C} = \mathcal{R} + \mathcal{M}$ because we have constantly $Pc = 1$. In fact, a sure and certain prediction of any random phenomenon is reached in the universe $\mathcal{C}$ because in this set, we eliminate and subtract from the measured degree of our knowledge the computed chaotic factor. Consequently, this will lead to in the universe $\mathcal{C}$ a probability permanently equal to one as it is shown in the following equation: $Pc^2 = DOK - Chf = DOK + MChf = 1 = Pc$ deduced from the complex probability paradigm. Moreover, various discrete and continuous stochastic distributions illustrate in my seventeen previous research works this hypothesis and innovative and original model. The figure that follows shows and summarizes the Extended Kolmogorov Axioms (*EKA*) or the Complex Probability Paradigm (*CPP*) (**Figure 2**).

4. The complex probability paradigm applied to Bayes' Theorem

4.1 The case of a discrete binary random variable

4.1.1 The probabilities and the conditional probabilities

We define the probabilities for the binary random variable A as follows [23–37]:
A is an event occurring in the real probabilities set $\mathcal{R}$ such that: $P_{rob}(A) = P_r$.

The corresponding associated imaginary complementary event to the event A in the probabilities set $\mathcal{M}$ is the event B such that: $P_{rob}(B) = P_m = i(1 - P_r)$.

The real complementary event to the event A in $\mathcal{R}$ is the event $\overline{A}$ such that:
$A \cup \overline{A} = \mathcal{R}$ and $A \cap \overline{A} = \varnothing$ (mutually exclusive events)

$$P_{rob}(\overline{A}) = 1 - P_{rob}(A) = 1 - P_r = P_m / i = P_{rob}(B)/i$$

$$\Rightarrow P_{rob}(B) = iP_{rob}(\overline{A})$$

$$P_{rob}(\mathcal{R}) = P_{rob}(A \cup \overline{A}) = P_{rob}(A) + P_{rob}(\overline{A}) = P_r + (1 - P_r) = 1$$

The imaginary complementary event to the event B in $\mathcal{M}$ is the event $\overline{B}$ such that:

$B \cup \overline{B} = \mathcal{M}$ and $B \cap \overline{B} = \emptyset$ (mutually exclusive events)

$$P_{rob}(\overline{B}) = i - P_{rob}(B) = i - P_m = i - i(1 - P_r) = i - i + iP_r = iP_r = iP_{rob}(A)$$

$\Rightarrow P_{rob}(A) = P_{rob}(\overline{B})/i = -iP_{rob}(\overline{B})$ since $1/i = -i$.

$$P_{rob}(\mathcal{M}) = P_{rob}(B \cup \overline{B}) = P_{rob}(B) + P_{rob}(\overline{B}) = P_m + (i - P_m) = i$$

$\Rightarrow P_{rob}(\mathcal{R}) = P_{rob}(\mathcal{M})/i = 1$, just as predicted by *CPP*.

We have also, as derived from *CPP* that:

$P_{rob}(A/B) = P_{rob}(A) = P_r$, that means if the event B occurs in $\mathcal{M}$ then the event A, which is its real complementary event, occurs in $\mathcal{R}$.

$P_{rob}(B/A) = P_{rob}(B) = P_m$, that means if the event A occurs in $\mathcal{R}$ then the event B, which is its imaginary complementary event, occurs in $\mathcal{M}$.

Furthermore, we can deduce from *CPP* the following:

$P_{rob}(A/\overline{B}) = iP_r/i = P_r = P_{rob}(A)$, that means if the event $\overline{B}$ occurs in $\mathcal{M}$ then the event A, which is its real correspondent and associated event, occurs in $\mathcal{R}$.

$P_{rob}(B/\overline{A}) = i(1 - P_r) = P_m = P_{rob}(B)$, that means if the event $\overline{A}$ occurs in $\mathcal{R}$ then the event B, which is its imaginary correspondent and associated event, occurs in $\mathcal{M}$.

$P_{rob}(\overline{A}/B) = i(1 - P_r)/i = 1 - P_r = P_{rob}(\overline{A})$, that means if the event B occurs in $\mathcal{M}$ then the event $\overline{A}$, which is its real correspondent and associated event, occurs in $\mathcal{R}$.

$P_{rob}(\overline{B}/A) = iP_r = i - P_m = P_{rob}(\overline{B})$, that means if the event A occurs in $\mathcal{R}$ then the event $\overline{B}$, which is its imaginary correspondent and associated event, occurs in $\mathcal{M}$.

$P_{rob}(\overline{A}/\overline{B}) = 1 - iP_r/i = 1 - P_r = P_{rob}(\overline{A})$, that means if the event $\overline{B}$ occurs in $\mathcal{M}$ then the event $\overline{A}$, which is its real complementary event, occurs in $\mathcal{R}$.

$P_{rob}(\overline{B}/\overline{A}) = i - i(1 - P_r) = iP_r = P_{rob}(\overline{B})$, that means if the event $\overline{A}$ occurs in $\mathcal{R}$ then the event $\overline{B}$, which is its imaginary complementary event, occurs in $\mathcal{M}$.

4.1.2 The relations to Bayes' theorem

Another form of Bayes' theorem for two competing statements or hypotheses that is, a binary random variable, is in the probability set $\mathcal{R}$ equal to:

$$P_{rob}(A/B) = \frac{P_{rob}(B/A)P_{rob}(A)}{P_{rob}(B)} = \frac{P_{rob}(B/A)P_{rob}(A)}{P_{rob}(B/A)P_{rob}(A) + P_{rob}(B/\overline{A})P_{rob}(\overline{A})}$$

For an epistemological interpretation:

For proposition A and evidence or background B,

- $P_{rob}(A)$ is the prior probability, the initial degree of belief in A.
- $P_{rob}(\overline{A})$ is the corresponding initial degree of belief in not-A, that A is false
- $P_{rob}(B/A)$ is the conditional probability or likelihood, the degree of belief in B given that proposition A is true.

- $P_{rob}(B/\overline{A})$ is the conditional probability or likelihood, the degree of belief in B given that proposition A is false.

- $P_{rob}(A/B)$ is the posterior probability, the probability of A after taking into account B.

Therefore, in *CPP* and hence in $\mathcal{C} = \mathcal{R} + \mathcal{M}$, we can deduce the new forms of Bayes' theorem for the case considered as follows:

$$\begin{aligned}
P_{rob}(A/B) &= \frac{P_{rob}(B/A)P_{rob}(A)}{P_{rob}(B)} = \frac{P_{rob}(B)P_{rob}(A)}{P_{rob}(B)} = \frac{P_m P_r}{P_m} = P_r = P_{rob}(A) \\
&= \frac{P_{rob}(B/A)P_{rob}(A)}{P_{rob}(B/A)P_{rob}(A) + P_{rob}(B/\overline{A})P_{rob}(\overline{A})} \\
&= \frac{P_{rob}(B)P_{rob}(A)}{P_{rob}(B)P_{rob}(A) + P_{rob}(B)P_{rob}(\overline{A})} \\
&= \frac{P_m P_r}{P_m P_r + P_m(1-P_r)} = \frac{P_m P_r}{P_m P_r + P_m - P_m P_r} = \frac{P_m P_r}{P_m} = P_r = P_{rob}(A)
\end{aligned}$$

and this independently of the distribution of the binary random variables A in $\mathcal{R}$ and correspondingly of B in $\mathcal{M}$.

And, its corresponding Bayes' relation in $\mathcal{M}$ is:

$$\begin{aligned}
P_{rob}(B/A) &= \frac{P_{rob}(A/B)P_{rob}(B)}{P_{rob}(A)} = \frac{P_{rob}(A)P_{rob}(B)}{P_{rob}(A)} = \frac{P_r P_m}{P_r} = P_m = P_{rob}(B) \\
&= i(N-1)\left[\frac{P_{rob}(A/B)P_{rob}(B)}{P_{rob}(A/B)P_{rob}(B) + P_{rob}(A/\overline{B})P_{rob}(\overline{B})}\right] \\
&= i(2-1)\left[\frac{P_{rob}(A)P_{rob}(B)}{P_{rob}(A)P_{rob}(B) + P_{rob}(A)P_{rob}(\overline{B})}\right] \\
&= i\left[\frac{P_r P_m}{P_r P_m + P_r(i-P_m)}\right] = i\left[\frac{P_r P_m}{P_r P_m + iP_r - P_r P_m}\right] = i\left[\frac{P_r P_m}{iP_r}\right] = i\left[\frac{P_m}{i}\right] \\
&= P_m = P_{rob}(B)
\end{aligned}$$

and this independently of the distribution of the binary random variables A in $\mathcal{R}$ and correspondingly of B in $\mathcal{M}$. Note that $N = 2$ corresponds to the binary random variable considered in this case.

Similarly,

$$\begin{aligned}
P_{rob}(\overline{A}/\overline{B}) &= \frac{P_{rob}(\overline{B}/\overline{A})P_{rob}(\overline{A})}{P_{rob}(\overline{B})} = \frac{P_{rob}(\overline{B})P_{rob}(\overline{A})}{P_{rob}(\overline{B})} = \frac{iP_r(1-P_r)}{iP_r} = 1 - P_r = P_{rob}(\overline{A}) \\
&= (N-1)\left[\frac{P_{rob}(\overline{B}/\overline{A})P_{rob}(\overline{A})}{P_{rob}(\overline{B}/\overline{A})P_{rob}(\overline{A}) + P_{rob}(\overline{B}/A)P_{rob}(A)}\right] \\
&= (2-1)\left[\frac{P_{rob}(\overline{B})P_{rob}(\overline{A})}{P_{rob}(\overline{B})P_{rob}(\overline{A}) + P_{rob}(\overline{B})P_{rob}(A)}\right] \\
&= \frac{iP_r(1-P_r)}{iP_r(1-P_r) + iP_r P_r} = \frac{iP_r(1-P_r)}{iP_r - iP_r^2 + iP_r^2} = \frac{iP_r(1-P_r)}{iP_r} = 1 - P_r = P_{rob}(\overline{A})
\end{aligned}$$

and this independently of the distribution of the binary random variables A in $\mathcal{R}$ and correspondingly of B in $\mathcal{M}$.

And, its corresponding Bayes' relation in $\mathcal{M}$ is:

$$P_{rob}(\overline{B}/\overline{A}) = \frac{P_{rob}(\overline{A}/\overline{B})P_{rob}(\overline{B})}{P_{rob}(\overline{A})} = \frac{P_{rob}(\overline{A})P_{rob}(\overline{B})}{P_{rob}(\overline{A})} = \frac{(1-P_r)iP_r}{(1-P_r)} = iP_r = i - P_m = P(\overline{B})$$

$$= i\left[\frac{P_{rob}(\overline{A}/\overline{B})P_{rob}(\overline{B})}{P_{rob}(\overline{A}/\overline{B})P_{rob}(\overline{B}) + P_{rob}(\overline{A}/B)P_{rob}(B)}\right]$$

$$= i\left[\frac{P_{rob}(\overline{A})P_{rob}(\overline{B})}{P_{rob}(\overline{A})P_{rob}(\overline{B}) + P_{rob}(\overline{A})P_{rob}(B)}\right]$$

$$= i\left[\frac{(1-P_r)iP_r}{(1-P_r)iP_r + (1-P_r)i(1-P_r)}\right] = i\left[\frac{iP_r}{iP_r + i(1-P_r)}\right]$$

$$= i\left[\frac{iP_r}{iP_r + i - iP_r}\right] = i\left[\frac{iP_r}{i}\right] = iP_r = i - P_m = P(\overline{B})$$

and this independently of the distribution of the binary random variables A in $\mathcal{R}$ and correspondingly of B in $\mathcal{M}$.

Moreover,

$$P_{rob}(A/\overline{B}) = \frac{P_{rob}(\overline{B}/A)P_{rob}(A)}{P_{rob}(\overline{B})} = \frac{P_{rob}(\overline{B})P_{rob}(A)}{P_{rob}(\overline{B})} = \frac{iP_rP_r}{iP_r} = P_r = P_{rob}(A)$$

$$= \frac{P_{rob}(\overline{B}/A)P_{rob}(A)}{P_{rob}(\overline{B}/A)P_{rob}(A) + P_{rob}(\overline{B}/\overline{A})P_{rob}(\overline{A})}$$

$$= \frac{P_{rob}(\overline{B})P_{rob}(A)}{P_{rob}(\overline{B})P_{rob}(A) + P_{rob}(\overline{B})P_{rob}(\overline{A})}$$

$$= \frac{iP_rP_r}{iP_rP_r + iP_r(1-P_r)} = \frac{iP_rP_r}{iP_r^2 + iP_r - iP_r^2} = \frac{iP_rP_r}{iP_r} = P_r = P_{rob}(A)$$

and this independently of the distribution of the binary random variables A in $\mathcal{R}$ and correspondingly of B in $\mathcal{M}$.

And, its corresponding Bayes' relation in $\mathcal{M}$ is:

$$P_{rob}(B/\overline{A}) = \frac{P_{rob}(\overline{A}/B)P_{rob}(B)}{P_{rob}(\overline{A})} = \frac{P_{rob}(\overline{A})P_{rob}(B)}{P_{rob}(\overline{A})} = \frac{(1-P_r)i(1-P_r)}{(1-P_r)} = i(1-P_r) = P_{rob}(B)$$

$$= i(N-1)\left[\frac{P_{rob}(\overline{A}/B)P_{rob}(B)}{P_{rob}(\overline{A}/B)P_{rob}(B) + P_{rob}(\overline{A}/\overline{B})P_{rob}(\overline{B})}\right]$$

$$= i(2-1)\left[\frac{P_{rob}(\overline{A})P_{rob}(B)}{P_{rob}(\overline{A})P_{rob}(B) + P_{rob}(\overline{A})P_{rob}(\overline{B})}\right]$$

$$= i\left[\frac{(1-P_r)i(1-P_r)}{(1-P_r)i(1-P_r) + (1-P_r)iP_r}\right] = i\left[\frac{i(1-P_r)}{i(1-P_r) + iP_r}\right]$$

$$= i\left[\frac{i(1-P_r)}{i - iP_r + iP_r}\right] = i\left[\frac{i(1-P_r)}{i}\right] = i(1-P_r) = P_{rob}(B)$$

and this independently of the distribution of the binary random variables A in $\mathcal{R}$ and correspondingly of B in $\mathcal{M}$.

Furthermore,

$$P_{rob}(\overline{A}/B) = \frac{P_{rob}(B/\overline{A})P_{rob}(\overline{A})}{P_{rob}(B)} = \frac{P_{rob}(B)P_{rob}(\overline{A})}{P_{rob}(B)} = \frac{P_m(1-P_r)}{P_m} = 1 - P_r = P_{rob}(\overline{A})$$
$$= (N-1)\left[\frac{P_{rob}(B/\overline{A})P_{rob}(\overline{A})}{P_{rob}(B/\overline{A})P_{rob}(\overline{A}) + P_{rob}(B/A)P_{rob}(A)}\right]$$
$$= (2-1)\left[\frac{P_{rob}(B)P_{rob}(\overline{A})}{P_{rob}(B)P_{rob}(\overline{A}) + P_{rob}(B)P_{rob}(A)}\right]$$
$$= \frac{P_m(1-P_r)}{P_m(1-P_r)+P_mP_r} = \frac{P_m(1-P_r)}{P_m - P_mP_r + P_mP_r} = \frac{P_m(1-P_r)}{P_m} = 1 - P_r = P_{rob}(\overline{A})$$

and this independently of the distribution of the binary random variables A in $\mathcal{R}$ and correspondingly of B in $\mathcal{M}$.

And, its corresponding Bayes' relation in $\mathcal{M}$ is:

$$P_{rob}(\overline{B}/A) = \frac{P_{rob}(A/\overline{B})P_{rob}(\overline{B})}{P_{rob}(A)} = \frac{P_{rob}(A)P_{rob}(\overline{B})}{P_{rob}(A)} = \frac{P_r iP_r}{P_r} = iP_r = i - P_m = P_{rob}(\overline{B})$$
$$= i\left[\frac{P_{rob}(A/\overline{B})P_{rob}(\overline{B})}{P_{rob}(A/\overline{B})P_{rob}(\overline{B}) + P_{rob}(A/B)P_{rob}(B)}\right]$$
$$= i\left[\frac{P_{rob}(A)P_{rob}(\overline{B})}{P_{rob}(A)P_{rob}(\overline{B}) + P_{rob}(A)P_{rob}(B)}\right]$$
$$= i\left[\frac{P_r iP_r}{P_r iP_r + P_r i(1-P_r)}\right] = i\left[\frac{P_r iP_r}{iP_r^2 + iP_r - iP_r^2}\right] = i\left[\frac{P_r iP_r}{iP_r}\right] = iP_r = i - P_m = P_{rob}(\overline{B})$$

and this independently of the distribution of the binary random variables A in $\mathcal{R}$ and correspondingly of B in $\mathcal{M}$.

Since the complex random vector in CPP is $z = P_r + P_m = P_r + i(1-P_r)$ then:

$$\Rightarrow P_{rob}(A/B) + P_{rob}(B/A) = P_{rob}(A) + P_{rob}(B) = P_r + P_m = z_1$$

$$\text{And } P_{rob}(A/\overline{B}) + P_{rob}(B/\overline{A}) = P_{rob}(A) + P_{rob}(B) = P_r + P_m = z_1$$

$$\Rightarrow P_{rob}(\overline{A}/\overline{B}) + P_{rob}(\overline{B}/\overline{A}) = P_{rob}(\overline{A}) + P_{rob}(\overline{B}) = (1-P_r) + (i - P_m) = z_2$$

$$\text{And } P_{rob}(\overline{A}/B) + P_{rob}(\overline{B}/A) = P_{rob}(\overline{A}) + P_{rob}(\overline{B}) = (1-P_r) + (i - P_m) = z_2$$

Therefore, the resultant complex random vector in CPP is:

$Z = \sum_{j=1}^{2} z_j = z_1 + z_2 = [P_r + (1-P_r)] + [P_m + (i - P_m)] = 1 + i = 1 + (N-1)i,$

where N = 2 corresponds to the binary random variable considered in this case. And, $\frac{Z}{N} = \frac{\sum_{j=1}^{2} z_j}{N} = \frac{z_1+z_2}{N} = \frac{1+(N-1)i}{N} = \frac{1}{N} + \left(1 - \frac{1}{N}\right)i = P_{rZ} + P_{mZ} = 0.5 + 0.5i$ for N = 2 in this case. Thus,

$Pc_Z = P_{rZ} + \frac{P_{mZ}}{i} = 0.5 + \frac{0.5i}{i} = 0.5 + 0.5 = 1$, just as predicted by CPP.

$$\Rightarrow P_{rZ} = P_{mZ}/i = 0.5$$

$\Rightarrow P_{rob}(Z/N \text{ in} \mathcal{R}) = P_{rob}(Z/N \text{ in} \mathcal{M})/i = 0.5.$

To interpret the results obtained, that means that the two probabilities sets $\mathcal{R}$ and $\mathcal{M}$ are not only associated and complementary and dependent but also equiprobable, which means that there is no preference of considering one probability set on another. Both $\mathcal{R}$ and $\mathcal{M}$ have the same chance of 0.5 = 1/2 to be chosen in the complex probabilities set $\mathcal{C} = \mathcal{R} + \mathcal{M}$.

Since $\mathcal{C} = \mathcal{R} + \mathcal{M}$ and $Pc^2 = (P_r + P_m/i)^2 = 1 = Pc$ in *CPP* then:

$$P_{rob}(A/B) + P_{rob}(B/A)/i = P_{rob}(A) + P_{rob}(B)/i = P_r + P_m/i = 1 = Pc_{z1}$$

$$P_{rob}(A/\overline{B}) + P_{rob}(B/\overline{A})/i = P_{rob}(A) + P_{rob}(B)/i = P_r + P_m/i = 1 = Pc_{z1}$$

$$P_{rob}(\overline{A}/\overline{B}) + P_{rob}(\overline{B}/\overline{A})/i = P_{rob}(\overline{A}) + P_{rob}(\overline{B})/i = (1 - P_r) + (i - P_m)/i = 1 = Pc_{z2}$$

$$P_{rob}(\overline{A}/B) + P_{rob}(\overline{B}/A)/i = P_{rob}(\overline{A}) + P_{rob}(\overline{B})/i = (1 - P_r) + (i - P_m)/i = 1 = Pc_{z2}$$

That means that the probability in the set $\mathcal{C} = \mathcal{R} + \mathcal{M}$ is equal to 1, just as predicted by *CPP* (**Table 1**).

4.1.3 The probabilities of dependent and of joint events in $\mathcal{C} = \mathcal{R} + \mathcal{M}$

Additionally, we have:

$$\begin{aligned} P_{rob}(A \cap B) &= P_{rob}(A)P_{rob}(B/A) = P_{rob}(A)P_{rob}(B) \\ &= P_{rob}(B)P_{rob}(A/B) = P_{rob}(B)P_{rob}(A) \\ &= P_rP_m = P_mP_r = iP_r(1 - P_r) \end{aligned}$$

And,

$$\begin{aligned} P_{rob}(A \cup B) &= P_{rob}(A) + P_{rob}(B) - P_{rob}(A \cap B) \\ &= P_r + P_m - P_rP_m \\ \Rightarrow P_{rob}(A \cup B) &= P_r + i(1 - P_r) - P_r[i(1 - P_r)] = P_r + i - iP_r - iP_r + iP_r^2 = P_r + i - 2iP_r + iP_r^2 \\ &= P_r + i(1 - 2P_r + P_r^2) = P_r + i(1 - P_r)^2 \end{aligned}$$

So, if $P_r = 1 \Rightarrow A = \mathcal{R}$ and $\overline{A} = \varnothing$ and $B = \varnothing$ and $\overline{B} = \mathcal{M} \Rightarrow P_{rob}(A \cup B) = 1 = P_r = P_{rob}(\mathcal{R})$, that means we have a 100% deterministic certain experiment A in $\mathcal{R}$.

And if $P_r = 0 \Rightarrow A = \varnothing$ and $\overline{A} = \mathcal{R}$ and $B = \mathcal{M}$ and $\overline{B} = \varnothing \Rightarrow P_{rob}(A \cup B) = i = P_{rob}(\mathcal{M})$, that means we have a 100% deterministic impossible experiment A in $\mathcal{R}$.

Moreover,

$$\begin{aligned} P_{rob}(\overline{A} \cap B) &= P_{rob}(\overline{A})P_{rob}(B/\overline{A}) = P_{rob}(\overline{A})P_{rob}(B) = (1 - P_r) \times i(1 - P_r) \\ &= P_{rob}(B)P_{rob}(\overline{A}/B) = P_{rob}(B)P_{rob}(\overline{A}) = i(1 - P_r) \times (1 - P_r) \\ &= i(1 - P_r)^2 \end{aligned}$$

Probability Sets	Event Probability	Complementary Event Probability
In $\mathcal{R}$	$P_{rob}(A) = P_r$	$P_{rob}(\overline{A}) = 1 - P_{rob}(A) = 1 - P_r$
In $\mathcal{M}$	$P_{rob}(B) = P_m = i(1 - P_r)$	$P_{rob}(\overline{B}) = i - P_m = iP_{rob}(A) = iP_r$
In $\mathcal{C} = \mathcal{R} + \mathcal{M}$	$z_1 = P_{rob}(A) + P_{rob}(B) = P_r + P_m$	$z_2 = P_{rob}(\overline{A}) + P_{rob}(\overline{B}) = (1 - P_r) + (i - P_m)$
Deterministic Probabilities in $\mathcal{C}$	$Pc_{z1} = P_{rob}(A) + P_{rob}(B)/i = P_r + P_m/i = 1$	$Pc_{z2} = P_{rob}(\overline{A}) + P_{rob}(\overline{B})/i = (1 - P_r) + (i - P_m)/i = 1$

Table 1.
The table of the probabilities in $\mathcal{R}$, $\mathcal{M}$, and $\mathcal{C}$.

And,

$$\begin{aligned} P_{rob}\left(\overline{A} \cup B\right) &= P_{rob}\left(\overline{A}\right) + P_{rob}(B) - P_{rob}\left(\overline{A} \cap B\right) \\ &= (1 - P_r) + P_m - i(1 - P_r)^2 \\ &\Rightarrow P_{rob}\left(\overline{A} \cup B\right) = 1 - P_r + i(1 - P_r) - i(1 - P_r)^2 \\ &= (1 - P_r)[1 + i - i(1 - P_r)] = (1 - P_r)(1 + iP_r) \end{aligned}$$

So, if $P_r = 1 \Rightarrow A = \mathcal{R}$ and $\overline{A} = \varnothing$ and $B = \varnothing$ and $\overline{B} = \mathcal{M} \Rightarrow P_{rob}\left(\overline{A} \cup B\right) = P_{rob}(\varnothing) = 0$, that means we have a 100% deterministic certain experiment A in $\mathcal{R}$.

And if $P_r = 0 \Rightarrow A = \varnothing$ and $\overline{A} = \mathcal{R}$ and $B = \mathcal{M}$ and $\overline{B} = \varnothing$.

$\Rightarrow P_{rob}\left(\overline{A} \cup B\right) = P_{rob}(\mathcal{R} \cup \mathcal{M}) = P_{rob}(\mathcal{C}) = 1$, that means we have a 100% deterministic impossible experiment A in $\mathcal{R}$.

In addition,

$$\begin{aligned} P_{rob}\left(A \cap \overline{B}\right) &= P_{rob}(A)P_{rob}\left(\overline{B}/A\right) = P_{rob}(A)P_{rob}\left(\overline{B}\right) = P_r \times iP_r \\ &= P_{rob}\left(\overline{B}\right)P_{rob}\left(A/\overline{B}\right) = P_{rob}\left(\overline{B}\right)P_{rob}(A) = iP_r \times P_r \\ &= iP_r^2 \end{aligned}$$

And,

$$\begin{aligned} P_{rob}\left(A \cup \overline{B}\right) &= P_{rob}(A) + P_{rob}\left(\overline{B}\right) - P_{rob}\left(A \cap \overline{B}\right) \\ &= P_r + iP_r - iP_r^2 \\ &= P_r[1 + i(1 - P_r)] \end{aligned}$$

So, if $P_r = 1 \Rightarrow A = \mathcal{R}$ and $\overline{A} = \varnothing$ and $B = \varnothing$ and $\overline{B} = \mathcal{M}$.

$\Rightarrow P_{rob}\left(A \cup \overline{B}\right) = P_{rob}(\mathcal{R} \cup \mathcal{M}) = P_{rob}(\mathcal{C}) = 1$, that means we have a 100% deterministic certain experiment A in $\mathcal{R}$.

And if $P_r = 0 \Rightarrow A = \varnothing$ and $\overline{A} = \mathcal{R}$ and $B = \mathcal{M}$ and $\overline{B} = \varnothing \Rightarrow P_{rob}\left(A \cup \overline{B}\right) = P_{rob}(\varnothing) = 0$, that means we have a 100% deterministic impossible experiment A in $\mathcal{R}$.

Furthermore,

$$\begin{aligned} P_{rob}\left(\overline{A} \cap \overline{B}\right) &= P_{rob}\left(\overline{A}\right)P_{rob}\left(\overline{B}/\overline{A}\right) = P_{rob}\left(\overline{A}\right)P_{rob}\left(\overline{B}\right) = (1 - P_r) \times iP_r = P_r \times i(1 - P_r) \\ &= P_{rob}\left(\overline{B}\right)P_{rob}\left(\overline{A}/\overline{B}\right) = P_{rob}\left(\overline{B}\right)P_{rob}\left(\overline{A}\right) = iP_r \times (1 - P_r) = P_r \times i(1 - P_r) \\ &= P_rP_m = P_mP_r = iP_r(1 - P_r) \end{aligned}$$

And,

$$\begin{aligned} P_{rob}\left(\overline{A} \cup \overline{B}\right) &= P_{rob}\left(\overline{A}\right) + P_{rob}\left(\overline{B}\right) - P_{rob}\left(\overline{A} \cap \overline{B}\right) \\ &= 1 - P_r + (i - P_m) - P_rP_m \\ &= 1 - P_r + iP_r - P_rP_m \\ &\Rightarrow P_{rob}\left(\overline{A} \cup \overline{B}\right) = 1 - P_r + iP_r - P_r[i(1 - P_r)] \\ &= 1 - P_r + iP_r - iP_r + iP_r^2 = (1 - P_r) + iP_r^2 \end{aligned}$$

So, if $P_r = 1 \Rightarrow A = \mathcal{R}$ and $\overline{A} = \varnothing$ and $B = \varnothing$ and $\overline{B} = \mathcal{M} \Rightarrow P_{rob}\left(\overline{A} \cup \overline{B}\right) = i = P_{rob}(\mathcal{M})$, that means we have a 100% deterministic certain experiment A in $\mathcal{R}$.

And if $P_r = 0 \Rightarrow A = \varnothing$ and $\overline{A} = \mathcal{R}$ and $B = \mathcal{M}$ and $\overline{B} = \varnothing \Rightarrow P_{rob}\left(\overline{A} \cup \overline{B}\right) = 1 = P_{rob}(\mathcal{R})$, that means we have a 100% deterministic impossible experiment A in $\mathcal{R}$ (**Table 2**).

Sets and Events	Sets Intersection	Sets Union
A, B	$P_{rob}(A\cap B) = P_rP_m$	$P_{rob}(A\cup B) = P_r + P_m - P_rP_m = P_r + i(1-P_r)^2$
$\overline{A}, B$	$P_{rob}(\overline{A}\cap B) = i(1-P_r)^2$	$P_{rob}(\overline{A}\cup B) = (1-P_r)(1+iP_r)$
$A,\overline{B}$	$P_{rob}(A\cap\overline{B}) = iP_r^2$	$P_{rob}(A\cup\overline{B}) = P_r[1+i(1-P_r)]$
$\overline{A}, \overline{B}$	$P_{rob}(\overline{A}\cap\overline{B}) = P_rP_m$	$P_{rob}(\overline{A}\cup\overline{B}) = 1 - P_r + iP_r - P_rP_m = (1-P_r) + iP_r^2$

Table 2.
The table of the probabilities of dependent and of joint events in $\mathcal{C} = \mathcal{R} + \mathcal{M}$.

Finally, we can directly notice that:

$$\begin{aligned} P_{rob}(A\cap B) &= P_{rob}(\overline{A}\cap\overline{B}) \\ &= P_{rob}(A)P_{rob}(B) \\ &= P_{rob}(\overline{A})P_{rob}(\overline{B}) \\ &= P_rP_m = P_mP_r = iP_r(1-P_r) \end{aligned}$$

4.1.4 The relations to CPP *parameters*

The complex random vector $z_1 = P_r + P_m$.
The complex random vector $z_2 = (1-P_r) + (i-P_m)$.
Therefore, the resultant complex random vector is:
$Z = \sum_{j=1}^{2} z_j = z_1 + z_2 = 1 + i = 1 + (2-1)i = 1 + (N-1)i$, where $N = 2$ corresponds to the binary random variable that we have studied in this case. Thus,

$$\frac{Z}{N} = P_{rZ} + P_{mZ} = \frac{1}{N} + \left(1 - \frac{1}{N}\right)i = \frac{1}{2} + \left(1 - \frac{1}{2}\right)i = 0.5 + 0.5i$$
$$\Rightarrow P_{rZ} = 0.5 \text{ and } P_{mZ} = 0.5i$$

The Degree of our knowledge or DOK_{z1} of z_1 is: $DOK_{z1} = |z_1|^2 = P_r^2 + (P_m/i)^2$.
The Degree of our knowledge or DOK_{z2} of z_2 is: $DOK_{z2} = |z_2|^2 = (1-P_r)^2 + ([i-P_m]/i)^2$.
The Degree of our knowledge or DOK_Z of $\frac{Z}{N}$ is:

$$DOK_Z = \frac{|Z|^2}{N^2} = \frac{|Z|^2}{2^2} = \frac{|1+i|^2}{4} = \frac{1^2+1^2}{4} = P_{rZ}^2 + (P_{mZ}/i)^2 = (0.5)^2 + (0.5i/i)^2$$
$$= 0.25 + 0.25 = 0.5$$

The Chaotic Factor or Chf_{z1} of z_1 is: $Chf_{z1} = 2iP_rP_m$.
The Chaotic Factor or Chf_{z2} of z_2 is: $Chf_{z2} = 2i(1-P_r)(i-P_m)$.
The Chaotic Factor or Chf_Z of $\frac{Z}{N}$ is: $Chf_Z = 2iP_{rZ}P_{mZ} = 2i(0.5)(0.5i) = -0.5$.
The Magnitude of the Chaotic Factor or $MChf_{z1}$ of z_1 is: $MChf_{z1} = |Chf_{z1}| = |2iP_rP_m|$.
The Magnitude of the Chaotic Factor or $MChf_{z2}$ of z_2 is: $MChf_{z2} = |Chf_{z2}| = |2i(1-P_r)(i-P_m)|$.
The Magnitude of the Chaotic Factor or $MChf_Z$ of $\frac{Z}{N}$ is:

$$MChf_Z = |Chf_Z| = |2iP_{rZ}P_{mZ}| = |2i(0.5)(0.5i)| = |-0.5| = 0.5$$

The probability Pc_{z1} in $\mathcal{C} = \mathcal{R} + \mathcal{M}$ of z_1 is:

$$Pc_{z1}^2 = (P_r + P_m/i)^2 = (P_r + 1 - P_r)^2 = 1^2 = 1 = Pc_{z1}$$

The probability Pc_{z2} in $\mathcal{C} = \mathcal{R} + \mathcal{M}$ of z_2 is:

$$Pc_{z2}^2 = [(1 - P_r) + (i - P_m)/i]^2 = [(1 - P_r) + iP_r/i]^2 = [(1 - P_r) + P_r]^2 = 1^2 = 1 = Pc_{z2}$$

The probability Pc_Z in $\mathcal{C} = \mathcal{R} + \mathcal{M}$ of $\frac{Z}{N}$ is:

$$Pc_Z^2 = (P_{rZ} + P_{mZ}/i)^2 = (0.5 + 0.5i/i)^2 = 1^2 = 1 = Pc_Z$$

It is important to note here that all the results of the calculations done above confirm the predictions made by *CPP*.

4.1.5 Bayes' theorem and CPP *and the contingency tables*

See **Tables 3–7**.

Intersection	A	$\overline{A}$	**Total**
B	$P_{rob}(A \cap B) = P_{rob}(A)P_{rob}(B/A) = P_{rob}(B)P_{rob}(A/B)$	$P_{rob}(\overline{A} \cap B) = P_{rob}(\overline{A})P_{rob}(B/\overline{A}) = P_{rob}(B)P_{rob}(\overline{A}/B)$	$P_{rob}(B)$
$\overline{B}$	$P_{rob}(A \cap \overline{B}) = P_{rob}(A)P_{rob}(\overline{B}/A) = P_{rob}(\overline{B})P_{rob}(A/\overline{B})$	$P_{rob}(\overline{A} \cap \overline{B}) = P_{rob}(\overline{A})P_{rob}(\overline{B}/\overline{A}) = P_{rob}(\overline{B})P_{rob}(\overline{A}/\overline{B})$	$P_{rob}(\overline{B}) = i - P_{rob}(B)$
Total	$iP_{rob}(A) = P_{rob}(\overline{B})$	$iP_{rob}(\overline{A}) = i[1 - P_{rob}(A)] = P_{rob}(B)$	i

Table 3.
The table of Bayes' theorem and CPP.

Probabilities in $\mathcal{R}$	B	$\overline{B}$
A	$P_{rob}(A/B) = P_{rob}(A) = P_r$	$P_{rob}(A/\overline{B}) = P_{rob}(A) = P_r$
$\overline{A}$	$P_{rob}(\overline{A}/B) = P_{rob}(\overline{A}) = 1 - P_r$	$P_{rob}(\overline{A}/\overline{B}) = P_{rob}(\overline{A}) = 1 - P_r$
Total	1	1

Table 4.
The table of the real probabilities in $\mathcal{R}$.

Probabilities in $\mathcal{M}$	A	$\overline{A}$
B	$P_{rob}(B/A) = P_{rob}(B) = P_m$	$P_{rob}(B/\overline{A}) = P_{rob}(B) = P_m$
$\overline{B}$	$P_{rob}(\overline{B}/A) = P_{rob}(\overline{B}) = i - P_m$	$P_{rob}(\overline{B}/\overline{A}) = P_{rob}(\overline{B}) = i - P_m$
Total	i	i

Table 5.
The table of the imaginary probabilities in $\mathcal{M}$.

Complex probabilities in $\mathcal{C} = \mathcal{R} + \mathcal{M}$	A	$\overline{A}$
B	$z_1 = P_{rob}(A/B) + P_{rob}(B/A) = P_{rob}(A) + P_{rob}(B) = P_r + P_m$	$z_1 = P_{rob}(A/\overline{B}) + P_{rob}(B/\overline{A}) = P_{rob}(A) + P_{rob}(B) = P_r + P_m$
$\overline{B}$	$z_2 = P_{rob}(\overline{A}/B) + P_{rob}(\overline{B}/A) = P_{rob}(\overline{A}) + P_{rob}(\overline{B}) = (1 - P_r) + (i - P_m)$	$z_2 = P_{rob}(\overline{A}/\overline{B}) + P_{rob}(\overline{B}/\overline{A}) = P_{rob}(\overline{A}) + P_{rob}(\overline{B}) = (1 - P_r) + (i - P_m)$
Total = Resultant Complex Random Vector	$Z = z_1 + z_2 = 1 + i$	$Z = z_1 + z_2 = 1 + i$

Table 6.
The table of the complex probabilities in $\mathcal{C} = \mathcal{R} + \mathcal{M}$.

Probability Pc **in** $\mathcal{C} = \mathcal{R} + \mathcal{M}$	A	$\overline{A}$
B	$P_{rob}(A/B) + P_{rob}(B/A)/i = 1 = Pc_{z1}$	$P_{rob}(\overline{A}/B) + P_{rob}(\overline{B}/A)/i = 1 = Pc_{z2}$
$\overline{B}$	$P_{rob}(A/\overline{B}) + P_{rob}(B/\overline{A})/i = 1 = Pc_{z1}$	$P_{rob}(\overline{A}/\overline{B}) + P_{rob}(\overline{B}/\overline{A})/i = 1 = Pc_{z2}$

Table 7.
The table of the deterministic real probabilities in $\mathcal{C} = \mathcal{R} + \mathcal{M}$.

4.2 The case of a general discrete uniform random variable

4.2.1 The probabilities and the conditional probabilities

Let us consider here a discrete uniform random distribution in the probability set $\mathcal{R}$ to illustrate the results obtained for the new Bayes' theorem when related to *CPP*.

A_j is an event occurring in the real probabilities set $\mathcal{R}$ such that:

$$P_{rob}(A_j) = P_{rj} = \frac{1}{N}, \quad \forall j : 1 \le j \le N$$

The corresponding associated imaginary complementary event to the event A_j in the probabilities set $\mathcal{M}$ is the event B_j such that:

$$P_{rob}(B_j) = P_{mj} = i(1 - P_{rj}) = i\left(1 - \frac{1}{N}\right), \quad \forall j : 1 \le j \le N$$

The real complementary event to the event A_j in $\mathcal{R}$ is the event $\overline{A}_j$ such that:

$$A_j \cup \overline{A}_j = A_1 \cup A_2 \cup \cdots \cup A_j \cup \ldots \cup A_N = \mathcal{R}$$

and $A_j \cap A_k = \emptyset, \forall j \neq k$ (pairwise mutually exclusive events)

$$\begin{aligned}
P_{rob}(\overline{A}_j) &= 1 - P_{rob}(A_j) = 1 - P_{rj} = P_{mj}/i = P_{rob}(B_j)/i = 1 - \frac{1}{N} \\
P_{rob}(\mathcal{R}) &= P_{rob}(A_j \cup \overline{A}_j) = P_{rob}(A_1 \cup A_2 \cup \cdots \cup A_j \cup \ldots \cup A_N) \\
&= P_{rob}(A_1) + P_{rob}(A_2) + \cdots + P_{rob}(A_j) + \cdots + P_{rob}(A_N) \\
&= N \times P_{rob}(A_j) = N \times \frac{1}{N} = 1
\end{aligned}$$

The imaginary complementary event to the event B_j in $\mathcal{M}$ is the event $\overline{B}_j$ such that:

$$B_j \cup \overline{B}_j = B_1 \cup B_2 \cup \cdots \cup B_j \cup \ldots \cup B_N = \mathcal{M}$$

and $B_j \cap B_k = \varnothing, \forall j \neq k$ (pairwise mutually exclusive events)

$$P_{rob}(\overline{B}_j) = i - P_{rob}(B_j) = i - P_{mj} = i - i(1 - P_{rj}) = i - i + iP_{rj} = iP_{rj} = iP_{rob}(A_j) = \frac{i}{N}$$

$$\begin{aligned} P_{rob}(\mathcal{M}) &= P_{rob}(B_j \cup \overline{B}_j) = P_{rob}(B_1 \cup B_2 \cup \cdots \cup B_j \cup \ldots \cup B_N) \\ &= P_{rob}(B_1) + P_{rob}(B_2) + \cdots + P_{rob}(B_j) + \cdots + P_{rob}(B_N) \\ &= N \times P_{rob}(B_j) = N \times i\left(1 - \frac{1}{N}\right) = i(N-1) \end{aligned}$$

We have also, as derived from *CPP* that:

$P_{rob}(A_j/B_j) = P_{rob}(A_j) = P_{rj} = \frac{1}{N}$, that means if the event B_j occurs in $\mathcal{M}$ then the event A_j, which is its real complementary event, occurs in $\mathcal{R}$.

$P_{rob}(B_j/A_j) = P_{rob}(B_j) = P_{mj} = i\left(1 - \frac{1}{N}\right)$, that means if the event A_j occurs in $\mathcal{R}$ then the event B_j, which is its imaginary complementary event, occurs in $\mathcal{M}$.

$P_{rob}(\overline{A}_j/\overline{B}_j) = P_{rob}(\overline{A}_j) = 1 - P_{rob}(A_j) = 1 - P_{rj} = 1 - \frac{1}{N}$, that means if the event $\overline{B}_j$ occurs in $\mathcal{M}$ then the event $\overline{A}_j$, which is its real complementary event, occurs in $\mathcal{R}$.

$P_{rob}(\overline{B}_j/\overline{A}_j) = P_{rob}(\overline{B}_j) = i - P_{rob}(B_j) = i - P_{mj} = iP_{rj} = \frac{i}{N}$, that means if the event $\overline{A}_j$ occurs in $\mathcal{R}$ then the event $\overline{B}_j$, which is its imaginary complementary event, occurs in $\mathcal{M}$.

4.2.2 The relations to Bayes' theorem

Bayes' theorem for N competing statements or hypotheses that is, for N random variables, is in the probability set $\mathcal{R}$ equal to:

$$P_{rob}(A_j/B) = \frac{P_{rob}(B/A_j)P_{rob}(A_j)}{P_{rob}(B)} = \frac{P_{rob}(B/A_j)P_{rob}(A_j)}{\sum_{k=1}^{N} P_{rob}(B/A_k)P_{rob}(A_k)}$$

Therefore, in *CPP* and hence in $\mathcal{C} = \mathcal{R} + \mathcal{M}$, we can deduce the new forms of Bayes' theorem for the case considered as follows:

$$\begin{aligned} P_{rob}(A_j/B_j) &= \frac{P_{rob}(B_j/A_j)P_{rob}(A_j)}{P_{rob}(B_j)} = \frac{P_{rob}(B_j)P_{rob}(A_j)}{P_{rob}(B_j)} = P_{rob}(A_j) \\ &= \frac{P_{rob}(B_j/A_j)P_{rob}(A_j)}{\sum_{k=1}^{N} P_{rob}(B_j/A_k)P_{rob}(A_k)} == \frac{P_{rob}(B_j)P_{rob}(A_j)}{\sum_{k=1}^{N} P_{rob}(B_j)P_{rob}(A_k)} = \frac{P_{rob}(B_j)P_{rob}(A_j)}{P_{rob}(B_j)\sum_{k=1}^{N} P_{rob}(A_k)} \\ &= \frac{P_{rob}(B_j)P_{rob}(A_j)}{P_{rob}(B_j) \times N\left(\frac{1}{N}\right)} = P_{rob}(A_j) = \frac{1}{N}, \quad \forall j : 1 \leq j \leq N \end{aligned}$$

And, its corresponding Bayes' relation in $\mathcal{M}$ is:

$$P_{rob}(B_j/A_j) = \frac{P_{rob}(A_j/B_j)P_{rob}(B_j)}{P_{rob}(A_j)} = \frac{P_{rob}(A_j)P_{rob}(B_j)}{P_{rob}(A_j)} = P_{rob}(B_j)$$

$$= i(N-1)\left[\frac{P_{rob}(A_j/B_j)P_{rob}(B_j)}{\sum_{k=1}^{N} P_{rob}(A_j/B_k)P_{rob}(B_k)}\right] = i(N-1)\left[\frac{P_{rob}(A_j)P_{rob}(B_j)}{\sum_{k=1}^{N} P_{rob}(A_j)P_{rob}(B_k)}\right]$$

$$= i(N-1)\left[\frac{P_{rob}(A_j)P_{rob}(B_j)}{P_{rob}(A_j)\sum_{k=1}^{N} P_{rob}(B_k)}\right] = i(N-1)\left[\frac{P_{rob}(A_j)P_{rob}(B_j)}{P_{rob}(A_j)\times i(N-1)}\right]$$

$$= P_{rob}(B_j) = i\left[1 - P_{rob}(A_j)\right] = i\left(1-\frac{1}{N}\right), \quad \forall j : 1 \le j \le N$$

Similarly,

$$P_{rob}(\overline{A}_j/\overline{B}_j) = \frac{P_{rob}(\overline{B}_j/\overline{A}_j)P_{rob}(\overline{A}_j)}{P_{rob}(\overline{B}_j)} = \frac{P_{rob}(\overline{B}_j)P_{rob}(\overline{A}_j)}{P_{rob}(\overline{B}_j)} = P_{rob}(\overline{A}_j)$$

$$= (N-1)\left[\frac{P_{rob}(\overline{B}_j/\overline{A}_j)P_{rob}(\overline{A}_j)}{\sum_{k=1}^{N} P_{rob}(\overline{B}_j/\overline{A}_k)P_{rob}(\overline{A}_k)}\right] = (N-1)\left[\frac{P_{rob}(\overline{B}_j)P_{rob}(\overline{A}_j)}{\sum_{k=1}^{N} P_{rob}(\overline{B}_j)P_{rob}(\overline{A}_k)}\right]$$

$$= (N-1)\left[\frac{P_{rob}(\overline{B}_j)P_{rob}(\overline{A}_j)}{P_{rob}(\overline{B}_j)\sum_{k=1}^{N} P_{rob}(\overline{A}_k)}\right] = (N-1)\left[\frac{P_{rob}(\overline{B}_j)P_{rob}(\overline{A}_j)}{P_{rob}(\overline{B}_j)\times N\left(1-\frac{1}{N}\right)}\right]$$

$$= (N-1)\left[\frac{P_{rob}(\overline{B}_j)P_{rob}(\overline{A}_j)}{P_{rob}(\overline{B}_j)\times (N-1)}\right]$$

$$= P_{rob}(\overline{A}_j) = 1 - P_{rob}(A_j) = 1 - \frac{1}{N}, \quad \forall j : 1 \le j \le N$$

And, its corresponding Bayes' relation in $\mathcal{M}$ is:

$$P_{rob}(\overline{B}_j/\overline{A}_j) = \frac{P_{rob}(\overline{A}_j/\overline{B}_j)P_{rob}(\overline{B}_j)}{P_{rob}(\overline{A}_j)} = \frac{P_{rob}(\overline{A}_j)P_{rob}(\overline{B}_j)}{P_{rob}(\overline{A}_j)} = P_{rob}(\overline{B}_j)$$

$$= i\left[\frac{P_{rob}(\overline{A}_j/\overline{B}_j)P_{rob}(\overline{B}_j)}{\sum_{k=1}^{N} P_{rob}(\overline{A}_j/\overline{B}_k)P_{rob}(\overline{B}_k)}\right] = i\left[\frac{P_{rob}(\overline{A}_j)P_{rob}(\overline{B}_j)}{\sum_{k=1}^{N} P_{rob}(\overline{A}_j)P_{rob}(\overline{B}_k)}\right]$$

$$= i\left[\frac{P_{rob}(\overline{A}_j)P_{rob}(\overline{B}_j)}{P_{rob}(\overline{A}_j)\sum_{k=1}^{N} P_{rob}(\overline{B}_k)}\right] = i\left[\frac{P_{rob}(\overline{A}_j)P_{rob}(\overline{B}_j)}{P_{rob}(\overline{A}_j)\times N\left(\frac{i}{N}\right)}\right] = i\left[\frac{P_{rob}(\overline{A}_j)P_{rob}(\overline{B}_j)}{P_{rob}(\overline{A}_j)\times i}\right]$$

$$= P_{rob}(\overline{B}_j) = i - P_{rob}(B_j) = i - i\left(1-\frac{1}{N}\right) = i\left[1 - P_{rob}(\overline{A}_j)\right]$$

$$= i\left[1 - \left(1-\frac{1}{N}\right)\right] = \frac{i}{N}, \quad \forall j : 1 \le j \le N$$

Furthermore,

$$P_{rob}(A_j/\overline{B}_j) = \frac{P_{rob}(\overline{B}_j/A_j)P_{rob}(A_j)}{P_{rob}(\overline{B}_j)} = \frac{P_{rob}(\overline{B}_j)P_{rob}(A_j)}{P_{rob}(\overline{B}_j)} = P_{rob}(A_j)$$

$$= \frac{P_{rob}(\overline{B}_j/A_j)P_{rob}(A_j)}{\sum_{k=1}^{N} P_{rob}(\overline{B}_j/A_k)P_{rob}(A_k)} = \frac{P_{rob}(\overline{B}_j)P_{rob}(A_j)}{\sum_{k=1}^{N} P_{rob}(\overline{B}_j)P_{rob}(A_k)} = \frac{P_{rob}(\overline{B}_j)P_{rob}(A_j)}{P_{rob}(\overline{B}_j)\sum_{k=1}^{N} P_{rob}(A_k)}$$

$$= \frac{P_{rob}(\overline{B}_j)P_{rob}(A_j)}{P_{rob}(\overline{B}_j) \times N\left(\frac{1}{N}\right)} = P_{rob}(A_j) = \frac{1}{N}, \quad \forall j : 1 \le j \le N$$

And, its corresponding Bayes' relation in $\mathcal{M}$ is:

$$P_{rob}(B_j/\overline{A}_j) = \frac{P_{rob}(\overline{A}_j/B_j)P_{rob}(B_j)}{P_{rob}(\overline{A}_j)} = \frac{P_{rob}(\overline{A}_j)P_{rob}(B_j)}{P_{rob}(\overline{A}_j)} = P_{rob}(B_j)$$

$$= i(N-1)\left[\frac{P_{rob}(\overline{A}_j/B_j)P_{rob}(B_j)}{\sum_{k=1}^{N} P_{rob}(\overline{A}_j/B_k)P_{rob}(B_k)}\right] = i(N-1)\left[\frac{P_{rob}(\overline{A}_j)P_{rob}(B_j)}{\sum_{k=1}^{N} P_{rob}(\overline{A}_j)P_{rob}(B_k)}\right]$$

$$= i(N-1)\left[\frac{P_{rob}(\overline{A}_j)P_{rob}(B_j)}{P_{rob}(\overline{A}_j)\sum_{k=1}^{N} P_{rob}(B_k)}\right] = i(N-1)\left[\frac{P_{rob}(\overline{A}_j)P_{rob}(B_j)}{P_{rob}(\overline{A}_j) \times i(N-1)}\right]$$

$$= P_{rob}(B_j) = i\left(1 - \frac{1}{N}\right), \quad \forall j : 1 \le j \le N$$

Moreover,

$$P_{rob}(\overline{A}_j/B_j) = \frac{P_{rob}(B_j/\overline{A}_j)P_{rob}(\overline{A}_j)}{P_{rob}(B_j)} = \frac{P_{rob}(B_j)P_{rob}(\overline{A}_j)}{P_{rob}(B_j)} = P_{rob}(\overline{A}_j)$$

$$= (N-1)\left[\frac{P_{rob}(B_j/\overline{A}_j)P_{rob}(\overline{A}_j)}{\sum_{k=1}^{N} P_{rob}(B_j/\overline{A}_k)P_{rob}(\overline{A}_k)}\right] = (N-1)\left[\frac{P_{rob}(B_j)P_{rob}(\overline{A}_j)}{\sum_{k=1}^{N} P_{rob}(B_j)P_{rob}(\overline{A}_k)}\right]$$

$$= (N-1)\left[\frac{P_{rob}(B_j)P_{rob}(\overline{A}_j)}{P_{rob}(B_j)\sum_{k=1}^{N} P_{rob}(\overline{A}_k)}\right] = (N-1)\left[\frac{P_{rob}(B_j)P_{rob}(\overline{A}_j)}{P_{rob}(B_j) \times N\left(1 - \frac{1}{N}\right)}\right]$$

$$= (N-1)\left[\frac{P_{rob}(B_j)P_{rob}(\overline{A}_j)}{P_{rob}(B_j) \times (N-1)}\right]$$

$$= P_{rob}(\overline{A}_j) = 1 - \frac{1}{N}, \quad \forall j : 1 \le j \le N$$

And, its corresponding Bayes' relation in $\mathcal{M}$ is:

$$P_{rob}(\overline{B}_j/A_j) = \frac{P_{rob}(A_j/\overline{B}_j)P_{rob}(\overline{B}_j)}{P_{rob}(A_j)} = \frac{P_{rob}(A_j)P_{rob}(\overline{B}_j)}{P_{rob}(A_j)} = P_{rob}(\overline{B}_j)$$

$$= i\left[\frac{P_{rob}(A_j/\overline{B}_j)P_{rob}(\overline{B}_j)}{\sum_{k=1}^{N} P_{rob}(A_j/\overline{B}_k)P_{rob}(\overline{B}_k)}\right] = i\left[\frac{P_{rob}(A_j)P_{rob}(\overline{B}_j)}{\sum_{k=1}^{N} P_{rob}(A_j)P_{rob}(\overline{B}_k)}\right]$$

$$= i\left[\frac{P_{rob}(A_j)P_{rob}(\overline{B}_j)}{P_{rob}(A_j)\sum_{k=1}^{N} P_{rob}(\overline{B}_k)}\right] = i\left[\frac{P_{rob}(A_j)P_{rob}(\overline{B}_j)}{P_{rob}(A_j) \times N\left(\frac{i}{N}\right)}\right] = i\left[\frac{P_{rob}(A_j)P_{rob}(\overline{B}_j)}{P_{rob}(A_j) \times i}\right]$$

$$= P_{rob}(\overline{B}_j) = \frac{i}{N}, \quad \forall j : 1 \le j \le N$$

Since the complex random vector in *CPP* is $z = P_r + P_m = P_r + i(1 - P_r)$ then:

$$\Rightarrow P_{rob}(A_j/B_j) + P_{rob}(B_j/A_j) = P_{rob}(A_j/\overline{B}_j) + P_{rob}(B_j/\overline{A}_j)$$

$$= P_{rob}(A_j) + P_{rob}(B_j) = P_{rj} + P_{mj}$$

$$= \frac{1}{N} + i\left(1 - \frac{1}{N}\right) = z_j, \quad \forall j : 1 \le j \le N$$

$$\Rightarrow P_{rob}(\overline{A}_j/\overline{B}_j) + P_{rob}(\overline{B}_j/\overline{A}_j) = P_{rob}(\overline{A}_j/B_j) + P_{rob}(\overline{B}_j/A_j)$$

$$= P_{rob}(\overline{A}_j) + P_{rob}(\overline{B}_j) = P_{rj}^* + P_{mj}^*$$

$$= \left(1 - \frac{1}{N}\right) + \frac{i}{N} = z_j^*, \quad \forall j : 1 \le j \le N$$

Therefore, the resultant complex random vectors in *CPP* of the uniform discrete random distribution are:

$$Z_U = \sum_{j=1}^{N} z_j = z_1 + z_2 + \cdots + z_N = Nz_j = N\left[\frac{1}{N} + i\left(1 - \frac{1}{N}\right)\right] = 1 + (N-1)i$$

$$Z_U^* = \sum_{j=1}^{N} z_j^* = z_1^* + z_2^* + \cdots + z_N^* = Nz_j^* = N\left[\left(1 - \frac{1}{N}\right) + \frac{i}{N}\right] = (N-1) + i$$

And,

$\frac{Z_U}{N} = \frac{\sum_{j=1}^{N} z_j}{N} = \frac{Nz_j}{N} = z_j = \frac{1}{N} + \left(1 - \frac{1}{N}\right)i = P_r|_{Z_U} + P_m|_{Z_U}$. Thus,

$Pc|_{Z_U} = P_r|_{Z_U} + \frac{P_m|_{Z_U}}{i} = \frac{1}{N} + \frac{\left(1-\frac{1}{N}\right)i}{i} = \frac{1}{N} + 1 - \frac{1}{N} = 1$, just as predicted by *CPP*.

Analogously, $\frac{Z_U^*}{N} = \frac{\sum_{j=1}^{N} z_j^*}{N} = \frac{Nz_j^*}{N} = z_j^* = \left(1 - \frac{1}{N}\right) + \frac{i}{N} = P_r^*|_{Z_U^*} + P_m^*|_{Z_U^*}$. Thus,

$Pc^*|_{Z_U^*} = P_r^*|_{Z_U^*} + \frac{P_m^*|_{Z_U^*}}{i} = \left(1 - \frac{1}{N}\right) + \frac{\frac{i}{N}}{i} = 1 - \frac{1}{N} + \frac{1}{N} = 1$, just as predicted by *CPP*.

Since $\mathcal{C} = \mathcal{R} + \mathcal{M}$ and $Pc^2 = (P_r + P_m/i)^2 = 1 = Pc$ in *CPP* then:

$$\Rightarrow P_{rob}(A_j/B_j) + P_{rob}(B_j/A_j)/i = P_{rob}(A_j) + P_{rob}(B_j)/i = P_{rj} + P_{mj}/i$$
$$= \frac{1}{N} + \frac{\left(1 - \frac{1}{N}\right)i}{i} = \frac{1}{N} + 1 - \frac{1}{N} = 1 = Pc_j, \quad \forall j : 1 \le j \le N$$

$$\Rightarrow P_{rob}(A_j/\overline{B}_j) + P_{rob}(B_j/\overline{A}_j)/i = P_{rob}(A_j) + P_{rob}(B_j)/i = P_{rj} + P_{mj}/i$$
$$= \frac{1}{N} + \frac{\left(1 - \frac{1}{N}\right)i}{i} = \frac{1}{N} + 1 - \frac{1}{N} = 1 = Pc_j, \quad \forall j : 1 \le j \le N$$

$$\Rightarrow P_{rob}(\overline{A}_j/\overline{B}_j) + P_{rob}(\overline{B}_j/\overline{A}_j)/i = P_{rob}(\overline{A}_j) + P_{rob}(\overline{B}_j)/i = P_{rj}^* + P_{mj}^*/i = (1 - P_{rj}) + (i - P_{mj})/i$$
$$= \left(1 - \frac{1}{N}\right) + \frac{\left(\frac{i}{N}\right)}{i} = 1 - \frac{1}{N} + \frac{1}{N} = 1 = Pc_j^*, \quad \forall j : 1 \le j \le N$$

$$\Rightarrow P_{rob}(\overline{A}_j/B_j) + P_{rob}(\overline{B}_j/A_j)/i = P_{rob}(\overline{A}_j) + P_{rob}(\overline{B}_j)/i = P_{rj}^* + P_{mj}^*/i = (1 - P_{rj}) + (i - P_{mj})/i$$
$$= \left(1 - \frac{1}{N}\right) + \frac{\left(\frac{i}{N}\right)}{i} = 1 - \frac{1}{N} + \frac{1}{N} = 1 = Pc_j^*, \quad \forall j : 1 \le j \le N$$

That means that the probability in the set $\mathcal{C} = \mathcal{R} + \mathcal{M}$ is equal to 1, just as predicted by *CPP*.

Additionally, we have:

$$P_{rob}(A_j \cap B_j) = P_{rob}(A_j)P_{rob}(B_j/A_j) = P_{rob}(A_j)P_{rob}(B_j)$$
$$= P_{rob}(B_j)P_{rob}(A_j/B_j) = P_{rob}(B_j)P_{rob}(A_j)$$
$$= P_{rj}P_{mj} = P_{mj}P_{rj}$$

Moreover, we have:

$$P_{rob}(A_j \cup B_j) = P_{rob}(A_j) + P_{rob}(B_j) - P_{rob}(A_j \cap B_j)$$
$$= P_{rj} + P_{mj} - P_{rj}P_{mj}$$
$$\Rightarrow P_{rob}(A_j \cup B_j) = P_{rj} + i(1 - P_{rj}) - P_{rj}[i(1 - P_{rj})] = P_{rj} + i - iP_{rj} - iP_{rj} + iP_{rj}^2 = P_{rj} + i - 2iP_{rj} + iP_{rj}^2$$
$$= P_{rj} + i\left(1 - 2P_{rj} + P_{rj}^2\right) = P_{rj} + i(1 - P_{rj})^2$$

So, if $P_{rj} = 1$ then $P_{rob}(A_j \cup B_j) = P_{rj} = 1 = P_{rob}(\mathcal{R})$, that means we have a 100% deterministic certain experiment A_j in $\mathcal{R}$.

And if $P_{rj} = 0$ then $P_{rob}(A_j \cup B_j) = i$, that means we have a 100% deterministic impossible experiment A_j in $\mathcal{R}$.

4.2.3 The relations to CPP *parameters*

The first complex random vector is: $z_j = P_{rj} + P_{mj} = \frac{1}{N} + \left(1 - \frac{1}{N}\right)i, \quad \forall j : 1 \le j \le N$.
Therefore, the first resultant complex random vector is:

$$Z_U = \sum_{j=1}^{N} z_j = z_1 + z_2 + \cdots + z_N = Nz_j = N\left[\frac{1}{N} + \left(1 - \frac{1}{N}\right)i\right] = 1 + (N-1)i$$

And, $\frac{Z_U}{N} = P_r|_{Z_U} + P_m|_{Z_U} = \frac{\sum_{j=1}^{N} z_j}{N} = \frac{Nz_j}{N} = z_j = \frac{1}{N} + \left(1 - \frac{1}{N}\right)i$.

The second complex random vector is: $z_j^* = P_{rj}^* + P_{mj}^* = \left(1 - \frac{1}{N}\right) + \frac{i}{N}$, $\forall j : 1 \le j \le N$.

Therefore, the second resultant complex random vector is:

$$Z_U^* = \sum_{j=1}^{N} z_j^* = z_1^* + z_2^* + \cdots + z_N^* = N z_j^* = N\left[\left(1 - \frac{1}{N}\right) + \frac{i}{N}\right] = (N-1) + i$$

And, $\frac{Z_U^*}{N} = P_r^*\big|_{Z_U^*} + P_m^*\big|_{Z_U^*} = \frac{\sum_{j=1}^{N} z_j^*}{N} = \frac{N z_j^*}{N} = z_j^* = \left(1 - \frac{1}{N}\right) + \frac{i}{N}$.

The Degree of our knowledge or DOK_{z_j} of z_j is:

$$DOK_{z_j} = |z_j|^2 = P_{rj}^2 + (P_{mj}/i)^2 = \left(\frac{1}{N}\right)^2 + \left(1 - \frac{1}{N}\right)^2 = \frac{1 + (N-1)^2}{N^2}, \quad \forall j : 1 \le j \le N$$

The Degree of our knowledge or $DOK_{z_j^*}$ of z_j^* is:

$$DOK_{z_j^*} = \left|z_j^*\right|^2 = \left(P_{rj}^*\right)^2 + \left(P_{mj}^*/i\right)^2 = \left(1 - \frac{1}{N}\right)^2 + \left(\frac{1}{N}\right)^2 = \frac{1 + (N-1)^2}{N^2}, \quad \forall j : 1 \le j \le N$$

The Degree of our knowledge or DOK_{Z_U} of $\frac{Z_U}{N}$ is:

$$\begin{aligned} DOK_{Z_U} &= \frac{|Z_U|^2}{N^2} = \frac{|1 + (N-1)i|^2}{N^2} = P_r^2\big|_{Z_U} + \left(\frac{P_m|_{Z_U}}{i}\right)^2 = \left(\frac{1}{N}\right)^2 + \left(1 - \frac{1}{N}\right)^2 \\ &= \frac{1 + (N-1)^2}{N^2} \end{aligned}$$

The Degree of our knowledge or $DOK_{Z_U^*}$ of $\frac{Z_U^*}{N}$ is:

$$\begin{aligned} DOK_{Z_U^*} &= \frac{|Z_U^*|^2}{N^2} = \frac{|(N-1) + i|^2}{N^2} = P_r^*\big|_{Z_U^*}^2 + \left(\frac{P_m^*|_{Z_U^*}}{i}\right)^2 \\ &= \left(1 - \frac{1}{N}\right)^2 + \left(\frac{1}{N}\right)^2 = \frac{1 + (N-1)^2}{N^2} \end{aligned}$$

$$\Leftrightarrow DOK_{z_j} = DOK_{z_j^*} = DOK_{Z_U} = DOK_{Z_U^*}$$

The Chaotic Factor or Chf_{z_j} of z_j is:

$Chf_{z_j} = 2iP_{rj}P_{mj} = 2i\left(\frac{1}{N}\right)i\left(1 - \frac{1}{N}\right) = \frac{-2(N-1)}{N^2}$ since $i^2 = -1$, $\forall j : 1 \le j \le N$.

The Chaotic Factor or $Chf_{z_j^*}$ of z_j^* is:

$Chf_{z_j^*} = 2iP_{rj}^*P_{mj}^* = 2i\left(1 - \frac{1}{N}\right)i\left(\frac{1}{N}\right) = \frac{-2(N-1)}{N^2}$ since $i^2 = -1$, $\forall j : 1 \le j \le N$.

The Chaotic Factor or Chf_{Z_U} of $\frac{Z_U}{N}$ is:

$$Chf_{Z_U} = 2iP_r|_{Z_U}P_m|_{Z_U} = 2i\left(\frac{1}{N}\right)i\left(1 - \frac{1}{N}\right) = \frac{-2(N-1)}{N^2}$$

The Chaotic Factor or $Chf_{Z_U^*}$ of $\frac{Z_U^*}{N}$ is:

It is important to note here that all the results of the calculations done above confirm the predictions made by *CPP*.

5. Flowchart of the complex probability and Bayes' theorem prognostic model

The following flowchart summarizes all the procedures of the proposed complex probability prognostic model where X is between the lower bound L_b and the upper bound U_b:

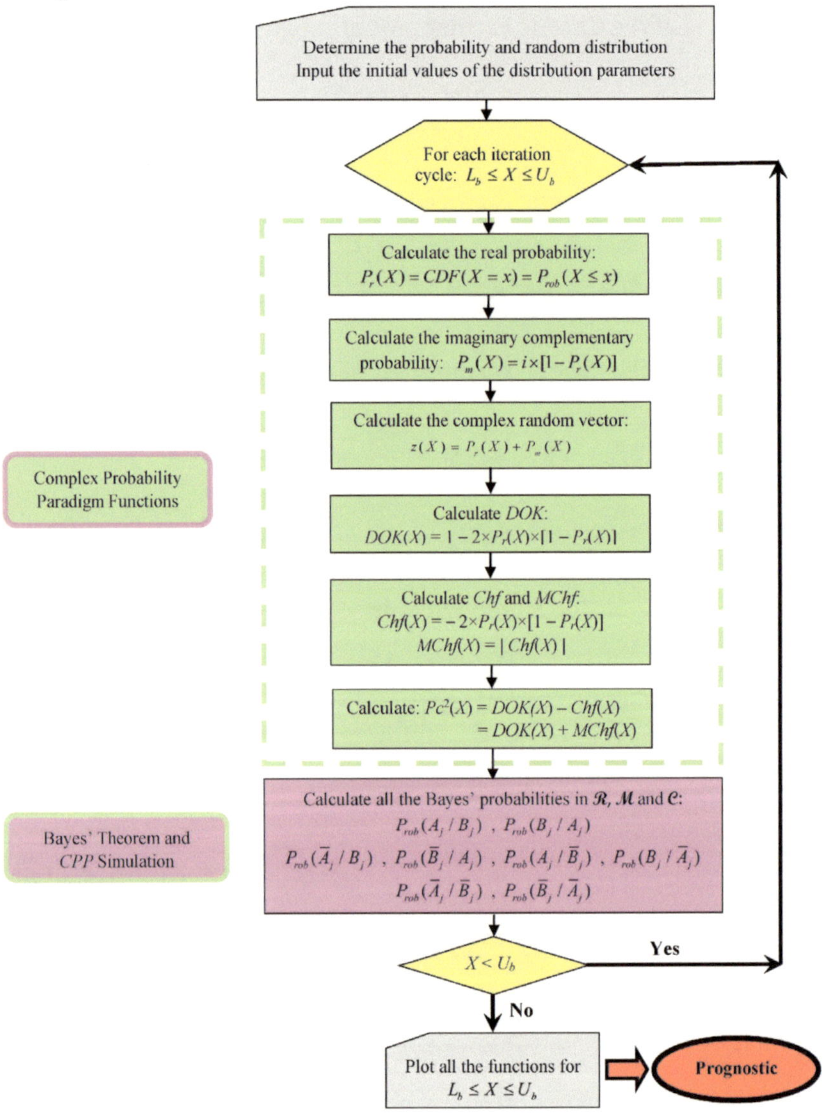

6. The new paradigm applied to discrete and continuous stochastic distributions

In this section, the simulation of the novel *CPP* model for a discrete and a continuous random distribution will be done. Note that all the numerical values found in the paradigm functions analysis for all the simulations were computed using the 64-Bit MATLAB version 2021 software. It is important to mention here that two important and well-known probability distributions were considered although the original *CPP* model can be applied to any stochastic distribution beside the studied random cases below. This will lead to similar results and conclusions. Hence, the new paradigm is successful with any discrete or continuous random case.

6.1 Simulation of the discrete binomial probability distribution

The probability density function (*PDF*) of this discrete stochastic distribution is:

$$f(x) = {}_N C_x p^x q^{N-x} = \binom{N}{x} p^x q^{N-x}, \quad \text{for } (L_b = 0) \le x \le (U_b = N)$$

I have taken the domain for the binomial random variable to be: $x \in [L_b = 0, U_b = N = 10]$ and $\forall k : 1 \le k \le 10$ we have $\Delta x_k = x_k - x_{k-1} = 1$, then: $x = 0, 1, 2, \ldots, 10$.

Taking in our simulation $N = 10$ and $p + q = 1, p = q = 0.5$ then:

The mean of this binomial discrete random distribution is: $\mu = Np = 10 \times 0.5 = 5$.

The standard deviation is: $\sigma = \sqrt{Npq} = \sqrt{10 \times 0.5 \times 0.5} = \sqrt{2.5} = 1.58113883\ldots$.

The median is $Md = \mu = 5$.

The mode for this symmetric distribution is = 5 = Md = μ.

The cumulative distribution function (*CDF*) is:

$$CDF(x) = P_{rob}(X \le x) = \sum_{k=0}^{x} f(k; N) = \sum_{k=0}^{x} {}_N C_k p^k q^{N-k} = \sum_{k=0}^{x} {}_{10} C_k p^k q^{10-k}, \quad \forall x : 0 \le x \le (N = 10)$$

Note that:

If $x = 0 \Rightarrow X = L_b \Rightarrow CDF(x) = P_{rob}(X \le 0) = f(X = L_b; N) = {}_N C_0 p^0 q^{N-0} = q^N = 0.5^{10} \cong 0$.

If $x = N = 10 \Rightarrow X = U_b \Rightarrow CDF(x) = P_{rob}(X \le x) = \sum_{k=0}^{x=N} {}_N C_k p^k q^{N-k} = (p+q)^N = 1^N = 1^{10} = 1$ by the binomial theorem.

The real probability $P_{rj}(x)$ is:

$$P_{rj}(x) = CDF(x) = \sum_{k=0}^{x} f(k; N) = \sum_{k=0}^{x} {}_N C_k p^k q^{N-k} = \sum_{k=0}^{x} {}_{10} C_k p^k q^{10-k}, \quad \forall x : 0 \le x \le (N = 10)$$

$$\Rightarrow P_{rob}(A_j/B_j) = P_{rob}(A_j/\overline{B}_j) = P_{rob}(A_j) = P_{rj}(x) = \sum_{k=0}^{x} {}_{10} C_k p^k q^{10-k}$$

The imaginary complementary probability $P_{mj}(x)$ to $P_{rj}(x)$ is:

$$P_{mj}(x) = i[1 - P_{rj}(x)] = i[1 - CDF(x)] = i\left[1 - \sum_{k=0}^{x} f(k; N)\right]$$

$$= i\left(1 - \sum_{k=0}^{x} {}_N C_k p^k q^{N-k}\right) = i \sum_{k=x+1}^{N} {}_N C_k p^k q^{N-k} = i \sum_{k=x+1}^{10} {}_{10} C_k p^k q^{10-k}, \quad \forall x : 0 \le x \le (N = 10)$$

$$\Rightarrow P_{rob}(B_j/A_j) = P_{rob}(B_j/\overline{A}_j) = P_{rob}(B_j) = P_{mj}(x) = i\left(\sum_{k=x+1}^{10} {}_{10}C_k p^k q^{10-k}\right)$$

The real complementary probability $P_{rj}^*(x)$ to $P_{rj}(x)$ is:

$$P_{rj}^*(x) = 1 - P_{rj}(x) = P_{mj}(x)/i = 1 - CDF(x) = 1 - \sum_{k=0}^{x} f(k;N) = \sum_{k=x+1}^{N} {}_N C_k p^k q^{N-k}$$

$$= \sum_{k=x+1}^{10} {}_{10}C_k p^k q^{10-k}, \quad \forall x : 0 \le x \le (N = 10)$$

$$\Rightarrow P_{rob}(\overline{A}_j/B_j) = P_{rob}(\overline{A}_j/\overline{B}_j) = P_{rob}(\overline{A}_j) = P_{rj}^*(x) = \sum_{k=x+1}^{10} {}_{10}C_k p^k q^{10-k}$$

The imaginary complementary probability $P_{mj}^*(x)$ to $P_{mj}(x)$ is:

$$P_{mj}^*(x) = i - P_{mj}(x) = i - i[1 - P_{rj}(x)] = iP_{rj}(x) = iCDF(x) = i\left[\sum_{k=0}^{x} f(k;N)\right]$$

$$= i\sum_{k=0}^{x} {}_{10}C_k p^k q^{10-k}, \quad \forall x : 0 \le x \le (N = 10)$$

$$\Rightarrow P_{rob}(\overline{B}_j/A_j) = P_{rob}(\overline{B}_j/\overline{A}_j) = P_{rob}(\overline{B}_j) = P_{mj}^*(x) = i\sum_{k=0}^{x} {}_{10}C_k p^k q^{10-k}$$

The complex probability or random vectors are:

$$z_j(x) = P_{rj}(x) + P_{mj}(x) = \left(\sum_{k=0}^{x} {}_{10}C_k p^k q^{10-k}\right) + i\left(1 - \sum_{k=0}^{x} {}_{10}C_k p^k q^{10-k}\right)$$

$$= \sum_{k=0}^{x} {}_{10}C_k p^k q^{10-k} + i\left(\sum_{k=x+1}^{10} {}_{10}C_k p^k q^{10-k}\right), \quad \forall x : 0 \le x \le (N = 10)$$

$$z_j^*(x) = P_{rj}^*(x) + P_{mj}^*(x) = [1 - P_{rj}(x)] + [i - P_{mj}(x)] = [1 - P_{rj}(x)] + iP_{rj}(x)$$

$$= \left(1 - \sum_{k=0}^{x} {}_{10}C_k p^k q^{10-k}\right) + i\left(\sum_{k=0}^{x} {}_{10}C_k p^k q^{10-k}\right)$$

$$= \left(\sum_{k=x+1}^{10} {}_{10}C_k p^k q^{10-k}\right) + i\left(\sum_{k=0}^{x} {}_{10}C_k p^k q^{10-k}\right), \quad \forall x : 0 \le x \le (N = 10)$$

The Degree of Our Knowledge of $z_j(x)$:

$$DOK_j(x) = |z_j(x)|^2 = P_{rj}^2(x) + [P_{mj}(x)/i]^2 = \left(\sum_{k=0}^{x} {}_N C_k p^k q^{N-k}\right)^2 + \left(1 - \sum_{k=0}^{x} {}_N C_k p^k q^{N-k}\right)^2$$

$$= 1 + 2iP_{rj}(x)P_{mj}(x) = 1 - 2P_{rj}(x)[1 - P_{rj}(x)] = 1 - 2P_{rj}(x) + 2P_{rj}^2(x)$$

$$= 1 - 2\left(\sum_{k=0}^{x} {}_N C_k p^k q^{N-k}\right) + 2\left(\sum_{k=0}^{x} {}_N C_k p^k q^{N-k}\right)^2$$

$$= 1 - 2\left(\sum_{k=0}^{x} {}_{10}C_k p^k q^{10-k}\right) + 2\left(\sum_{k=0}^{x} {}_{10}C_k p^k q^{10-k}\right)^2, \quad \forall x : 0 \le x \le (N = 10)$$

$DOK_j(x)$ is equal to 1 when $P_{rj}(x) = P_{rj}(L_b = 0) = 0$ and when $P_{rj}(x) = P_{rj}(U_b = 10) = 1$.

The Degree of Our Knowledge of $z_j^*(x)$:

$$
\begin{aligned}
DOK_j^*(x) &= \left|z_j^*(x)\right|^2 = \left[P_{rj}^*(x)\right]^2 + \left[P_{mj}^*(x)/i\right]^2 \\
&= \left[1 - P_{rj}(x)\right]^2 + \left[\frac{i - P_{mj}(x)}{i}\right]^2 = \left(1 - \sum_{k=0}^{x} {}_N C_k p^k q^{N-k}\right)^2 + \left(\sum_{k=0}^{x} {}_N C_k p^k q^{N-k}\right)^2 \\
&= 1 + 2iP_{rj}(x)P_{mj}(x) = 1 - 2P_{rj}(x)\left[1 - P_{rj}(x)\right] = 1 - 2P_{rj}(x) + 2P_{rj}^2(x) \\
&= 1 - 2\left(\sum_{k=0}^{x} {}_N C_k p^k q^{N-k}\right) + 2\left(\sum_{k=0}^{x} {}_N C_k p^k q^{N-k}\right)^2 \\
&= 1 - 2\left(\sum_{k=0}^{x} {}_{10} C_k p^k q^{10-k}\right) + 2\left(\sum_{k=0}^{x} {}_{10} C_k p^k q^{10-k}\right)^2, \quad \forall x : 0 \leq x \leq (N = 10) \\
&= DOK_j(x)
\end{aligned}
$$

$DOK_j^*(x)$ is equal to 1 when $P_{rj}(x) = P_{rj}(L_b = 0) = 0$ and when $P_{rj}(x) = P_{rj}(U_b = 10) = 1$.

The Chaotic Factor of $z_j(x)$:

$$
\begin{aligned}
Chf_j(x) &= 2iP_{rj}(x)P_{mj}(x) = -2P_{rj}(x)\left[1 - P_{rj}(x)\right] = -2P_{rj}(x) + 2P_{rj}^2(x) \\
&= -2\left(\sum_{k=0}^{x} {}_N C_k p^k q^{N-k}\right) + 2\left(\sum_{k=0}^{x} {}_N C_k p^k q^{N-k}\right)^2 \\
&= -2\left(\sum_{k=0}^{x} {}_{10} C_k p^k q^{10-k}\right) + 2\left(\sum_{k=0}^{x} {}_{10} C_k p^k q^{10-k}\right)^2, \quad \forall x : 0 \leq x \leq (N = 10)
\end{aligned}
$$

$Chf_j(x)$ is null when $P_{rj}(x) = P_{rj}(L_b = 0) = 0$ and when $P_{rj}(x) = P_{rj}(U_b = 10) = 1$.

The Chaotic Factor of $z_j^*(x)$:

$$
\begin{aligned}
Chf_j^*(x) &= 2iP_{rj}^*(x)P_{mj}^*(x) = 2i\left[1 - P_{rj}(x)\right]\left[i - P_{mj}(x)\right] = -2\left[1 - P_{rj}(x)\right]P_{rj}(x) \\
&= -2P_{rj}(x) + 2P_{rj}^2(x) \\
&= -2\left(\sum_{k=0}^{x} {}_N C_k p^k q^{N-k}\right) + 2\left(\sum_{k=0}^{x} {}_N C_k p^k q^{N-k}\right)^2 \\
&= -2\left(\sum_{k=0}^{x} {}_{10} C_k p^k q^{10-k}\right) + 2\left(\sum_{k=0}^{x} {}_{10} C_k p^k q^{10-k}\right)^2, \quad \forall x : 0 \leq x \leq (N = 10) \\
&= Chf_j(x)
\end{aligned}
$$

$Chf_j^*(x)$ is null when $P_{rj}(x) = P_{rj}(L_b = 0) = 0$ and when $P_{rj}(x) = P_{rj}(U_b = 10) = 1$.

The Magnitude of the Chaotic Factor of $z_j(x)$:

$$MChf_j(x) = \left|Chf_j(x)\right| = -2iP_{rj}(x)P_{mj}(x) = 2P_{rj}(x)\left[1 - P_{rj}(x)\right] = 2P_{rj}(x) - 2P_{rj}^2(x)$$
$$= 2\left(\sum_{k=0}^{x} {}_NC_k p^k q^{N-k}\right) - 2\left(\sum_{k=0}^{x} {}_NC_k p^k q^{N-k}\right)^2$$
$$= 2\left(\sum_{k=0}^{x} {}_{10}C_k p^k q^{10-k}\right) - 2\left(\sum_{k=0}^{x} {}_{10}C_k p^k q^{10-k}\right)^2, \quad \forall x : 0 \le x \le (N = 10)$$

$MChf_j(x)$ is null when $P_{rj}(x) = P_{rj}(L_b = 0) = 0$ and when $P_{rj}(x) = P_{rj}(U_b = 10) = 1$.

The Magnitude of the Chaotic Factor of $z_j^*(x)$:

$$MChf_j^*(x) = \left|Chf_j^*(x)\right| = -2iP_{rj}^*(x)P_{mj}^*(x)$$
$$= -2i\left[1 - P_{rj}(x)\right]\left[i - P_{mj}(x)\right] = 2\left[1 - P_{rj}(x)\right]P_{rj}(x) = 2P_{rj}(x) - 2P_{rj}^2(x)$$
$$= 2\left(\sum_{k=0}^{x} {}_NC_k p^k q^{N-k}\right) - 2\left(\sum_{k=0}^{x} {}_NC_k p^k q^{N-k}\right)^2$$
$$= 2\left(\sum_{k=0}^{x} {}_{10}C_k p^k q^{10-k}\right) - 2\left(\sum_{k=0}^{x} {}_{10}C_k p^k q^{10-k}\right)^2, \quad \forall x : 0 \le x \le (N = 10)$$
$$= MChf_j(x)$$

$MChf_j^*(x)$ is null when $P_{rj}(x) = P_{rj}(L_b = 0) = 0$ and when $P_{rj}(x) = P_{rj}(U_b = 10) = 1$.

At any value of x: $\forall x : (L_b = 0) \le x \le (U_b = N = 10)$, the probability expressed in the complex probability set $\mathcal{C} = \mathcal{R} + \mathcal{M}$ is the following:

$$Pc_j^2(x) = \left[P_{rj}(x) + P_{mj}(x)/i\right]^2 = \left|z_j(x)\right|^2 - 2iP_{rj}(x)P_{mj}(x)$$
$$= DOK_j(x) - Chf_j(x)$$
$$= DOK_j(x) + MChf_j(x)$$
$$= 1$$

then,

$$Pc_j^2(x) = \left[P_{rj}(x) + P_{mj}(x)/i\right]^2 = \left\{P_{rj}(x) + \left[1 - P_{rj}(x)\right]\right\}^2 = 1^2 = 1$$
$$\Leftrightarrow Pc_j(x) = 1 \text{ always.}$$

And

$$Pc_j^*(x)\Big|^2 = \left[P_{rj}^*(x) + P_{mj}^*(x)/i\right]^2 = \left\{\left[1 - P_{rj}(x)\right] + \left[\frac{i - P_{mj}(x)}{i}\right]\right\}^2$$
$$= \left|z_j^*(x)\right|^2 - 2i\left[1 - P_{rj}(x)\right]\left[i - P_{mj}(x)\right] = \left|z_j^*(x)\right|^2 - 2iP_{rj}^*(x)P_{mj}^*(x)$$
$$= DOK_j^*(x) - Chf_j^*(x)$$
$$= DOK_j^*(x) + MChf_j^*(x)$$
$$= 1$$

then,

$$\begin{aligned}
\left|Pc_j^*(x)\right|^2 &= \left[P_{rj}^*(x) + P_{mj}^*(x)/i\right]^2 \\
&= \left\{\left[1 - P_{rj}(x)\right] + \left[\frac{i - P_{mj}(x)}{i}\right]\right\}^2 = \left\{\left[1 - P_{rj}(x)\right] + \left[\frac{i - i\left[1 - P_{rj}(x)\right]}{i}\right]\right\}^2 \\
&= \left\{\left[1 - P_{rj}(x)\right] + \left[\frac{iP_{rj}(x)}{i}\right]\right\}^2 \\
&= \left\{\left[1 - P_{rj}(x)\right] + P_{rj}(x)\right\}^2 = 1^2 = 1 \Leftrightarrow Pc_j^*(x) = 1 \text{ always}
\end{aligned}$$

Hence, the prediction of all the probabilities and of Bayes' theorem in the universe $\mathcal{C} = \mathcal{R} + \mathcal{M}$ is permanently certain and perfectly deterministic (**Figure 3**).

6.1.1 The Complex Probability Cubes.

In the first cube (**Figure 4**), the simulation of *DOK* and *Chf* as functions of each other and of the random variable *X* for the binomial probability distribution can be seen. The thick line in cyan is the projection of the plane $Pc^2(X) = DOK(X) - Chf(X) = 1 = Pc(X)$ on the plane $X = L_b$ = lower bound of $X = 0$. This thick line starts at the point J ($DOK = 1$, $Chf = 0$) when $X = L_b = 0$, reaches the point ($DOK = 0.5$, $Chf = -0.5$) when $X = 5$, and returns at the end to J ($DOK = 1$, $Chf = 0$) when $X = U_b$ = upper bound of $X = 10$. The other curves are the graphs of *DOK*(*X*) (red) and *Chf*(*X*) (green, blue, pink) in different simulation planes. Notice that they all

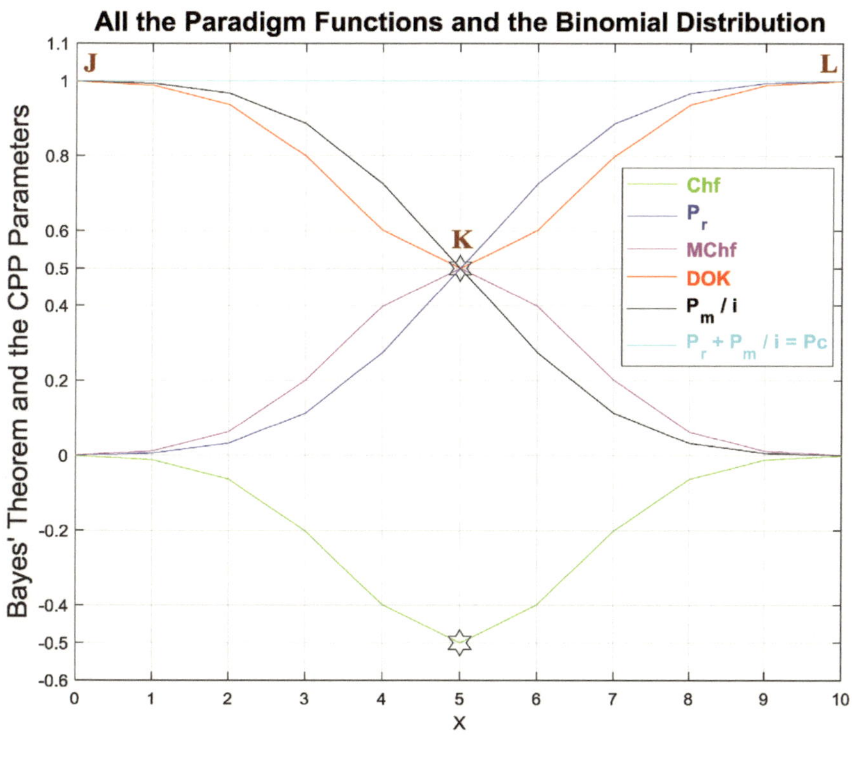

Figure 3.
The graphs of all the CPP *parameters as functions of the random variable* X *for this discrete binomial probability distribution.*

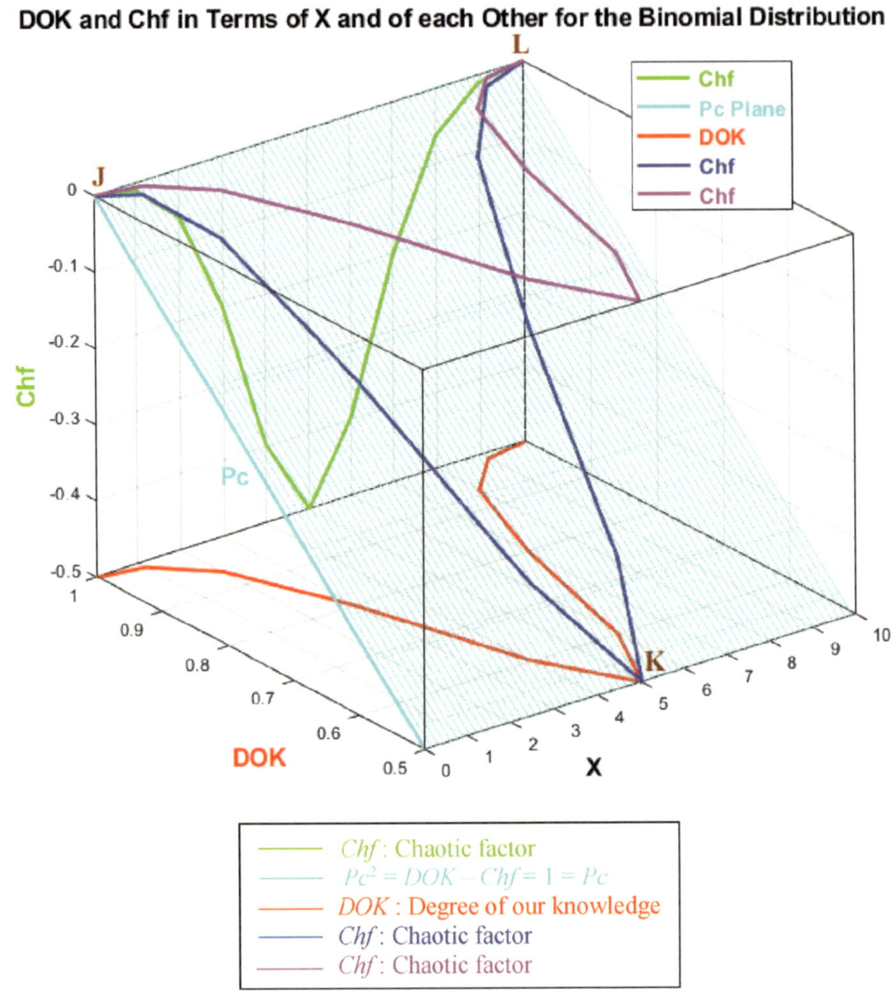

Figure 4.
The graphs of DOK *and of* Chf *and of* Pc *in terms of* X *and of each other for this binomial probability distribution.*

have a minimum at the point K ($DOK = 0.5$, $Chf = -0.5$, $X = 5$). The point L corresponds to ($DOK = 1$, $Chf = 0$, $X = U_b = 10$). The three points J, K, L are the same as in **Figure 3**.

In the second cube (**Figure 5**), we can notice the simulation of the real probability $P_r(X)$ in $\mathcal{R}$ and its complementary real probability $P_m(X)/i$ in $\mathcal{R}$ also in terms of the random variable X for the binomial probability distribution. The thick line in cyan is the projection of the plane $Pc^2(X) = P_r(X) + P_m(X)/i = 1 = Pc(X)$ on the plane $X = L_b$ = lower bound of $X = 0$. This thick line starts at the point ($P_r = 0$, $P_m/i = 1$) and ends at the point ($P_r = 1$, $P_m/i = 0$). The red curve represents $P_r(X)$ in the plane $P_r(X) = P_m(X)/i$ in light grey. This curve starts at the point J ($P_r = 0$, $P_m/i = 1$, $X = L_b$ = lower bound of $X = 0$), reaches the point K ($P_r = 0.5$, $P_m/i = 0.5$, $X = 5$), and gets at the end to L ($P_r = 1$, $P_m/i = 0$, $X = U_b$ = upper bound of $X = 10$). The blue curve represents $P_m(X)/i$ in the plane in cyan $P_r(X) + P_m(X)/i = 1 = Pc(X)$. Notice the importance of the point K which is the intersection of the red and blue curves at $X = 5$ and when $P_r(X) = P_m(X)/i = 0.5$. The three points J, K, L are the same as in **Figure 3**.

In the third cube (**Figure 6**), we can notice the simulation of the complex probability $z(X)$ in $\mathcal{C} = \mathcal{R} + \mathcal{M}$ as a function of the real probability $P_r(X) = \mathrm{Re}(z)$

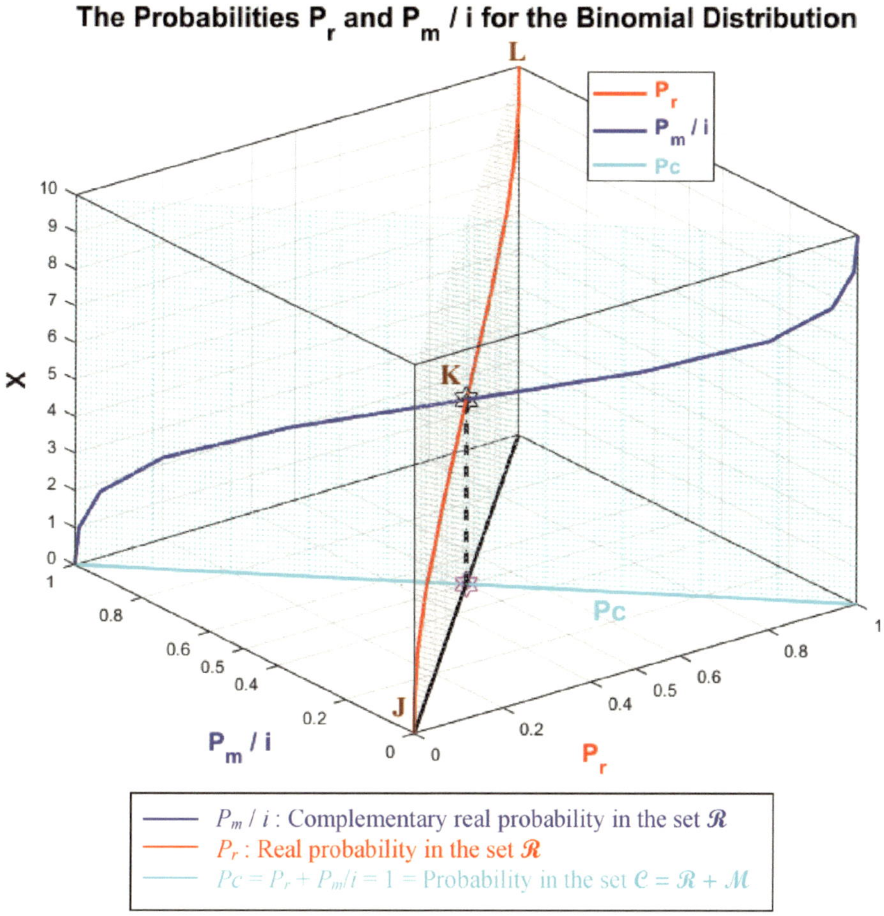

Figure 5.
The graphs of P_r *and of* P_m/i *and of* Pc *in terms of* X *and of each other for this binomial probability distribution.*

in $\mathcal{R}$ and of its complementary imaginary probability $P_m(X) = i \times \text{Im}(z)$ in $\mathcal{M}$, and this in terms of the random variable X for the binomial probability distribution. The red curve represents $P_r(X)$ in the plane $P_m(X) = 0$ and the blue curve represents $P_m(X)$ in the plane $P_r(X) = 0$. The green curve represents the complex probability $z(X) = P_r(X) + P_m(X) = \text{Re}(z) + i \times \text{Im}(z)$ in the plane $P_r(X) = iP_m(X) + 1$ or $z(X)$ plane in cyan. The curve of $z(X)$ starts at the point J ($P_r = 0, P_m = i, X = L_b =$ lower bound of $X = 0$) and ends at the point L ($P_r = 1, P_m = 0, X = U_b =$ upper bound of $X = 10$). The thick line in cyan is $P_r(X = L_b = 0) = iP_m(X = L_b = 0) + 1$ and it is the projection of the $z(X)$ curve on the complex probability plane whose equation is $X = L_b = 0$. This projected thick line starts at the point J ($P_r = 0, P_m = i, X = L_b = 0$) and ends at the point ($P_r = 1, P_m = 0, X = L_b = 0$). Notice the importance of the point K corresponding to $X = 5$ and $z = 0.5 + 0.5i$ when $P_r = 0.5$ and $P_m = 0.5i$. The three points J, K, L are the same as in **Figure 3**.

6.2 Simulation of the continuous standard Gaussian normal probability distribution

The probability density function (*PDF*) of this continuous stochastic distribution is:

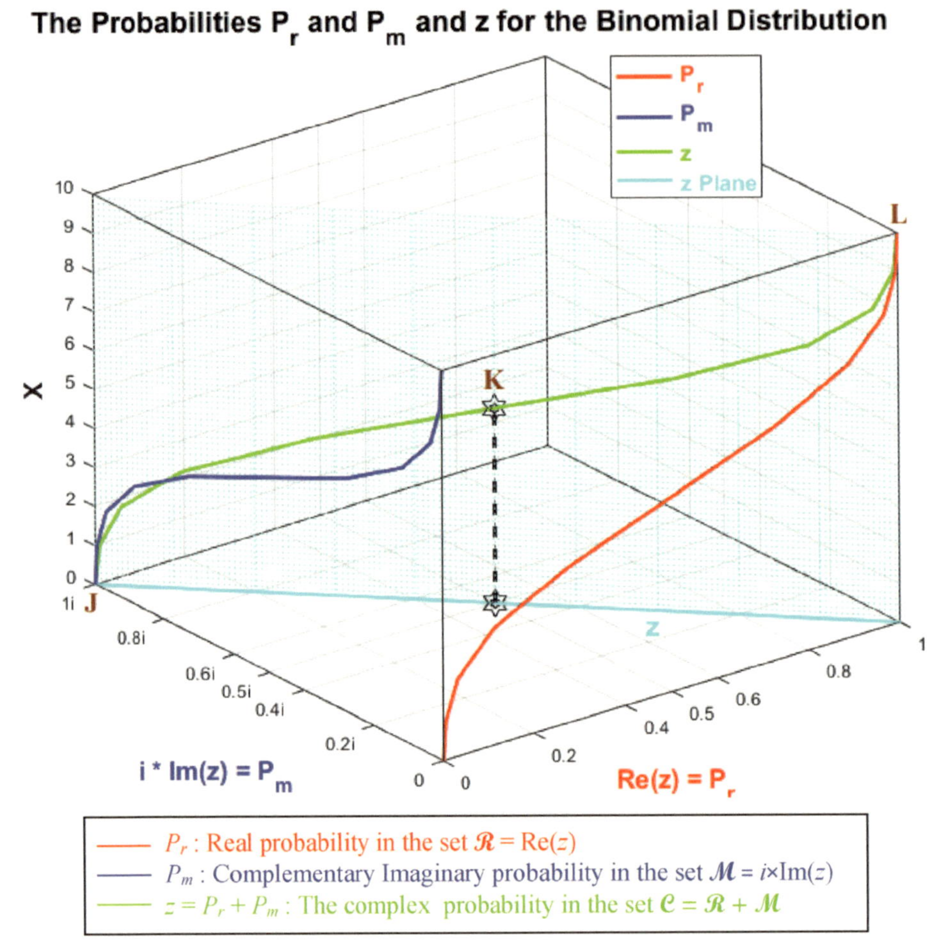

Figure 6.
The graphs of P_r *and of* P_m *and of* z *in terms of* X *for this binomial probability distribution.*

$$f(x) = \frac{d\,[CDF(x)]}{dx} = \frac{1}{\sqrt{2\pi}}\exp\left(-\frac{x^2}{2}\right), \text{for } -\infty < x < \infty$$

and the cumulative distribution function (CDF) is:

$$CDF(x) = P_{rob}(X \le x) = \int_{-\infty}^{x} f(t)dt = \int_{-\infty}^{x} \frac{1}{\sqrt{2\pi}}\exp\left(-\frac{t^2}{2}\right) dt$$

The domain for this standard Gaussian normal variable is considered in the simulations to be equal to: $x \in [L_b = -4, U_b = 4]$ and I have taken $dx = 0.01$.

In the simulations, the mean of this standard normal random distribution is $\mu = 0$.

The variance is $\sigma^2 = 1$.

The standard deviation is $\sigma = 1$.

The median is $Md = 0$.

The mode for this symmetric distribution is = 0 = Md = μ.

The real probability $P_{rj}(x)$ is:

$$P_{rj}(x) = CDF(x) = \int_{-\infty}^{x} \frac{1}{\sqrt{2\pi}} \exp\left(-\frac{t^2}{2}\right) dt = \int_{-4}^{x} \frac{1}{\sqrt{2\pi}} \exp\left(-\frac{t^2}{2}\right) dt, \quad \forall x : -4 \leq x \leq 4$$

$$\Rightarrow P_{rob}\left(A_j/B_j\right) = P_{rob}\left(A_j/\overline{B}_j\right) = P_{rob}\left(A_j\right) = P_{rj}(x) = \int_{-4}^{x} \frac{1}{\sqrt{2\pi}} \exp\left(-\frac{t^2}{2}\right) dt$$

The imaginary complementary probability $P_{mj}(x)$ to $P_{rj}(x)$ is:

$$P_{mj}(x) = i\left[1 - P_{rj}(x)\right] = i[1 - CDF(x)] = i\left[1 - \int_{-\infty}^{x} f(t)dt\right]$$

$$= i\left[\int_{x}^{+\infty} f(t)dt\right] = i\left[\int_{x}^{+\infty} \frac{1}{\sqrt{2\pi}} \exp\left(-\frac{t^2}{2}\right) dt\right] = i\left[\int_{x}^{4} \frac{1}{\sqrt{2\pi}} \exp\left(-\frac{t^2}{2}\right) dt\right],$$
$$\forall x : -4 \leq x \leq 4$$

$$\Rightarrow P_{rob}\left(B_j/A_j\right) = P_{rob}\left(B_j/\overline{A}_j\right) = P_{rob}\left(B_j\right) = P_{mj}(x) = i\left[\int_{x}^{4} \frac{1}{\sqrt{2\pi}} \exp\left(-\frac{t^2}{2}\right) dt\right]$$

The real complementary probability $P^*_{rj}(x)$ to $P_{rj}(x)$ is:

$$P^*_{rj}(x) = 1 - P_{rj}(x) = P_{mj}(x)/i = 1 - CDF(x) = 1 - \int_{-\infty}^{x} f(t)dt = \int_{x}^{+\infty} f(t)dt$$
$$= \int_{x}^{4} \frac{1}{\sqrt{2\pi}} \exp\left(-\frac{t^2}{2}\right) dt, \ \forall x : -4 \leq x \leq 4$$

$$\Rightarrow P_{rob}\left(\overline{A}_j/B_j\right) = P_{rob}\left(\overline{A}_j/\overline{B}_j\right) = P_{rob}\left(\overline{A}_j\right) = P^*_{rj}(x) = \int_{x}^{4} \frac{1}{\sqrt{2\pi}} \exp\left(-\frac{t^2}{2}\right) dt$$

The imaginary complementary probability $P^*_{mj}(x)$ to $P_{mj}(x)$ is:

$$P^*_{mj}(x) = i - P_{mj}(x) = i - i\left[1 - P_{rj}(x)\right] = iP_{rj}(x) = iCDF(x) = i\int_{-\infty}^{x} f(t)dt$$
$$= i\left[\int_{-4}^{x} f(t)dt\right] = i\left[\int_{-4}^{x} \frac{1}{\sqrt{2\pi}} \exp\left(-\frac{t^2}{2}\right) dt\right], \quad \forall x : -4 \leq x \leq 4$$

$$\Rightarrow P_{rob}\left(\overline{B}_j/A_j\right) = P_{rob}\left(\overline{B}_j/\overline{A}_j\right) = P_{rob}\left(\overline{B}_j\right) = P^*_{mj}(x) = i\left[\int_{-4}^{x} \frac{1}{\sqrt{2\pi}} \exp\left(-\frac{t^2}{2}\right) dt\right]$$

The complex probability or random vectors are:

$$z_j(x) = P_{rj}(x) + P_{mj}(x) = \int_{-4}^{x} \frac{1}{\sqrt{2\pi}} \exp\left(-\frac{t^2}{2}\right) dt + i\left[1 - \int_{-4}^{x} \frac{1}{\sqrt{2\pi}} \exp\left(-\frac{t^2}{2}\right) dt\right]$$

$$= \left[\int_{-4}^{x} \frac{1}{\sqrt{2\pi}} \exp\left(-\frac{t^2}{2}\right) dt\right] + i\left[\int_{x}^{4} \frac{1}{\sqrt{2\pi}} \exp\left(-\frac{t^2}{2}\right) dt\right], \quad \forall x : -4 \leq x \leq 4$$

$$z_j^*(x) = P_{rj}^*(x) + P_{mj}^*(x) = \left[1 - P_{rj}(x)\right] + \left[i - P_{mj}(x)\right]$$

$$= \left[1 - \int_{-4}^{x} \frac{1}{\sqrt{2\pi}} \exp\left(-\frac{t^2}{2}\right) dt\right] + i\left[\int_{-4}^{x} \frac{1}{\sqrt{2\pi}} \exp\left(-\frac{t^2}{2}\right) dt\right]$$

$$= \left[\int_{x}^{4} \frac{1}{\sqrt{2\pi}} \exp\left(-\frac{t^2}{2}\right) dt\right] + i\left[\int_{-4}^{x} \frac{1}{\sqrt{2\pi}} \exp\left(-\frac{t^2}{2}\right) dt\right], \quad \forall x : -4 \leq x \leq 4$$

The Degree of Our Knowledge of $z_j(x)$:

$$DOK_j(x) = |z_j(x)|^2 = P_{rj}^2(x) + \left[P_{mj}(x)/i\right]^2 = \left[\int_{-4}^{x} \frac{1}{\sqrt{2\pi}} \exp\left(-\frac{t^2}{2}\right) dt\right]^2$$

$$+ \left[1 - \int_{-4}^{x} \frac{1}{\sqrt{2\pi}} \exp\left(-\frac{t^2}{2}\right) dt\right]^2$$

$$= 1 + 2iP_{rj}(x)P_{mj}(x) = 1 - 2P_{rj}(x)\left[1 - P_{rj}(x)\right] = 1 - 2P_{rj}(x) + 2P_{rj}^2(x)$$

$$= 1 - 2\left[\int_{-4}^{x} \frac{1}{\sqrt{2\pi}} \exp\left(-\frac{t^2}{2}\right) dt\right] + 2\left[\int_{-4}^{x} \frac{1}{\sqrt{2\pi}} \exp\left(-\frac{t^2}{2}\right) dt\right]^2, \quad \forall x : -4 \leq x \leq 4$$

$DOK_j(x)$ is equal to 1 when $P_{rj}(x) = P_{rj}(L_b = -4) = 0$ and when $P_{rj}(x) = P_{rj}(U_b = 4) = 1$.

The Degree of Our Knowledge of $z_j^*(x)$:

$$DOK_j^*(x) = \left|z_j^*(x)\right|^2 = \left[P_{rj}^*(x)\right]^2 + \left[P_{mj}^*(x)/i\right]^2 = \left[1 - P_{rj}(x)\right]^2 + \left[\frac{i - P_{mj}(x)}{i}\right]^2$$

$$= \left[1 - \int_{-4}^{x} \frac{1}{\sqrt{2\pi}} \exp\left(-\frac{t^2}{2}\right) dt\right]^2 + \left[\int_{-4}^{x} \frac{1}{\sqrt{2\pi}} \exp\left(-\frac{t^2}{2}\right) dt\right]^2$$

$$= 1 + 2iP_{rj}(x)P_{mj}(x) = 1 - 2P_{rj}(x)\left[1 - P_{rj}(x)\right] = 1 - 2P_{rj}(x) + 2P_{rj}^2(x)$$

$$= 1 - 2\left[\int_{-4}^{x} \frac{1}{\sqrt{2\pi}} \exp\left(-\frac{t^2}{2}\right) dt\right] + 2\left[\int_{-4}^{x} \frac{1}{\sqrt{2\pi}} \exp\left(-\frac{t^2}{2}\right) dt\right]^2, \quad \forall x : -4 \leq x \leq 4$$

$$= DOK_j(x)$$

$DOK_j^*(x)$ is equal to 1 when $P_{rj}(x) = P_{rj}(L_b = -4) = 0$ and when $P_{rj}(x) = P_{rj}(U_b = 4) = 1$.

The Chaotic Factor of $z_j(x)$:

$$\begin{aligned}Chf_j(x) &= 2iP_{rj}(x)P_{mj}(x) = -2P_{rj}(x)\left[1 - P_{rj}(x)\right] = -2P_{rj}(x) + 2P_{rj}^2(x)\\ &= -2\left[\int_{-4}^{x} f(t)dt\right] + 2\left[\int_{-4}^{x} f(t)dt\right]^2\\ &= -2\left[\int_{-4}^{x} \frac{1}{\sqrt{2\pi}}\exp\left(-\frac{t^2}{2}\right)dt\right] + 2\left[\int_{-4}^{x} \frac{1}{\sqrt{2\pi}}\exp\left(-\frac{t^2}{2}\right)dt\right]^2, \quad \forall x : -4 \le x \le 4\end{aligned}$$

$Chf_j(x)$ is null when $P_{rj}(x) = P_{rj}(L_b = -4) = 0$ and when $P_{rj}(x) = P_{rj}(U_b = 4) = 1$.

The Chaotic Factor of $z_j^*(x)$:

$$\begin{aligned}Chf_j^*(x) &= 2iP_{rj}^*(x)P_{mj}^*(x) = 2i\left[1 - P_{rj}(x)\right]\left[i - P_{mj}(x)\right] = -2\left[1 - P_{rj}(x)\right]P_{rj}(x) = -2P_{rj}(x) + 2P_{rj}^2(x)\\ &= -2\left[\int_{-4}^{x} f(t)dt\right] + 2\left[\int_{-4}^{x} f(t)dt\right]^2\\ &= -2\left[\int_{-4}^{x} \frac{1}{\sqrt{2\pi}}\exp\left(-\frac{t^2}{2}\right)dt\right] + 2\left[\int_{-4}^{x} \frac{1}{\sqrt{2\pi}}\exp\left(-\frac{t^2}{2}\right)dt\right]^2, \quad \forall x : -4 \le x \le 4\\ &= Chf_j(x)\end{aligned}$$

$Chf_j^*(x)$ is null when $P_{rj}(x) = P_{rj}(L_b = -4) = 0$ and when $P_{rj}(x) = P_{rj}(U_b = 4) = 1$.

The Magnitude of the Chaotic Factor of $z_j(x)$:

$$\begin{aligned}MChf_j(x) &= \left|Chf_j(x)\right| = -2iP_{rj}(x)P_{mj}(x) = 2P_{rj}(x)\left[1 - P_{rj}(x)\right] = 2P_{rj}(x) - 2P_{rj}^2(x)\\ &= 2\left[\int_{-4}^{x} f(t)dt\right] - 2\left[\int_{-4}^{x} f(t)dt\right]^2\\ &= 2\left[\int_{-4}^{x} \frac{1}{\sqrt{2\pi}}\exp\left(-\frac{t^2}{2}\right)dt\right] - 2\left[\int_{-4}^{x} \frac{1}{\sqrt{2\pi}}\exp\left(-\frac{t^2}{2}\right)dt\right]^2, \quad \forall x : -4 \le x \le 4\end{aligned}$$

$MChf_j(x)$ is null when $P_{rj}(x) = P_{rj}(L_b = -4) = 0$ and when $P_{rj}(x) = P_{rj}(U_b = 4) = 1$.

The Magnitude of the Chaotic Factor of $z_j^*(x)$:

$$\begin{aligned}MChf_j^*(x) &= \left|Chf_j^*(x)\right| = -2iP_{rj}^*(x)P_{mj}^*(x)\\ &= -2i\left[1 - P_{rj}(x)\right]\left[i - P_{mj}(x)\right] = 2\left[1 - P_{rj}(x)\right]P_{rj}(x) = 2P_{rj}(x) - 2P_{rj}^2(x)\\ &= 2\left[\int_{-4}^{x} f(t)dt\right] - 2\left[\int_{-4}^{x} f(t)dt\right]^2\\ &= 2\left[\int_{-4}^{x} \frac{1}{\sqrt{2\pi}}\exp\left(-\frac{t^2}{2}\right)dt\right] - 2\left[\int_{-4}^{x} \frac{1}{\sqrt{2\pi}}\exp\left(-\frac{t^2}{2}\right)dt\right]^2, \quad \forall x : -4 \le x \le 4\\ &= MChf_j(x)\end{aligned}$$

$MChf_j^*(x)$ is null when $P_{rj}(x) = P_{rj}(L_b = -4) = 0$ and when $P_{rj}(x) = P_{rj}(U_b = 4) = 1$.

At any value of x: $\forall x : (L_b = -4) \leq x \leq (U_b = 4)$, the probability expressed in the complex probability set $\mathcal{C} = \mathcal{R} + \mathcal{M}$ is the following:

$$\begin{aligned} Pc_j^2(x) &= \left[P_{rj}(x) + P_{mj}(x)/i\right]^2 = \left|z_j(x)\right|^2 - 2iP_{rj}(x)P_{mj}(x) \\ &= DOK_j(x) - Chf_j(x) \\ &= DOK_j(x) + MChf_j(x) \\ &= 1 \end{aligned}$$

then,

$$\begin{aligned} &Pc_j^2(x) = \left[P_{rj}(x) + P_{mj}(x)/i\right]^2 = \left\{P_{rj}(x) + \left[1 - P_{rj}(x)\right]\right\}^2 = 1^2 = 1 \\ &\Leftrightarrow Pc_j(x) = 1 \text{ always.} \end{aligned}$$

And

$$\begin{aligned} Pc_j^*(x)\Big|^2 &= \left[P_{rj}^*(x) + P_{mj}^*(x)/i\right]^2 = \left\{\left[1 - P_{rj}(x)\right] + \left[\frac{i - P_{mj}(x)}{i}\right]\right\}^2 \\ &= \left|z_j^*(x)\right|^2 - 2i\left[1 - P_{rj}(x)\right]\left[i - P_{mj}(x)\right] = \left|z_j^*(x)\right|^2 - 2iP_{rj}^*(x)P_{mj}^*(x) \\ &= DOK_j^*(x) - Chf_j^*(x) \\ &= DOK_j^*(x) + MChf_j^*(x) \\ &= 1 \end{aligned}$$

then,

$$\begin{aligned} Pc_j^*(x)\Big|^2 &= \left[P_{rj}^*(x) + P_{mj}^*(x)/i\right]^2 \\ &= \left\{\left[1 - P_{rj}(x)\right] + \left[\frac{i - P_{mj}(x)}{i}\right]\right\}^2 = \left\{\left[1 - P_{rj}(x)\right] + \left[\frac{i - i\left[1 - P_{rj}(x)\right]}{i}\right]\right\}^2 \\ &= \left\{\left[1 - P_{rj}(x)\right] + \left[\frac{iP_{rj}(x)}{i}\right]\right\}^2 \\ &= \left\{\left[1 - P_{rj}(x)\right] + P_{rj}(x)\right\}^2 = 1^2 = 1 \Leftrightarrow Pc_j^*(x) = 1 \text{ always} \end{aligned}$$

Hence, the prediction of all the probabilities and of Bayes' theorem in the universe $\mathcal{C} = \mathcal{R} + \mathcal{M}$ is permanently certain and perfectly deterministic (**Figure 7**).

6.2.1 The complex probability cubes

In the first cube (**Figure 8**), the simulation of *DOK* and *Chf* as functions of each other and of the random variable *X* for the standard Gaussian normal probability distribution can be seen. The thick line in cyan is the projection of the plane $Pc^2(X) = DOK(X) - Chf(X) = 1 = Pc(X)$ on the plane $X = L_b$ = lower bound of $X = -4$. This thick line starts at the point J ($DOK = 1$, $Chf = 0$) when $X = L_b = -4$, reaches the point ($DOK = 0.5$, $Chf = -0.5$) when $X = 0$, and returns at the end to J ($DOK = 1$, $Chf = 0$) when $X = U_b$ = upper bound of $X = 4$. The other curves are the graphs of

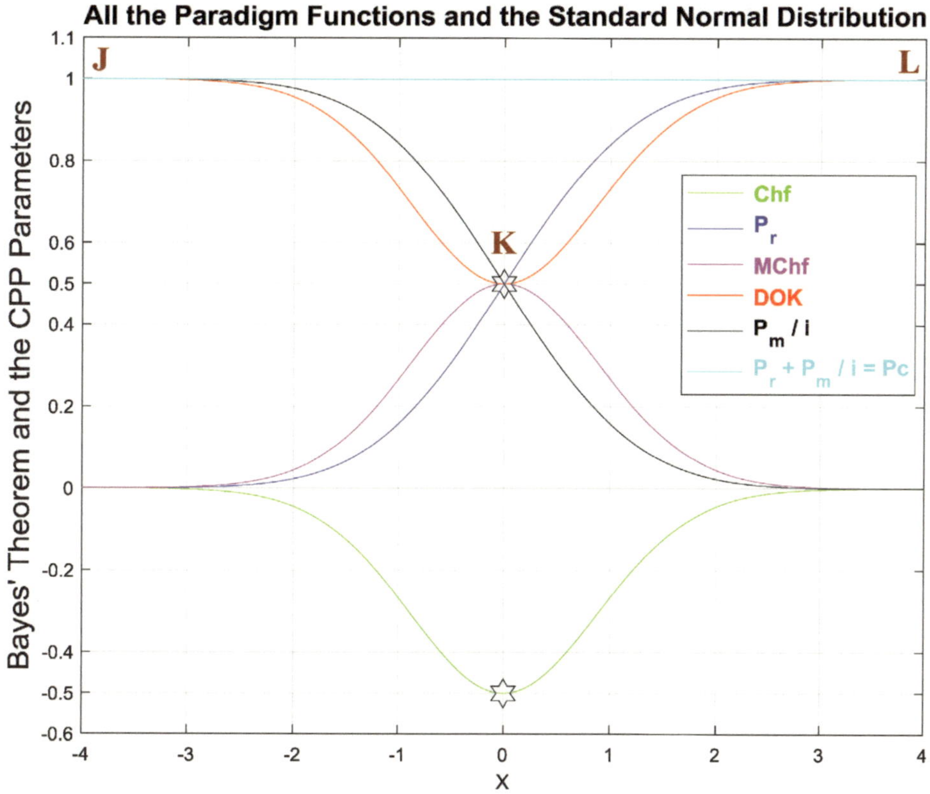

Figure 7.
The graphs of all the CPP *parameters as functions of the random variable* X *for the continuous standard Gaussian normal distribution.*

$DOK(X)$ (red) and $Chf(X)$ (green, blue, pink) in different simulation planes. Notice that they all have a minimum at the point K ($DOK = 0.5$, $Chf = -0.5$, $X = 0$). The point L corresponds to ($DOK = 1$, $Chf = 0$, $X = U_b = 4$). The three points J, K, L are the same as in **Figure 7**.

In the second cube (**Figure 9**), we can notice the simulation of the real probability $P_r(X)$ in $\mathcal{R}$ and its complementary real probability $P_m(X)/i$ in $\mathcal{R}$ also in terms of the random variable X for the standard Gaussian normal probability distribution. The thick line in cyan is the projection of the plane $Pc^2(X) = P_r(X) + P_m(X)/i = 1 = Pc(X)$ on the plane $X = L_b$ = lower bound of $X = -4$. This thick line starts at the point ($P_r = 0$, $P_m/i = 1$) and ends at the point ($P_r = 1$, $P_m/i = 0$). The red curve represents $P_r(X)$ in the plane $P_r(X) = P_m(X)/i$ in light grey. This curve starts at the point J ($P_r = 0$, $P_m/i = 1$, $X = L_b$ = lower bound of $X = -4$), reaches the point K ($P_r = 0.5$, $P_m/i = 0.5$, $X = 0$), and gets at the end to L ($P_r = 1$, $P_m/i = 0$, $X = U_b$ = upper bound of $X = 4$). The blue curve represents $P_m(X)/i$ in the plane in cyan $P_r(X) + P_m(X)/i = 1 = Pc(X)$. Notice the importance of the point K which is the intersection of the red and blue curves at $X = 0$ and when $P_r(X) = P_m(X)/i = 0.5$. The three points J, K, L are the same as in **Figure 7**.

In the third cube (**Figure 10**), we can notice the simulation of the complex probability $z(X)$ in $\mathcal{C} = \mathcal{R} + \mathcal{M}$ as a function of the real probability $P_r(X) = \text{Re}(z)$ in $\mathcal{R}$ and of its complementary imaginary probability $P_m(X) = i \times \text{Im}(z)$ in $\mathcal{M}$, and this in terms of the random variable X for the standard Gaussian normal probability distribution. The red curve represents $P_r(X)$ in the plane $P_m(X) = 0$ and the blue curve represents $P_m(X)$ in the plane $P_r(X) = 0$. The green curve represents the complex probability $z(X) = P_r(X) + P_m(X) = \text{Re}(z) + i \times \text{Im}(z)$ in the plane

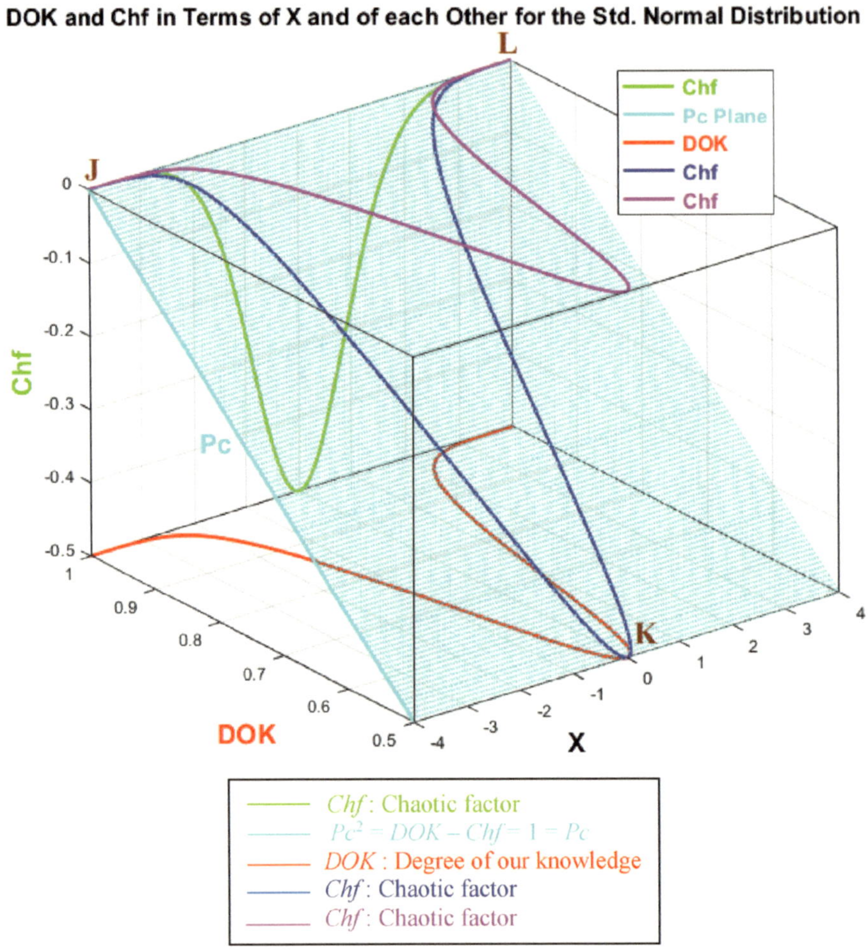

Figure 8.
The graphs of DOK *and of* Chf *and of* Pc *in terms of* X *and of each other for the standard Gaussian normal probability distribution.*

$P_r(X) = iP_m(X) + 1$ or $z(X)$ plane in cyan. The curve of $z(X)$ starts at the point J ($P_r = 0$, $P_m = i$, $X = L_b$ = lower bound of $X = -4$) and ends at the point L ($P_r = 1$, $P_m = 0$, $X = U_b$ = upper bound of $X = 4$). The thick line in cyan is $P_r(X = L_b = -4) = iP_m(X = L_b = -4) + 1$ and it is the projection of the $z(X)$ curve on the complex probability plane whose equation is $X = L_b = -4$. This projected thick line starts at the point J ($P_r = 0$, $P_m = i$, $X = L_b = -4$) and ends at the point ($P_r = 1$, $P_m = 0$, $X = L_b = -4$). Notice the importance of the point K corresponding to $X = 0$ and $z = 0.5 + 0.5i$ when $P_r = 0.5$ and $P_m = 0.5i$. The three points J, K, L are the same as in **Figure** 7.

7. Conclusion and perspectives

In the current research work, the original extended model of eight axioms (EKA) of A. N. Kolmogorov was connected and applied to the classical Bayes' theorem. Thus, a tight link between this theorem and the novel paradigm was achieved. Consequently, the model of "Complex Probability" was more developed beyond the scope of my seventeen previous research works on this topic.

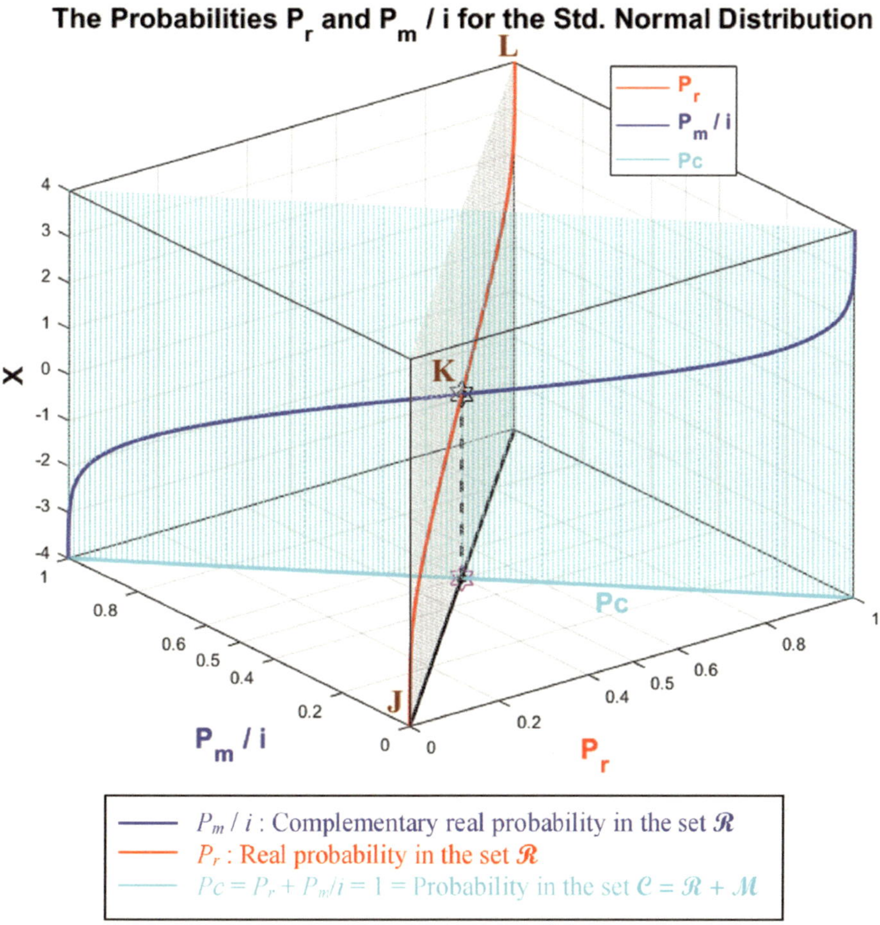

Figure 9.
The graphs of P_r *and of* P_m/i *and of* Pc *in terms of* X *and of each other for the standard Gaussian normal probability distribution.*

Additionally, as it was proved and verified in the novel model, before the beginning of the random phenomenon simulation and at its end we have the chaotic factor (*Chf* and *MChf*) is zero and the degree of our knowledge (*DOK*) is one since the stochastic fluctuations and effects have either not started yet or they have terminated and finished their task on the probabilistic phenomenon. During the execution of the nondeterministic phenomenon and experiment we also have: $0.5 \leq DOK < 1$, $-0.5 \leq Chf < 0$, and $0 < MChf \leq 0.5$. We can see that during this entire process we have incessantly and continually $Pc^2 = DOK - Chf = DOK + MChf = 1 = Pc$, that means that the simulation which behaved randomly and stochastically in the set $\mathcal{R}$ is now certain and deterministic in the probability set $\mathcal{C} = \mathcal{R} + \mathcal{M}$, and this after adding to the random experiment executed in $\mathcal{R}$ the contributions of the set $\mathcal{M}$ and hence after eliminating and subtracting the chaotic factor from the degree of our knowledge. Furthermore, the real, imaginary, complex, and deterministic probabilities that correspond to each value of the random variable X have been determined in the three probabilities sets which are $\mathcal{R}$, $\mathcal{M}$, and $\mathcal{C}$ by P_r, P_m, z and Pc respectively. Consequently, at each value of X, the novel Bayes' theorem and *CPP* parameters P_r, P_m, P_m/i, *DOK*, *Chf*, *MChf*, *Pc*, and z are surely and perfectly predicted in the complex probabilities set $\mathcal{C}$ with Pc maintained equal to one permanently and repeatedly. In addition, referring to all these obtained graphs and executed simulations throughout the whole research work, we are able

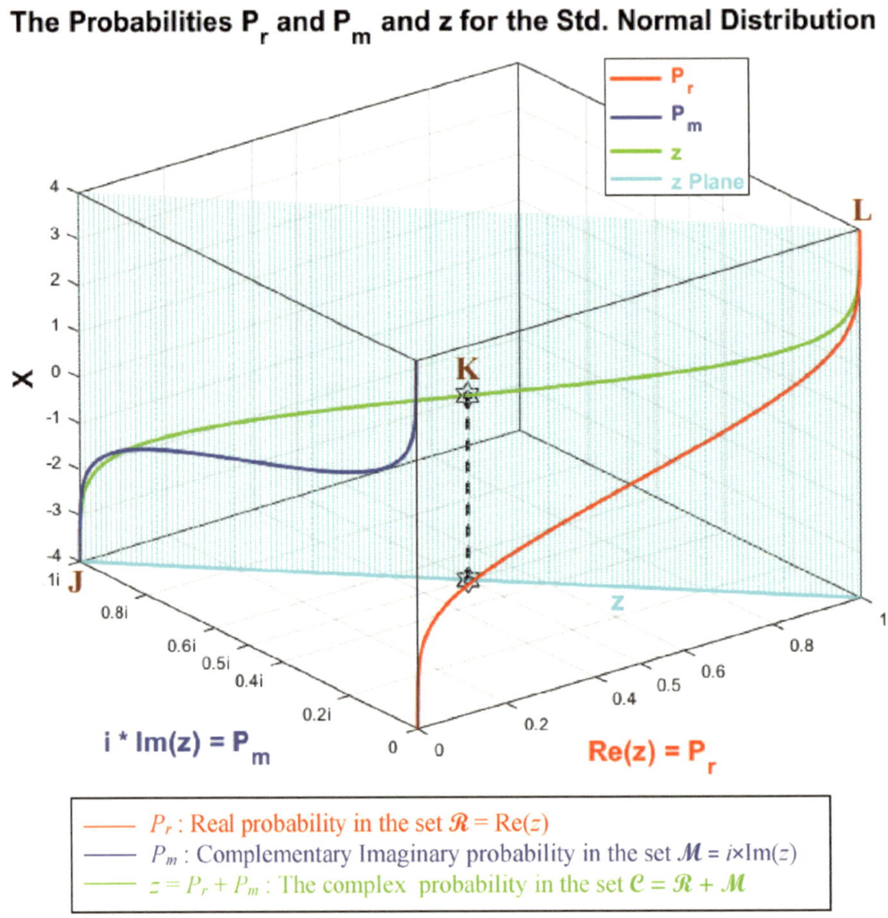

Figure 10.
The graphs of P_r *and of* P_m *and of* z *in terms of* X *for the standard Gaussian normal probability distribution.*

to quantify and to visualize both the system chaos and stochastic effects and influences (expressed and materialized by *Chf* and *MChf*) and the certain knowledge (expressed and materialized by *DOK* and *Pc*) of the new paradigm. This is without any doubt very fruitful, wonderful, and fascinating and proves and reveals once again the advantages of extending A. N. Kolmogorov's five axioms of probability and hence the novelty and benefits of this inventive and original model in the fields of prognostics and applied mathematics that can be called truly: "The Complex Probability Paradigm".

Furthermore, it is very crucial to state that using *CPP*, conditional probabilities, and Bayes' theorem, we have linked and joined and bonded the events probabilities sets $\mathcal{R}$ with $\mathcal{R}$, $\mathcal{M}$ with $\mathcal{M}$, $\mathcal{R}$ with $\mathcal{M}$, $\mathcal{M}$ with $\mathcal{R}$, $\mathcal{R}$ with $\mathcal{C}$, $\mathcal{M}$ with $\mathcal{C}$, and $\mathcal{C}$ with $\mathcal{C}$ using precise and exact mathematical relations and equations. Moreover, it is important to mention here that the novel *CPP* paradigm can be implemented to any probability distribution that exists in literature as it was shown in the simulation section. This will lead without any doubt to analogous and similar conclusions and results and will confirm certainly the success of my innovative and original model.

As a future and prospective research and challenges, we aim to more develop the novel prognostic paradigm conceived and to implement it to a large set of random and nondeterministic events like for other probabilistic phenomena as in stochastic processes and in the classical theory of probability. Additionally, we will apply *CPP* to the random walk problems which have huge and very interesting consequences when implemented to chemistry, to physics, to economics, to applied and pure mathematics.

Nomenclature

$\mathcal{R}$	real set of events
$\mathcal{M}$	imaginary set of events
$\mathcal{C}$	complex set of events
i	the imaginary number where $i = \sqrt{-1}$ or $i^2 = -1$
EKA	Extended Kolmogorov's Axioms
CPP	Complex Probability Paradigm
P_{rob}	probability of any event
P_r	probability in the real set $\mathcal{R}$
P_m	probability in the imaginary set $\mathcal{M}$ corresponding to the real probability in $\mathcal{R}$
Pc	probability of an event in $\mathcal{R}$ with its associated complementary event in $\mathcal{M}$
z	complex probability number = sum of P_r and P_m = complex random vector
$DOK = \|z\|^2$	the degree of our knowledge of the random system or experiment, it is the square of the norm of z
Chf	the chaotic factor of z
$MChf$	magnitude of the chaotic factor of z
N	number of random vectors
Z	the resultant complex random vector = $\sum_{j=1}^{N} z_j$
$DOK_Z = \frac{\|Z\|^2}{N^2}$	the degree of our knowledge of the whole stochastic system
$Chf_Z = \frac{Chf}{N^2}$	the chaotic factor of the whole stochastic system
$MChf_Z$	magnitude of the chaotic factor of the whole stochastic system
Z_U	the resultant complex random vector corresponding to a uniform random distribution
DOK_{Z_U}	the degree of our knowledge of the whole stochastic system corresponding to a uniform random distribution
Chf_{Z_U}	the chaotic factor of the whole stochastic system corresponding to a uniform random distribution
$MChf_{Z_U}$	the magnitude of the chaotic factor of the whole stochastic system corresponding to a uniform random distribution
$Pc\|_{Z_U}$	probability in the complex probability set $\mathcal{C}$ of the whole stochastic system corresponding to a uniform random distribution

Author details

Abdo Abou Jaoudé
Department of Mathematics and Statistics, Faculty of Natural and Applied Sciences, Notre Dame University-Louaize, Lebanon

*Address all correspondence to: abdoaj@idm.net.lb

References

[1] Abou Jaoude, A., El-Tawil, K., & Kadry, S. (2010). "*Prediction in Complex Dimension Using Kolmogorov's Set of Axioms*", Journal of Mathematics and Statistics, Science Publications, vol. 6 (2), pp. 116-124.

[2] Abou Jaoude, A. (2013)."*The Complex Statistics Paradigm and the Law of Large Numbers*", Journal of Mathematics and Statistics, Science Publications, vol. 9(4), pp. 289-304.

[3] Abou Jaoude, A. (2013). "*The Theory of Complex Probability and the First Order Reliability Method*", Journal of Mathematics and Statistics, Science Publications, vol. 9(4), pp. 310-324.

[4] Abou Jaoude, A. (2014). "*Complex Probability Theory and Prognostic*", Journal of Mathematics and Statistics, Science Publications, vol. 10(1), pp. 1-24.

[5] Abou Jaoude, A. (2015). "*The Complex Probability Paradigm and Analytic Linear Prognostic for Vehicle Suspension Systems*", American Journal of Engineering and Applied Sciences, Science Publications, vol. 8(1), pp. 147-175.

[6] Abou Jaoude, A. (2015). "*The Paradigm of Complex Probability and the Brownian Motion*", Systems Science and Control Engineering, Taylor and Francis Publishers, vol. 3(1), pp. 478-503.

[7] Abou Jaoude, A. (2016). "*The Paradigm of Complex Probability and Chebyshev's Inequality*", Systems Science and Control Engineering, Taylor and Francis Publishers, vol. 4(1), pp. 99-137.

[8] Abou Jaoude, A. (2016). "*The Paradigm of Complex Probability and Analytic Nonlinear Prognostic for Vehicle Suspension Systems*", Systems Science and Control Engineering, Taylor and Francis Publishers, vol. 4(1), pp. 99-137.

[9] Abou Jaoude, A. (2017). "*The Paradigm of Complex Probability and Analytic Linear Prognostic for Unburied Petrochemical Pipelines*", Systems Science and Control Engineering, Taylor and Francis Publishers, vol. 5(1), pp. 178-214.

[10] Abou Jaoude, A. (2017). "*The Paradigm of Complex Probability and Claude Shannon's Information Theory*", Systems Science and Control Engineering, Taylor and Francis Publishers, vol. 5(1), pp. 380-425.

[11] Abou Jaoude, A. (2017). "*The Paradigm of Complex Probability and Analytic Nonlinear Prognostic for Unburied Petrochemical Pipelines*", Systems Science and Control Engineering, Taylor and Francis Publishers, vol. 5(1), pp. 495-534.

[12] Abou Jaoude, A. (2018). "*The Paradigm of Complex Probability and Ludwig Boltzmann's Entropy*", Systems Science and Control Engineering, Taylor and Francis Publishers, vol. 6(1), pp. 108-149.

[13] Abou Jaoude, A. (2019). "*The Paradigm of Complex Probability and Monte Carlo Methods*", Systems Science and Control Engineering, Taylor and Francis Publishers, vol. 7(1), pp. 407-451.

[14] Abou Jaoude, A. (2020). "*Analytic Prognostic in the Linear Damage Case Applied to Buried Petrochemical Pipelines and the Complex Probability Paradigm*", Fault Detection, Diagnosis and Prognosis, IntechOpen. DOI: 10.5772/intechopen.90157.

[15] Abou Jaoude, A. (July 7th 2020). "*The Monte Carlo Techniques and The Complex Probability Paradigm*", Forecasting in Mathematics - Recent Advances, New Perspectives and Applications, Abdo Abou Jaoude,

IntechOpen. DOI: 10.5772/intechopen.93048.

[16] Abou Jaoude, A. (2020). "*The Paradigm of Complex Probability and Prognostic Using FORM*", London Journal of Research in Science: Natural and Formal (LJRS), London Journals Press, vol. 20(4), pp. 1-65. Print ISSN: 2631-8490, Online ISSN: 2631-8504, DOI: 10.17472/LJRS, 2020.

[17] Abou Jaoude, A. (2020). "*The Paradigm of Complex Probability and The Central Limit Theorem*", London Journal of Research in Science: Natural and Formal (LJRS), London Journals Press, vol. 20(5), pp. 1-57. Print ISSN: 2631-8490, Online ISSN: 2631-8504, DOI: 10.17472/LJRS, 2020.

[18] Benton, W. (1966). *Probability, Encyclopedia Britannica*. vol. 18, pp. 570-574, Chicago, Encyclopedia Britannica Inc.

[19] Benton, W. (1966). *Mathematical Probability, Encyclopedia Britannica*. vol. 18, pp. 574-579, Chicago, Encyclopedia Britannica Inc.

[20] Feller, W. (1968). *An Introduction to Probability Theory and Its Applications*. 3rd Edition. New York, Wiley.

[21] Walpole, R., Myers, R., Myers, S., & Ye, K. (2002). *Probability and Statistics for Engineers and Scientists*. 7th Edition, New Jersey, Prentice Hall.

[22] Freund, J. E. (1973). *Introduction to Probability*. New York: Dover Publications.

[23] Abou Jaoude, A. (2019). *The Computer Simulation of Monté Carlo Methods and Random Phenomena*. United Kingdom: Cambridge Scholars Publishing.

[24] Abou Jaoude, A. (2019). *The Analysis of Selected Algorithms for the Stochastic Paradigm*. United Kingdom: Cambridge Scholars Publishing.

[25] Abou Jaoude, A. (2020). *The Analysis of Selected Algorithms for the Statistical Paradigm*. The Republic of Moldova: Generis Publishing.

[26] Abou Jaoude, A. (August 1st 2004). Ph.D. Thesis in Applied Mathematics: *Numerical Methods and Algorithms for Applied Mathematicians*. Bircham International University. http://www.bircham.edu.

[27] Abou Jaoude, A. (October 2005). Ph.D. Thesis in Computer Science: *Computer Simulation of Monté Carlo Methods and Random Phenomena*. Bircham International University. http://www.bircham.edu.

[28] Abou Jaoude, A. (27 April 2007). Ph.D. Thesis in Applied Statistics and Probability: *Analysis and Algorithms for the Statistical and Stochastic Paradigm*. Bircham International University. http://www.bircham.edu.

[29] Stuart, A., Ord, K. (1994). *Kendall's Advanced Theory of Statistics: Volume I – Distribution Theory*, Edward Arnold, Section 8.7.

[30] Lee, P. M. (2012). *"Chapter 1". Bayesian Statistics*. Wiley. ISBN 978-1-1183-3257-3.

[31] Bayes, T., & Price, R. (1763). *"An Essay towards solving a Problem in the Doctrine of Chance. By the late Rev. Mr. Bayes, communicated by Mr. Price, in a letter to John Canton, A. M. F. R. S."* (PDF). Philosophical Transactions of the Royal Society of London. 53: pp. 370–418. doi:10.1098/rstl.1763.0053. Archived from the original (PDF) on 2011-04-10. Retrieved 2003-12-27.

[32] Daston, L. (1988). *Classical Probability in the Enlightenment*. Princeton University Press. pp. 268. ISBN 0-691-08497-1.

[33] Stigler, S. M. (1986). "*Inverse Probability*". *The History of Statistics: The*

Measurement of Uncertainty Before 1900. Harvard University Press. pp. 99–138. ISBN 978-0-674-40341-3.

[34] Jeffreys, H. (1973). *Scientific Inference* (3rd edition). Cambridge University Press. pp. 31. ISBN 978-0-521-18078-8.

[35] Stigler, S. M. (1983). "*Who Discovered Bayes' Theorem?*". The American Statistician. 37 (4): pp. 290–296. doi:10.1080/00031305.1983.10483122.

[36] Hooper, M. (2013). "*Richard Price, Bayes' theorem, and God*". Significance. 10 (1): pp. 36–39. doi:10.1111/j.1740-9713.2013.00638.x. S2CID 153704746.

[37] Wikipedia, the free encyclopedia, *Bayes' Theorem*. https://en.wikipedia.org/

Chapter 2

The Paradigm of Complex Probability and Isaac Newton's Classical Mechanics: On the Foundation of Statistical Physics

Abdo Abou Jaoudé

"Imagination is more important than knowledge. Knowledge is limited. Imagination encircles the world."

Albert Einstein.

"Our minds are finite, and yet even in these circumstances of finitude we are surrounded by possibilities that are infinite, and the purpose of life is to grasp as much as we can out of that infinitude."

Alfred North Whitehead.

"The important thing is not to stop questioning. Curiosity has its own reason for existence."

Albert Einstein.

"A theory with mathematical beauty is more likely to be correct than an ugly one that fits some experimental data. God is a mathematician of a very high order, and He used very advanced mathematics in constructing the universe."

Paul Adrien Maurice Dirac.

Abstract

The concept of mathematical probability was established in 1933 by *Andrey Nikolaevich Kolmogorov* by defining a system of five axioms. This system can be enhanced to encompass the imaginary numbers set after the addition of three novel axioms. As a result, any random experiment can be executed in the complex probabilities set $\mathcal{C}$ which is the sum of the real probabilities set $\mathcal{R}$ and the imaginary probabilities set $\mathcal{M}$. We aim here to incorporate supplementary imaginary dimensions to the random experiment occurring in the "real" laboratory in $\mathcal{R}$ and therefore to compute all the probabilities in the sets $\mathcal{R}$, $\mathcal{M}$, and $\mathcal{C}$. Accordingly, the probability in the whole set $\mathcal{C} = \mathcal{R} + \mathcal{M}$ is constantly equivalent to one independently of the distribution of the input random variable in $\mathcal{R}$, and subsequently the output of the stochastic experiment in $\mathcal{R}$ can be determined absolutely in $\mathcal{C}$. This is the consequence of the fact that the probability in $\mathcal{C}$ is computed after the subtraction of the chaotic factor from the degree of our knowledge of the nondeterministic experiment. We will apply this innovative paradigm to Isaac Newton's classical mechanics and to prove as well in an original way an important property at the foundation of statistical physics.

Keywords: Chaotic factor, degree of our knowledge, complex random vector, probability norm, complex probability set, random forces, complex force, resultant force

1. Introduction

Firstly, classical mechanics is a theory in physics studying the macroscopic objects motion whether they are parts of machinery or projectiles or objects in astronomy like for example planets or spacecrafts or galaxies or stars. As it was established, classical mechanics is deterministic that means that we can predict the motion of objects in the future when we know their present state. It is also reversible and that means we can know the motion of objects in the past when we know their present state also [1].

Since classical mechanics was developed at the beginning by Sir Isaac Newton therefore it is usually referred to as Newtonian mechanics. It comprises the mathematical methods and the employed physical concepts developed, as we have mentioned, by Newton, Gottfried Wilhelm Leibniz and others in the seventeenth century to study the bodies motion under the effect of a set of forces. The theory was more developed later on to embody more abstract methods which have led to the reformulations of classical mechanics and hence to the establishment of Hamiltonian mechanics and Lagrangian mechanics. These developments which were done in the eighteenth and nineteenth centuries are substantial extensions beyond the work of Newton because they used more particularly analytical mechanics. After doing some modifications, modern physics makes use of them in all its areas [2].

Moreover, exceptionally precise results are provided by classical mechanics when considering objects with velocities far from the speed of light and when they do not possess extreme masses. It is mandatory to make use of quantum mechanics which is a sub-field of mechanics when studying objects which have an atom diameter size. Additionally, we need Albert Einstein's special relativity when considering speeds near the velocity of light. Furthermore, Einstein's general relativity is applied when objects have huge masses. It is important to note that many modern sources include in classical physics the relativistic mechanics which represents according to them the most precise, developed, and complete form of classical mechanics [3].

Furthermore, we now present classical mechanics fundamental concepts. The theory assumes that the objects of the real world are of negligible size that means that they are point particles. And it also characterizes the point particle motion by few parameters which are: its mass, its position, and the applied forces to it. We will discuss each of these parameters in turn [4].

In fact, and in reality, classical mechanics can describe always the kind of objects that have a non-zero size. Whereas, very small particles like electrons are described more accurately by the physics of quantum mechanics. Additionally, hypothetical point particles have more simplified behavior than non-zero size objects like for example a baseball that can spin when it is in motion. Moreover, such non-zero objects are considered as composite objects constituted of a large number of point particles acting collectively; hence, the point particles results can be used in such large objects study [5].

Common sense notions are used by classical mechanics of how matter and forces interact and exist. Its basic assumption is that energy and matter have knowable and definite attributes such as speed and location in space. Additionally, it is assumed by non-relativistic mechanics the instantaneous action of forces or instantaneous action at a distance [6].

The bodies motion study is very ancient, this makes classical mechanics one of the largest and oldest subjects in engineering, technology, and science [7].

Aristotle, one among antiquity Greek philosophers and who is the founder of Aristotelian physics, may have been the first to postulate that theoretical principles can assist nature understanding and to assume that "everything happens for a reason". Many of these ideas preserved are considered as eminently reasonable by a modern reader but there is an obvious lack of controlled experiment and mathematical theory as we know it. In fact, modern science was formed by these later decisive factors and classical mechanics came to be known as their early application [8].

The medieval mathematician Jordanus de Nemore introduced in his Elementa demonstrationem ponderum the "positional gravity" concept and the component forces use [9].

Johannes Kepler published in 1609 Astronomia nova which was the first published causal explanation of the planets motion. Based on the observations made by Tycho Brahe on Mars orbit, he concluded that the orbits of the planet were ellipses. This epistemological revolution occurred at the same time when Galileo was proposing for objects motion abstract mathematical laws. Perhaps he may have performed the historical experiment of the two cannonballs of different weights dropping from Pisa tower. Hence, he showed that these two cannonballs hit the ground simultaneously. We doubt in fact the reality of that particular experiment, but Galileo conducted quantitative experiments which were to roll balls on an inclined plane. From such experiments results he derived his accelerated motion theory [10].

Sir Isaac Newton laid down classical mechanics foundations by founding his natural philosophy principles on three laws of motion proposed by him: the inertia first law, the acceleration second law, and the action and reaction third law. A proper mathematical and scientific treatment in Philosophiae Naturalis Principia Mathematica of Newton was given to his second and third laws. They are in fact different from the attempts laid earlier to explain similar phenomena and which were either incorrect, incomplete, or they lack a precise mathematical expression. Moreover, the principles of conservation of angular momentum and momentum were postulated by Newton. Additionally, the universal gravitational law of Newton was also provided by him to give the first accurate mathematical and scientific formulation of gravity. The most accurate and fullest description of classical mechanics was provided by the combination of the laws of motion and gravitation of Newton. Newton showed that his three laws can be applied to the objects of everyday as well to heavenly objects. Particularly, Newton derived a theoretical explanation of the planets' laws of motion of Kepler [11].

Newton performed the mathematical calculation by inventing previously the mathematical calculus. In fact, calculus eclipsed his book, the Principia, which was formulated totally in terms of geometric methods which were long established and to gain hence acceptability. Moreover, the notation of the integral and of the derivative which are preferred today were developed by Leibniz however [12].

All phenomena, including light in the form of geometric optic, can be explained by classical mechanics as it was assumed by Newton and most of his contemporaries, with the notable exception of Christiaan Huygens. Newton maintained his own corpuscular light theory even when they discovered the wave interference phenomenon or the so-called Newton's rings [13].

Classical mechanics became a major field of study in physics as well in mathematics and this after Newton. A far greater number to problems solutions were allowed by several progressive reformulations of his mechanics. Joseph Louis Lagrange was the first to reformulate in 1788 Newtons' mechanics. William Rowan Hamilton in his turn reformulated Lagrangian mechanics in 1833 [14].

More modern physics resolved some difficulties that were discovered in the late nineteenth century. Compatibility with the theory of electromagnetism and the famous Michelson-Morley experiment were some of these difficulties. Often still considered as a part of classical mechanics, the special relativity theory was led by the resolution of these problems [15].

Explaining all thermodynamics, raised another set of difficulties and problems with classical mechanics. Gibbs paradox of classical statistical mechanics was the result of the combination of classical mechanics with thermodynamics. In this paradox, entropy is not a quantity which was well defined. We introduced quanta to explain the black-body radiation otherwise this was not possible. Classical mechanics was unable to explain, not even approximately, such basic things as the sizes of the atoms, the photo-electric effect, and the energy levels and this when experiments delved into the atomic world. Quantum mechanics was the result of the efforts to resolve these problems [16].

Classical mechanics has no longer been considered as an independent theory since the end of the twentieth century. We consider classical mechanics now as an approximate theory to quantum mechanics which is a more general theory. The desire to understand the fundamental forces of nature has shifted our emphasis in our research and investigation and has led to the Standard Model and also has directed the studies to a unified theory of everything. For the study of the motion of low-energy, of non-quantum mechanical particles in weak gravitational fields, it is useful to make use of classical mechanics. Additionally, we were successful to extend classical mechanics to the complex domain. In fact, this extended complex classical mechanics behaves very similarly to quantum mechanics [17].

At the end, and to conclude, this research work is organized as follows: After the introduction in section 1, Newton's laws of classical mechanics are stated in section 2, then the purpose and the advantages of the present work are presented in section 3. Afterward, in section 4, the extended Kolmogorov's axioms and hence the complex probability paradigm with their original parameters and interpretation will be explained and summarized. Moreover, in section 5, the complex probability paradigm axioms are applied to classical mechanics which will be hence extended to the imaginary and complex sets. Additionally, in section 6, the resultant complex random vector Z of *CPP* will be applied to statistical physics to prove an important property at its foundation. Also, in section 7, the flowchart of the new paradigm will be shown. Furthermore, the simulations of the novel model for various discrete and continuous stochastic distributions are illustrated in section 8. Finally, we conclude the work by doing a comprehensive summary in section 9, and then present the list of references cited in the current research work.

2. Isaac Newton's laws of motion

The classical mechanics foundation was laid down by Isaac Newton's three physical laws of motion. These laws define and describe the forces acting upon a body as well as the response of the body to those forces. Moreover, and more precisely, the first law defines the force qualitatively, the second law measures the force quantitively. The third law states that an isolated single force does not exist [18–21]. Throughout nearly three centuries, these three laws have been stated in many different ways and we will summarize them as follows:

First law

In an inertial frame of reference, an object either remains at rest or continues to move at a constant velocity, unless acted upon by a force.

Second law

In an inertial frame of reference, the vector sum of the forces $\vec{F}$ on an object is equal to the mass m of that object multiplied by the acceleration $\vec{a}$ of the object: $\vec{F} = m\vec{a}$. (It is assumed here that the mass m is constant).

Third law

When one body exerts a force on a second body, the second body simultaneously exerts a force equal in magnitude and opposite in direction on the first body.

Isaac Newton was the first to state in his Mathematical Principles of Natural Philosophy (Philosophiae Naturalis Principia Mathematica), first published in 1687, the three laws of motion. Many systems and physical objects were investigated and explained by the three laws of motion of Newton. As an example, the planetary motion laws of Johannes Kepler were proved and demonstrated by Newton's laws when combined with the universal gravitational law, in the third volume of the text [22–25].

Fourth law

Some also describe a Fourth law which states that forces add up like vectors, that is, that forces obey the principle of superposition.

A single point masses idealize the objects to which we apply the laws of Newton, that means that the object body shape and size are to be ignored in order to concentrate on the body's motion more easily. This is achieved when the rotation and the deformation of the body are negligible and when the object is too small compared to the distances that the analysis involves. Hence, in the planet orbital motion around a star analysis, even a planet can be idealized as a particle [26–29].

Moreover, deformable bodies and the rigid bodies motion are not characterized by the original form of the laws of motion of Newton which reveal to be inadequate. Additionally, a generalization of the laws of motion of Newton for rigid bodies was introduced and achieved by Leonhard Euler in 1750 and they were called accordingly Euler's laws of motion. They were applied later on to deformable bodies which were postulated to be a continuum. Euler's laws can be derived from the laws of Newton if we represent a body as an assemblage of discrete particles where every particle is governed by the motion laws of Newton. Independently of the structure of any particle, the laws of Euler can be considered, however, as axioms that describe the motion laws of extended bodies [30–33].

Newtonian inertial reference frames are a certain set of frames that verify and confirm Newton's laws. The first law defines what an inertial frame of reference is and this according to some authors interpretation. Therefore, the first law cannot be demonstrated as special case of the second law since the second law is only valid when an inertial frame of reference is used in the observation. The second law is considered as a corollary of the first law by other authors. It was long after Newton's death that we have developed the inertial frame of reference explicit concept [34–37].

Furthermore, we assume that, momentum, acceleration, and most importantly force to be quantities defined externally in the given interpretation. This is not the only interpretation, but the most common way one can consider the definition of these quantities by Newton's laws [38–41].

Additionally, when the speeds considered are much closer to the speed of light, then Albert Einstein's special relativity replaces Newtonian mechanics which is still useful as an approximation of the studied phenomenon [42–44].

3. The purpose and the advantages of the current publication

The crucial job of the theory of classical probability is to compute and to assess probabilities. A deterministic expression of probability theory can be attained by adding supplementary dimensions to nondeterministic and stochastic experiments. This original and novel idea is at the foundations of my new paradigm of complex probability. In its core, probability theory is a nondeterministic system of axioms that means that the phenomena and experiments outputs are the products of chance and randomness. In fact, a deterministic expression of the stochastic experiment will be realized and achieved by the addition of imaginary new dimensions to the stochastic phenomenon taking place in the real probability set $\mathcal{R}$ and hence this will lead to a certain output in the set $\mathcal{C}$ of complex probabilities. Accordingly, we will be totally capable to foretell the random events outputs that occur in all probabilistic processes in the real world. This is possible because the chaotic phenomenon becomes completely predictable. Thus, the job that has been successfully completed here was to extend the set of real and random probabilities which is the set $\mathcal{R}$ to the complex and deterministic set of probabilities which is $\mathcal{C} = \mathcal{R} + \mathcal{M}$. This is achieved by taking into account the contributions of the imaginary and complementary set of probabilities to the set $\mathcal{R}$ and that we have called accordingly the set $\mathcal{M}$. This extension proved that it was effective and consequently we were successful to create an original paradigm dealing with prognostic and stochastic sciences in which we were able to express deterministically in $\mathcal{C}$ all the nondeterministic processes happening in the 'real' world $\mathcal{R}$. This innovative paradigm was coined by the term "The Complex Probability Paradigm" and was started and established in my seventeen earlier publications and research works [45–61].

The advantages and the purpose of this current work are to:

1. Extend the theory of classical probability to encompass the complex numbers set, hence to bond the theory of probability to the field of complex variables and analysis in mathematics. This mission was elaborated and initiated in my earlier seventeen papers.

2. Apply the novel probability axioms and paradigm to Newton's classical mechanics.

3. Show that all nondeterministic phenomena can be expressed deterministically in the complex probabilities set which is $\mathcal{C}$.

4. Compute and quantify both the degree of our knowledge and the chaotic factor of all the forces acting on a body in classical mechanics and *CPP* in the sets $\mathcal{R}$, $\mathcal{M}$, and $\mathcal{C}$.

5. Represent and show the graphs of the functions and parameters of the innovative paradigm related to Newton's mechanics.

6. Demonstrate that the classical concept of probability is permanently equal to one in the set of complex probabilities; hence, no randomness, no chaos, no ignorance, no uncertainty, no nondeterminism, no unpredictability, and no disorder exist in:

$$\mathcal{C}\text{ (complex set)} = \mathcal{R}\text{ (real set)} + \mathcal{M}\text{ (imaginary set)}.$$

7. Prove an important property at the foundation of statistical physics after applying *CPP* to classical mechanics.

8. Prepare to implement this creative model to other topics in prognostics and to the field of stochastic processes. These will be the job to be accomplished in my future research publications.

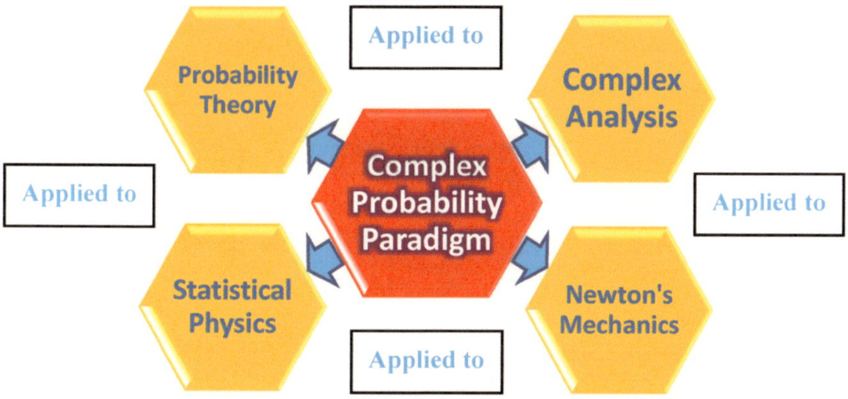

Figure 1.
The diagram of the complex probability paradigm major goals.

Concerning some applications of the novel founded paradigm and as a future work, it can be applied to any nondeterministic phenomenon using classical mechanics whether in the continuous or in the discrete cases. Moreover, compared with existing literature, the major contribution of the current research work is to apply the innovative paradigm of complex probability to Newton's classical mechanics and to statistical physics as well.

The next figure displays the major purposes and goals of the Complex Probability Paradigm (*CPP*) (**Figure 1**).

4. The complex probability paradigm

4.1 The original Andrey Nikolaevich Kolmogorov system of axioms

The simplicity of Kolmogorov's system of axioms may be surprising. Let E be a collection of elements $\{E_1, E_2, \ldots\}$ called elementary events and let F be a set of subsets of E called random events [62–66]. The five axioms for a finite set E are:

Axiom 1: F is a field of sets.

Axiom 2: F contains the set E.

Axiom 3: A non-negative real number $P_{rob}(A)$, called the probability of A, is assigned to each set A in F. We have always $0 \leq P_{rob}(A) \leq 1$.

Axiom 4: $P_{rob}(E)$ equals 1.

Axiom 5: If A and B have no elements in common, the number assigned to their union is:

$$P_{rob}(A \cup B) = P_{rob}(A) + P_{rob}(B)$$

hence, we say that A and B are disjoint; otherwise, we have:

$$P_{rob}(A \cup B) = P_{rob}(A) + P_{rob}(B) - P_{rob}(A \cap B)$$

And we say also that: $P_{rob}(A \cap B) = P_{rob}(A) \times P_{rob}(B/A) = P_{rob}(B) \times P_{rob}(A/B)$ which is the conditional probability. If both A and B are independent then: $P_{rob}(A \cap B) = P_{rob}(A) \times P_{rob}(B)$.

Moreover, we can generalize and say that for N disjoint (mutually exclusive) events $A_1, A_2, \ldots, A_j, \ldots, A_N$ (for $1 \leq j \leq N$), we have the following additivity rule:

$$P_{rob}\left(\bigcup_{j=1}^{N} A_j\right) = \sum_{j=1}^{N} P_{rob}(A_j)$$

And we say also that for N independent events $A_1, A_2, \ldots, A_j, \ldots, A_N$ (for $1 \leq j \leq N$), we have the following product rule:

$$P_{rob}\left(\bigcap_{j=1}^{N} A_j\right) = \prod_{j=1}^{N} P_{rob}(A_j)$$

4.2 Adding the imaginary part $\mathcal{M}$

Now, we can add to this system of axioms an imaginary part such that:

Axiom 6: Let $P_m = i \times (1 - P_r)$ be the probability of an associated complementary event in $\mathcal{M}$ (the imaginary part) to the event A in $\mathcal{R}$ (the real part). It follows that $P_r + P_m/i = 1$ where i is the imaginary number with $i = \sqrt{-1}$ or $i^2 = -1$.

Axiom 7: We construct the complex number or vector $z = P_r + P_m = P_r + i(1 - P_r)$ having a norm $|z|$ such that:

$$|z|^2 = P_r^2 + (P_m/i)^2.$$

Axiom 8: Let P_c denote the probability of an event in the complex probability universe $\mathcal{C}$ where $\mathcal{C} = \mathcal{R} + \mathcal{M}$. We say that P_c is the probability of an event A in $\mathcal{R}$ with its associated event in $\mathcal{M}$ such that:

$$P_c^{\,2} = (P_r + P_m/i)^2 = |z|^2 - 2iP_rP_m \text{ and is always equal to 1.}$$

We can see that by taking into consideration the set of imaginary probabilities we added three new and original axioms and consequently the system of axioms defined by Kolmogorov was hence expanded to encompass the set of imaginary numbers [45–61].

4.3 A Concise Interpretation of the Original Paradigm

As a summary of the new paradigm, we declare that in the universe $\mathcal{R}$ of real probabilities we have the degree of our certain knowledge is unfortunately incomplete and therefore insufficient and unsatisfactory, hence we encompass in our analysis the set $\mathcal{C}$ of complex numbers which integrates the contributions of both the real set $\mathcal{R}$ of probabilities and its complementary imaginary probabilities set that we have called accordingly $\mathcal{M}$. Subsequently, a perfect and an absolute degree of our knowledge is obtained and achieved in the universe of probabilities $\mathcal{C} = \mathcal{R} + \mathcal{M}$ because we have constantly $P_c = 1$. In fact, a sure and certain prediction of any random phenomenon is reached in the universe $\mathcal{C}$ because in this set, we eliminate and subtract from the measured degree of our knowledge the computed chaotic factor. Consequently, this will lead to in the universe $\mathcal{C}$ a probability permanently equal to one as it is shown in the following equation: $P_c^{\,2} = DOK - Chf = DOK + MChf = 1 = P_c$ deduced from the complex probability paradigm. Moreover, various discrete and continuous stochastic distributions illustrate in my seventeen previous research works this hypothesis and innovative and original model. The figure that follows shows and summarizes the Extended Kolmogorov Axioms (*EKA*) or the Complex Probability Paradigm (*CPP*) (**Figure 2**) [67–92]:

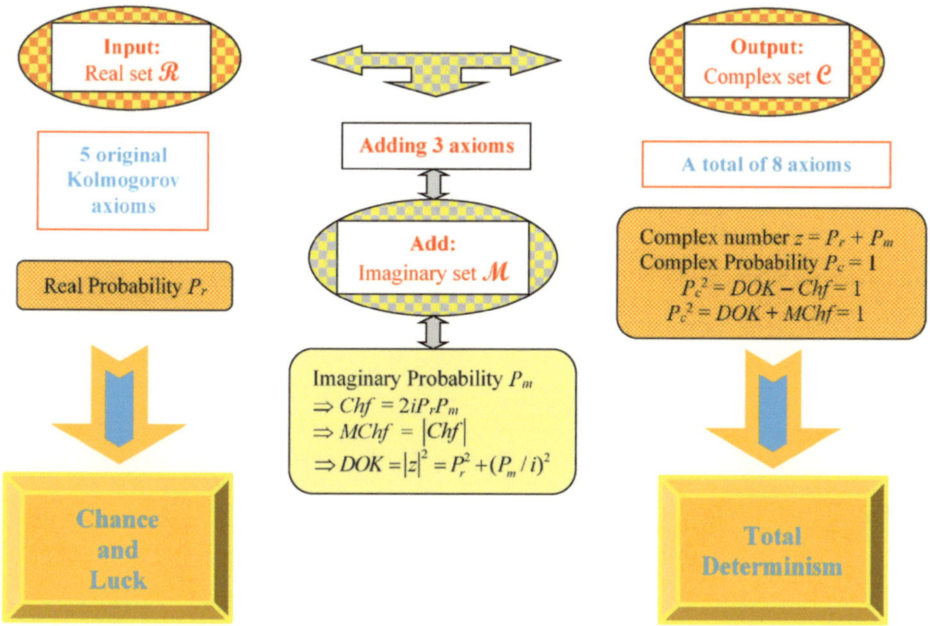

Figure 2.
The EKA *or the* CPP *diagram.*

5. The Newton's mechanics and the complex probability paradigm parameters

In this section we will relate and link Newton's mechanics to the complex probability paradigm with all its parameters by using four novel concepts which are: the real stochastic force $\vec{F}_r$ in the real probability set $\mathcal{R}$, the imaginary stochastic force $\vec{F}_m$ in the imaginary probability set $\mathcal{M}$, the complex resultant stochastic force $\vec{F}$ in the complex probability set $\mathcal{C} = \mathcal{R} + \mathcal{M}$, and the deterministic real force $\vec{F}_c$ also in the probability set $\mathcal{C}$ [45–61, 93–104].

5.1 The stochastic forces $\vec{F}_r$ in $\mathcal{R}$ and $\vec{F}_m$ in $\mathcal{M}$

The real stochastic force is defined by: $\vec{F}_r = P_r m\vec{a} \Leftrightarrow P_r = \frac{\vec{F}_r}{m\vec{a}}$.

Here P_r measures the probability that the real stochastic force $\vec{F}_r$ acting on a body in $\mathcal{R}$ will occur.

Since $0 \le P_r \le 1 \Leftrightarrow 0 \le \frac{\vec{F}_r}{m\vec{a}} \le 1 \Leftrightarrow \vec{0} \le \vec{F}_r \le m\vec{a}$

If $P_r = 0$ then $\vec{F}_r = \vec{0}$ that means that the real stochastic force in $\mathcal{R}$ is totally known and is equal to $\vec{0}$ or null in this case.

If $P_r = 1$ then $\vec{F}_r = m\vec{a}$ that means that the real stochastic force in $\mathcal{R}$ is totally known and totally deterministic and is equal to $m\vec{a}$ in this case.

The imaginary stochastic force is defined by:
$\vec{F}_m = P_m m\vec{a} = i(1 - P_r)m\vec{a} \Leftrightarrow P_m = \frac{\vec{F}_m}{m\vec{a}} = i(1 - P_r)$.

Here P_m measures the probability that the imaginary stochastic force $\vec{F}_m$ acting on a body in $\mathcal{M}$ will occur.

Since $0 \le P_r \le 1 \Leftrightarrow 0 \le P_m \le i \Leftrightarrow 0 \le \frac{\vec{F}_m}{m\vec{a}} \le i \Leftrightarrow \vec{0} \le \vec{F}_m \le im\vec{a}$

If $P_m = 0$ then $\vec{F}_m = \vec{0}$ that means that the imaginary stochastic force in $\mathcal{M}$ is totally known and is equal to $\vec{0}$ or null.

If $P_m = i$ then $\vec{F}_m = im\vec{a}$ that means that the imaginary stochastic force in $\mathcal{M}$ is totally known and totally deterministic and is equal to $im\vec{a}$.

5.1.1 The relation between the real and the imaginary stochastic forces

We have: $\vec{F}_m = P_m m\vec{a} = i(1 - P_r)m\vec{a} \Leftrightarrow P_m = \frac{\vec{F}_m}{m\vec{a}} = i(1 - P_r)$.

And since $P_r = \frac{\vec{F}_r}{m\vec{a}} \Leftrightarrow P_m = \frac{\vec{F}_m}{m\vec{a}} = i\left(1 - \frac{\vec{F}_r}{m\vec{a}}\right)$.

And we can deduce that: $P_r = 1 - P_m/i = 1 - \frac{\vec{F}_m}{im\vec{a}} \Leftrightarrow P_r = 1 + \frac{i\vec{F}_m}{m\vec{a}}$ since $i = -\frac{1}{i}$

Therefore, $\vec{F}_m = i\left(1 - \frac{\vec{F}_r}{m\vec{a}}\right)m\vec{a} = im\vec{a} - i\vec{F}_r$

$\Leftrightarrow \vec{F}_r = m\vec{a} - \frac{\vec{F}_m}{i} = m\vec{a} + i\vec{F}_m$ since $i = -\frac{1}{i}$ also.

5.2 The resultant complex stochastic force $\vec{F}$ in $\mathcal{C} = \mathcal{R} + \mathcal{M}$

We define the resultant complex stochastic force by: $\vec{F} = \vec{F}_r + \vec{F}_m = P_r m\vec{a} + P_m m\vec{a} = (P_r + P_m)m\vec{a} = zm\vec{a}$.

Here z measures here the complex probability that the resultant stochastic force $\vec{F} = \vec{F}_r + \vec{F}_m$ acting on a body in $\mathcal{C} = \mathcal{R} + \mathcal{M}$ will occur.

Since $z = P_r + P_m$ then:

If $P_r = 0 \Leftrightarrow P_m = i(1 - P_r) = i(1 - 0) = i \Leftrightarrow z = 0 + i = i \Leftrightarrow \vec{F} = zm\vec{a} = im\vec{a}$.

If $P_r = 1 \Leftrightarrow P_m = i(1 - P_r) = i(1 - 1) = 0 \Leftrightarrow z = 1 + 0 = 1 \Leftrightarrow \vec{F} = zm\vec{a} = m\vec{a}$.

5.2.1 The relations between the forces $\vec{F}_r$, $\vec{F}_m$, and $\vec{F}$

Since $\vec{F}_r = m\vec{a} + i\vec{F}_m \Leftrightarrow \vec{F} = \vec{F}_r + \vec{F}_m = m\vec{a} + i\vec{F}_m + \vec{F}_m = m\vec{a} + (1 + i)\vec{F}_m$.

where $\mathrm{Re}\left(\vec{F}\right) = m\vec{a} + i\vec{F}_m$ and $\mathrm{Im}\left(\vec{F}\right) = \vec{F}_m$.

Additionally, since $\vec{F}_m = im\vec{a} - i\vec{F}_r \Leftrightarrow \vec{F} = \vec{F}_r + \vec{F}_m = \vec{F}_r + im\vec{a} - i\vec{F}_r = im\vec{a} + (1 - i)\vec{F}_r$.

where $\mathrm{Re}\left(\vec{F}\right) = \vec{F}_r$ and $\mathrm{Im}\left(\vec{F}\right) = im\vec{a} - i\vec{F}_r = i\left(m\vec{a} - \vec{F}_r\right)$.

5.3 The deterministic real force $\vec{F}_c$ in the probability set $\mathcal{C} = \mathcal{R} + \mathcal{M}$

We define the deterministic real force by: $\vec{F}_c = P_c m\vec{a}$.

Since from *CPP* we have: $P_c = P_r + P_m/i = P_r + (1 - P_r) = 1 \Leftrightarrow \vec{F}_c = m\vec{a}$.

Here P_c measures the probability that the force $\vec{F}_c$ acting on a body in the probability universe $\mathcal{C} = \mathcal{R} + \mathcal{M}$ will occur. This means that the force acting on the body in the probability set $\mathcal{C}$ is totally known and is totally deterministic always $\forall P_r : 0 \le P_r \le 1$ and $\forall P_m : 0 \le P_m \le i$.

5.3.1 The relations between the forces $\vec{F}_r$, $\vec{F}_m$, and $\vec{F}_c$

Furthermore,

Since $\vec{F}_r = P_r m\vec{a} \Leftrightarrow \vec{F}_r = P_r \vec{F}_c$ and $P_r = \frac{\vec{F}_r}{\vec{F}_c}$.

Since $\vec{F}_m = P_m m\vec{a} \Leftrightarrow \vec{F}_m = P_m \vec{F}_c$ and $P_m = \frac{\vec{F}_m}{\vec{F}_c}$.

Since $P_m = i(1 - P_r) \Leftrightarrow P_r = 1 - \frac{P_m}{i} = 1 + iP_m$ because $i = -\frac{1}{i} \Leftrightarrow P_r = 1 + i\frac{\vec{F}_m}{\vec{F}_c}$.

Since $\vec{F} = zm\vec{a} \Leftrightarrow \vec{F} = z\vec{F}_c$, therefore:

If $P_r = 0 \Leftrightarrow P_m = i \Leftrightarrow z = 0 + i = i \Leftrightarrow \vec{F} = zm\vec{a} = im\vec{a} = i\vec{F}_c$.

If $P_r = 1 \Leftrightarrow P_m = 0 \Leftrightarrow z = 1 + 0 = 1 \Leftrightarrow \vec{F} = zm\vec{a} = m\vec{a} = \vec{F}_c$.

The second case shows and proves that if $P_r = 1$ then the complex resultant stochastic force will become equal to the real deterministic force that means that we will return directly to the classical deterministic Newtonian mechanics theory which is a special deterministic case of the stochastic complex probability paradigm general case.

Additionally, since $\vec{F}_m = im\vec{a} - i\vec{F}_r \Leftrightarrow i\vec{F}_r + \vec{F}_m = im\vec{a} = i\vec{F}_c$.

And $\vec{F}_r - i\vec{F}_m = m\vec{a} = \vec{F}_c$ since $i = -\frac{1}{i}$.

Since $\vec{F} = m\vec{a} + (1+i)\vec{F}_m \Leftrightarrow \vec{F} = \vec{F}_c + (1+i)\vec{F}_m$

And since $\vec{F} = im\vec{a} + (1-i)\vec{F}_r \Leftrightarrow \vec{F} = i\vec{F}_c + (1-i)\vec{F}_r$.

5.4 The relationships between the forces in $\mathcal{R}$, $\mathcal{M}$, and $\mathcal{C}$ and all the *CPP* parameters

5.4.1 The relationships between the real force in $\mathcal{R}$ and all the CPP *parameters*

Furthermore, according to *CPP*:

$$\begin{aligned} DOK &= |z|^2 = |P_r + P_m|^2 = P_r^2 + (P_m/i)^2 = P_r^2 + (1 - P_r)^2 \\ &= P_r^2 + 1 - 2P_r + P_r^2 \Leftrightarrow 2P_r^2 - 2P_r + 1 - DOK = 0 \end{aligned}$$

which is a second-degree equation in terms of P_r whose discriminant is:

$$\Delta = 4 - 8(1 - DOK) = 8DOK - 4.$$

Since $0.5 \le DOK \le 1 \Leftrightarrow 0 \le 8DOK - 4 \le 4 \Leftrightarrow 0 \le \Delta \le 4 \Leftrightarrow \Delta \ge 0, \forall DOK$. Therefore, the equation admits two real roots which are:

$$P_{r1} = \frac{2 - \sqrt{\Delta}}{4} = \frac{2 - \sqrt{8DOK - 4}}{4} = \frac{2 - 2\sqrt{2DOK - 1}}{4} = \frac{1 - \sqrt{2DOK - 1}}{2}$$

$$\text{and } P_{r2} = \frac{2 + \sqrt{\Delta}}{4} = \frac{2 + \sqrt{8DOK - 4}}{4} = \frac{2 + 2\sqrt{2DOK - 1}}{4} = \frac{1 + \sqrt{2DOK - 1}}{2}.$$

But according to *CPP*: $\forall P_r : 0 \le P_r \le 1 \Leftrightarrow 0.5 \le DOK \le 1$ and $-0.5 \le Chf \le 0$ and $0 \le MChf \le 0.5$.

And if $P_r = 0$ or $P_r = 1$ then $DOK = 1$ and $Chf = 0$ and $MChf = 0$.

And if $P_r = 0.5$ then $DOK = 0.5$ and $Chf = -0.5$ and $MChf = 0.5$.

Consequently,

$$P_r = \begin{cases} \dfrac{1-\sqrt{2DOK-1}}{2} & \text{if } 0 \le P_r \le 0.5 \\ \dfrac{1+\sqrt{2DOK-1}}{2} & \text{if } 0.5 \le P_r \le 1 \end{cases}$$

But $\vec{F}_r = P_r m\vec{a}$ hence (**Figure 3**):

$$\vec{F}_r = \begin{cases} \left(\dfrac{1-\sqrt{2DOK-1}}{2}\right)m\vec{a} & \text{if } 0 \le P_r \le 0.5 \\ \left(\dfrac{1+\sqrt{2DOK-1}}{2}\right)m\vec{a} & \text{if } 0.5 \le P_r \le 1 \end{cases}$$

We have $DOK = 1 + Chf \Leftrightarrow 2DOK - 1 = 1 + 2Chf$ thus (**Figure 4**):

$$\vec{F}_r = \begin{cases} \left(\dfrac{1-\sqrt{1+2Chf}}{2}\right)m\vec{a} & \text{if } 0 \le P_r \le 0.5 \\ \left(\dfrac{1+\sqrt{1+2Chf}}{2}\right)m\vec{a} & \text{if } 0.5 \le P_r \le 1 \end{cases}$$

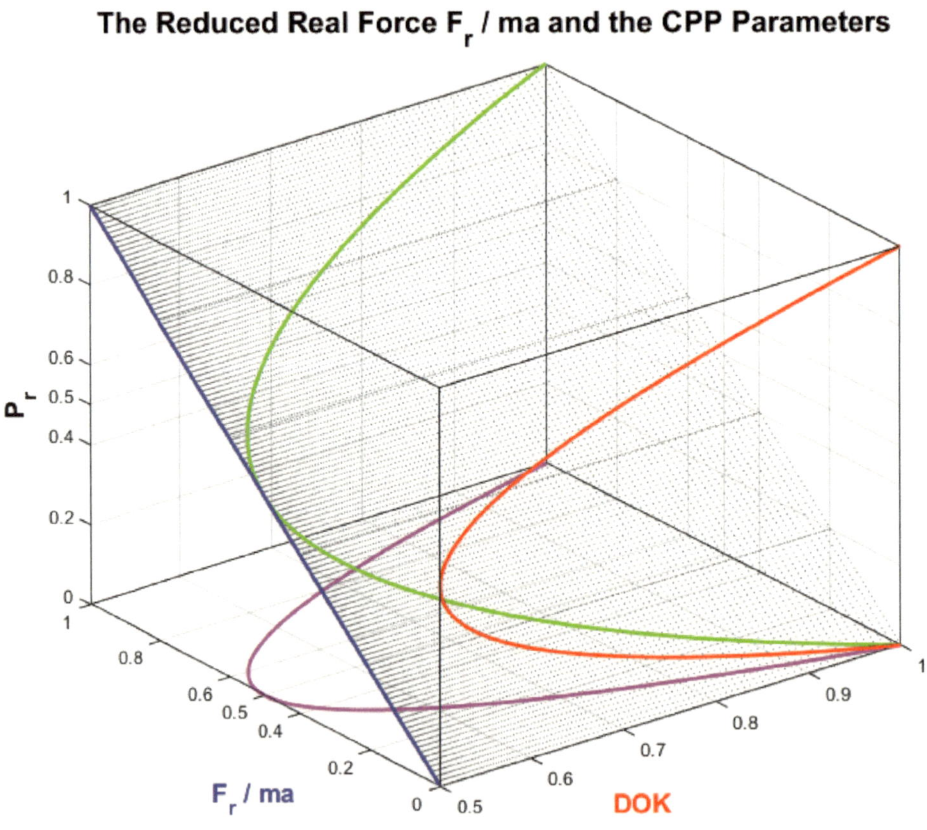

Figure 3.
The graphs of the reduced real force F_r (P_r) / ma *in blue and of* F_r (DOK) / ma *in pink and* DOK (P_r) *in red and of* F_r (DOK) / ma *in green in the* F_r (P_r) / ma *plane in light gray.*

We have $DOK = 1 - MChf \Leftrightarrow 2DOK - 1 = 1 - 2MChf$ thus (**Figure 5**):

$$\vec{F}_r = \begin{cases} \left(\dfrac{1 - \sqrt{1 - 2MChf}}{2}\right) m\vec{a} & \text{if } 0 \le P_r \le 0.5 \\ \left(\dfrac{1 + \sqrt{1 - 2MChf}}{2}\right) m\vec{a} & \text{if } 0.5 \le P_r \le 1 \end{cases}$$

We can deduce also from *CPP* that (**Figure 6**):

$$\vec{F}_r = \begin{cases} \left(\dfrac{1 - \sqrt{DOK + Chf}}{2}\right) m\vec{a} & \text{if } 0 \le P_r \le 0.5 \\ \left(\dfrac{1 + \sqrt{DOK + Chf}}{2}\right) m\vec{a} & \text{if } 0.5 \le P_r \le 1 \end{cases}$$

And we can infer using the fact that $MChf = -Chf$ that (**Figure 7**):

$$\vec{F}_r = \begin{cases} \left(\dfrac{1 - \sqrt{DOK - MChf}}{2}\right) m\vec{a} & \text{if } 0 \le P_r \le 0.5 \\ \left(\dfrac{1 + \sqrt{DOK - MChf}}{2}\right) m\vec{a} & \text{if } 0.5 \le P_r \le 1 \end{cases}$$

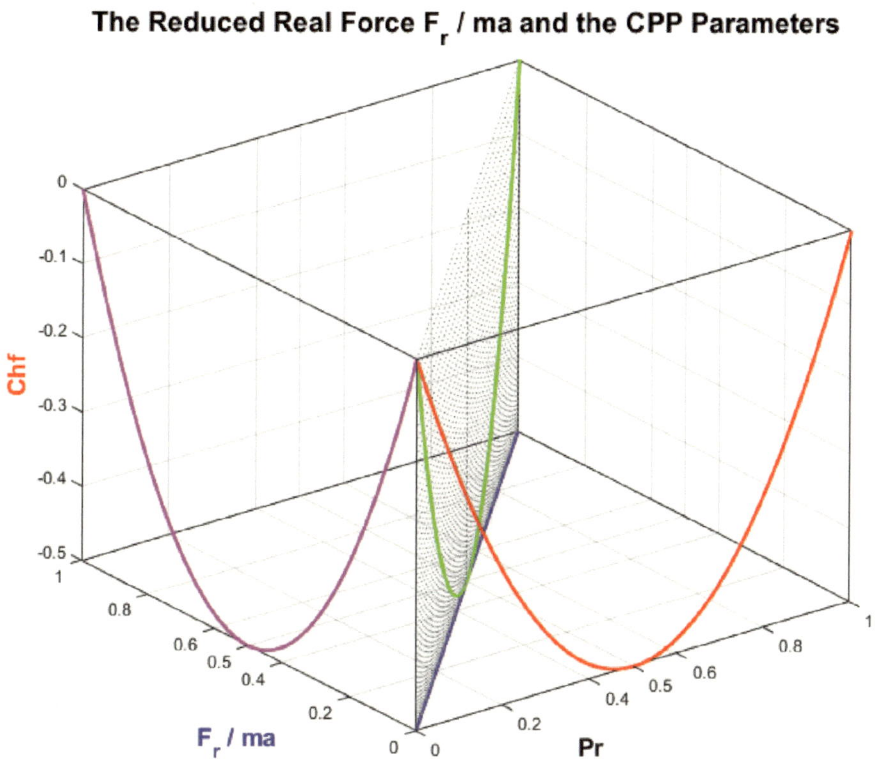

Figure 4.
The graphs of the reduced real force F_r *(*P_r*) / ma in blue and of* F_r *(Chf) / ma in pink and Chf (*P_r*) in red and of* F_r *(Chf) / ma in green in the* F_r *(*P_r*) / ma plane in light gray.*

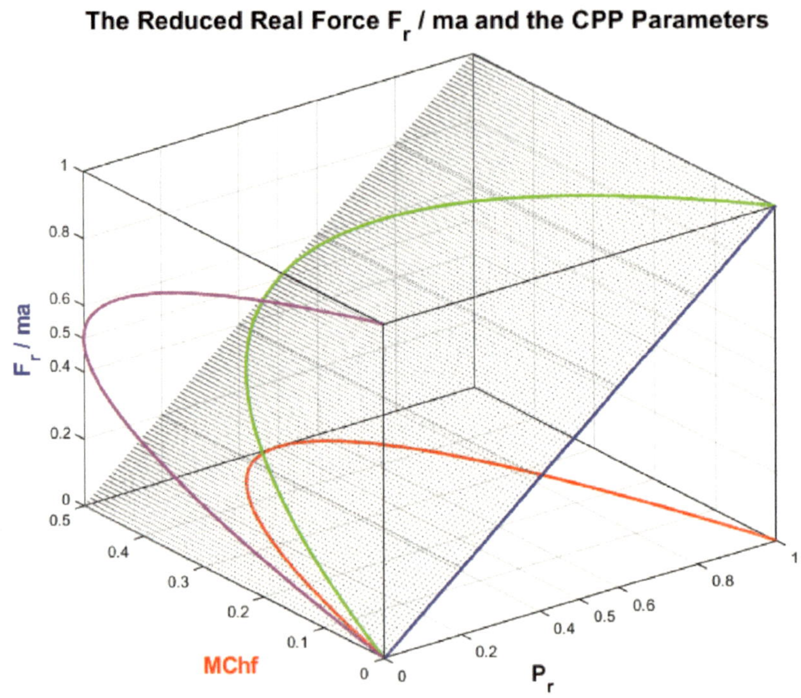

Figure 5.
The graphs of the reduced real force F_r (P_r) / ma *in blue and of* F_r (MChf) / ma *in pink and* MChf (P_r) *in red and of* F_r (MChf) / ma *in green in the* F_r (P_r) / ma *plane in light gray.*

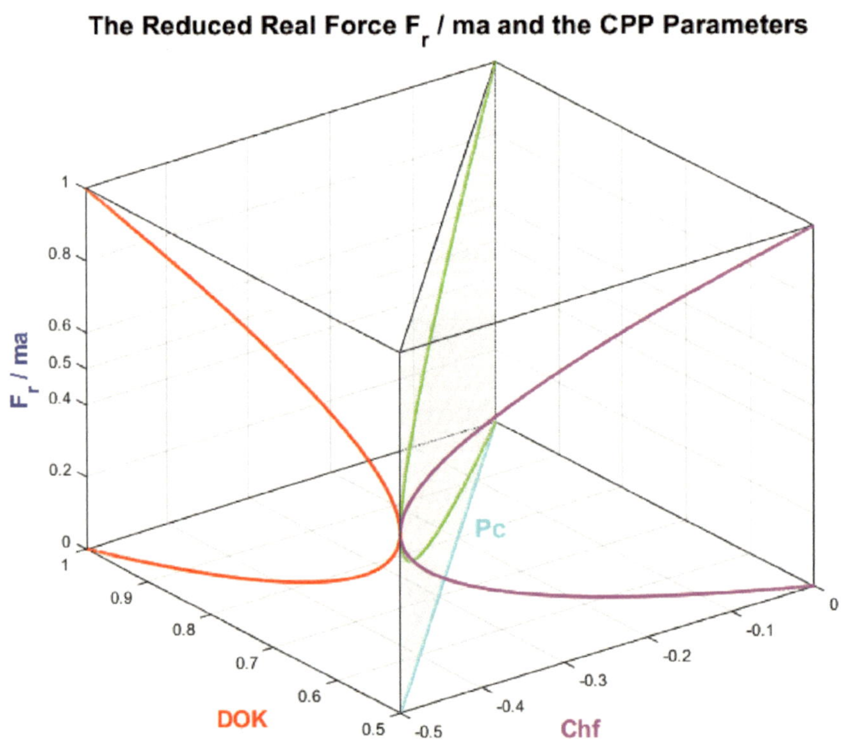

Figure 6.
The graphs of the reduced real force F_r (Chf) / ma *in pink and of* F_r (DOK) / ma *in red and of* P_c^2 = DOK – Chf = *1* = P_c (Chf, DOK) *in cyan and of* F_r (Chf, DOK) / ma *in green in the* P_c *plane in light gray.*

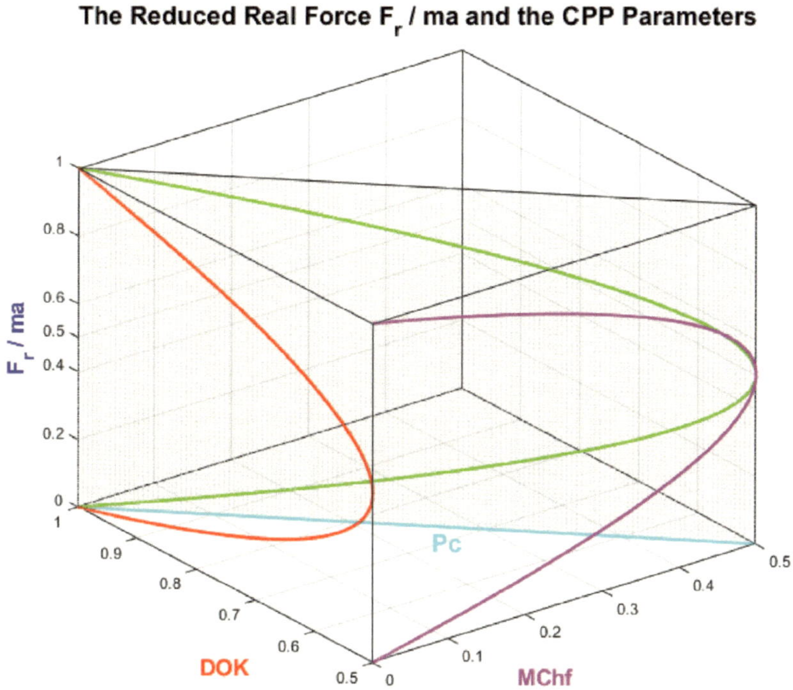

Figure 7.
The graphs of the reduced real force F_r (MChf) / ma *in pink and of* F_r (DOK) / ma *in red and of* P_c^2 = DOK + MChf = 1 = P_c (MChf, DOK) *in cyan and of* F_r (MChf, DOK) / ma *in green in the* P_c *plane in light gray.*

Also, we can calculate (**Figure 8**):

$$\vec{F}_r = \begin{cases} \left(\dfrac{1-\sqrt{1+Chf-MChf}}{2}\right) m\vec{a} & \text{if } 0 \le P_r \le 0.5 \\ \left(\dfrac{1+\sqrt{1+Chf-MChf}}{2}\right) m\vec{a} & \text{if } 0.5 \le P_r \le 1 \end{cases}$$

But according to *CPP*: $P_c^2 = DOK - Chf = DOK + MChf = 1 = P_c$ hence the real force $\vec{F}_r$ in $\mathcal{R}$ as a function of all the *CPP* parameters is the following:

$$\vec{F}_r = \begin{cases} \left(\dfrac{P_c-\sqrt{DOK-Chf-2MChf}}{2}\right) m\vec{a} & \text{if } 0 \le P_r \le 0.5 \\ \left(\dfrac{P_c+\sqrt{DOK-Chf-2MChf}}{2}\right) m\vec{a} & \text{if } 0.5 \le P_r \le 1 \end{cases}$$

5.4.2 The relationships between the imaginary force in $\mathcal{M}$ and all the CPP *parameters*

As we have computed:

$$P_r = \begin{cases} \dfrac{1-\sqrt{2DOK-1}}{2} & \text{if } 0 \le P_r \le 0.5 \\ \dfrac{1+\sqrt{2DOK-1}}{2} & \text{if } 0.5 \le P_r \le 1 \end{cases}$$

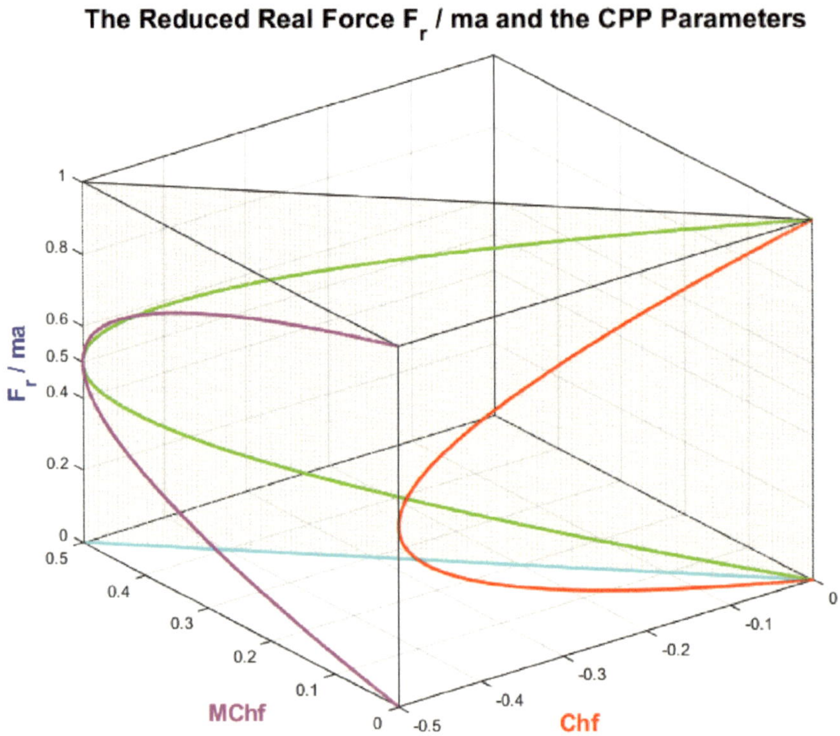

Figure 8.
The graphs of the reduced real force F_r *(MChf) / ma in pink and of* F_r *(Chf) / ma in red and of* Chf + MChf = 0 *in cyan and of* F_r *(Chf, MChf) / ma in green in the* Chf + MChf = 0 *plane in light gray.*

And since $P_m = i(1 - P_r)$ then:

$$P_m = \begin{cases} i\left(\dfrac{1+\sqrt{2DOK-1}}{2}\right) & \text{if } 0 \le P_r \le 0.5 \Leftrightarrow \text{if } 0.5i \le P_m \le i \\ i\left(\dfrac{1-\sqrt{2DOK-1}}{2}\right) & \text{if } 0.5 \le P_r \le 1 \Leftrightarrow \text{if } 0 \le P_m \le 0.5i \end{cases}$$

We have $\vec{F}_m = P_m m\vec{a}$, so similarly to the previous section we get (**Figure 9**):

$$\vec{F}_m = \begin{cases} i\left(\dfrac{1+\sqrt{2DOK-1}}{2}\right)m\vec{a} & \text{if } 0 \le P_r \le 0.5 \\ i\left(\dfrac{1-\sqrt{2DOK-1}}{2}\right)m\vec{a} & \text{if } 0.5 \le P_r \le 1 \end{cases}$$

And we can deduce that (**Figure 10**):

$$\vec{F}_m = \begin{cases} i\left(\dfrac{1+\sqrt{1+2Chf}}{2}\right)m\vec{a} & \text{if } 0 \le P_r \le 0.5 \\ i\left(\dfrac{1-\sqrt{1+2Chf}}{2}\right)m\vec{a} & \text{if } 0.5 \le P_r \le 1 \end{cases}$$

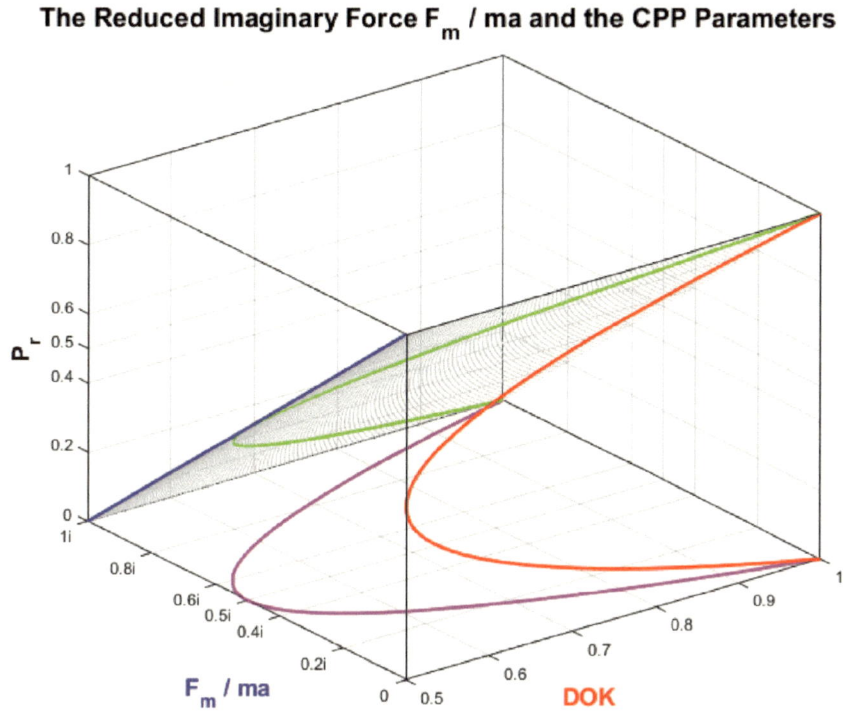

Figure 9.
The graphs of the reduced imaginary force F_m (P_r) / ma *in blue and of* F_m (DOK) / ma *in pink and* DOK (P_r) *in red and of* F_m (DOK) / ma *in green in the* F_m (P_r) / ma *plane in light gray.*

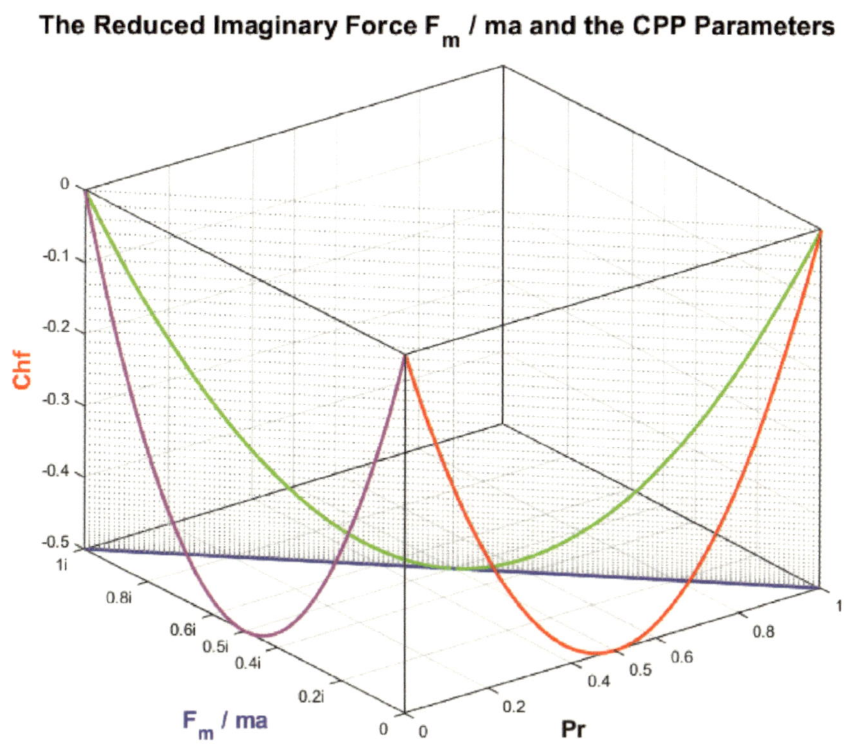

Figure 10.
The graphs of the reduced imaginary force F_m (P_r) / ma *in blue and of* F_m (Chf) / ma *in pink and* Chf (P_r) *in red and of* F_m (Chf) / ma *in green in the* F_m (P_r) / ma *plane in light gray.*

And we can infer that (**Figure 11**):

$$\vec{F}_m = \begin{cases} i\left(\dfrac{1+\sqrt{1-2MChf}}{2}\right)m\vec{a} & \text{if } 0 \le P_r \le 0.5 \\ i\left(\dfrac{1-\sqrt{1-2MChf}}{2}\right)m\vec{a} & \text{if } 0.5 \le P_r \le 1 \end{cases}$$

We can deduce also that (**Figure 12**):

$$\vec{F}_m = \begin{cases} i\left(\dfrac{1+\sqrt{DOK+Chf}}{2}\right)m\vec{a} & \text{if } 0 \le P_r \le 0.5 \\ i\left(\dfrac{1-\sqrt{DOK+Chf}}{2}\right)m\vec{a} & \text{if } 0.5 \le P_r \le 1 \end{cases}$$

And we can compute (**Figure 13**):

$$\vec{F}_m = \begin{cases} i\left(\dfrac{1+\sqrt{DOK-MChf}}{2}\right)m\vec{a} & \text{if } 0 \le P_r \le 0.5 \\ i\left(\dfrac{1-\sqrt{DOK-MChf}}{2}\right)m\vec{a} & \text{if } 0.5 \le P_r \le 1 \end{cases}$$

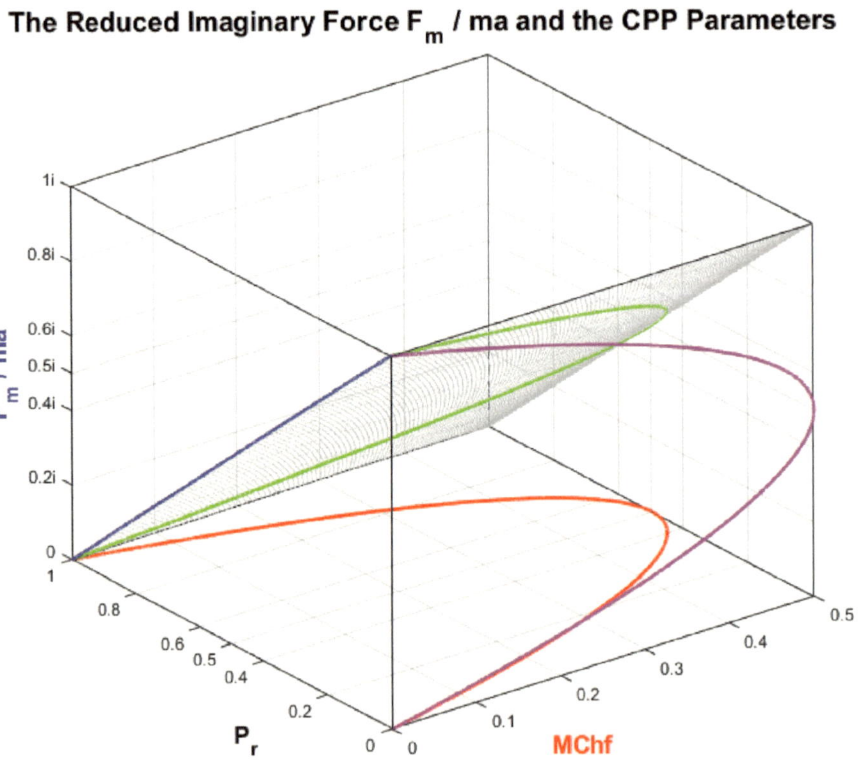

Figure 11.
The graphs of the reduced imaginary force F_m (P_r) / ma *in blue and of* F_m (MChf) / ma *in pink and* MChf (P_r) *in red and of* F_m (MChf) / ma *in green in the* F_m (P_r) / ma *plane in light gray.*

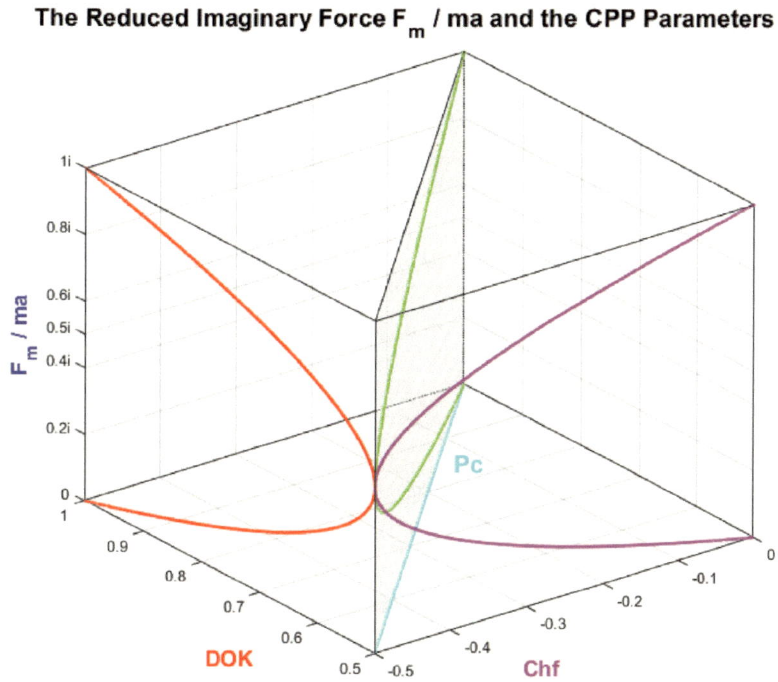

Figure 12.
The graphs of the reduced imaginary force F_m (Chf) / ma *in pink and of* F_m (DOK) / ma *in red and of* P_c^2 = DOK – Chf = *1* = P_c (Chf, DOK) *in cyan and of* F_m (Chf, DOK) / ma *in green in the* P_c *plane in light gray.*

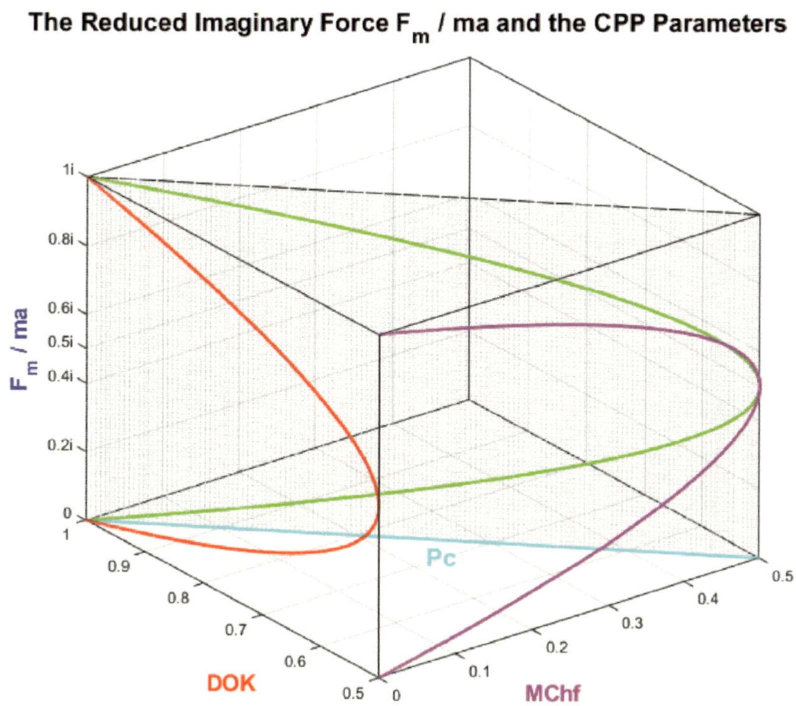

Figure 13.
The graphs of the reduced imaginary force F_m (MChf) / ma *in pink and of* F_m (DOK) / ma *in red and of* P_c^2 = DOK + MChf = *1* = P_c (MChf, DOK) *in cyan and of* F_m (MChf, DOK) / ma *in green in the* P_c *plane in light gray.*

And we can calculate (**Figure 14**):

$$\vec{F}_m = \begin{cases} i\left(\dfrac{1+\sqrt{1+Chf-MChf}}{2}\right)m\vec{a} & \text{if } 0 \le P_r \le 0.5 \\ i\left(\dfrac{1-\sqrt{1+Chf-MChf}}{2}\right)m\vec{a} & \text{if } 0.5 \le P_r \le 1 \end{cases}$$

But according to *CPP*: $P_c^2 = DOK - Chf = DOK + MChf = 1 = P_c$ hence the imaginary force $\vec{F}_m$ in $\mathcal{M}$ as a function of all the *CPP* parameters is the following:

$$\vec{F}_m = \begin{cases} i\left(\dfrac{P_c+\sqrt{DOK-Chf-2MChf}}{2}\right)m\vec{a} & \text{if } 0 \le P_r \le 0.5 \\ i\left(\dfrac{P_c-\sqrt{DOK-Chf-2MChf}}{2}\right)m\vec{a} & \text{if } 0.5 \le P_r \le 1 \end{cases}$$

5.4.3 The relationships between the resultant complex force in **C** *and all the* CPP *parameters*

Analogously, and since $\vec{F} = \vec{F}_r + \vec{F}_m$ then:

$$\vec{F} = \begin{cases} \left[\left(\dfrac{1-\sqrt{2DOK-1}}{2}\right)+i\left(\dfrac{1+\sqrt{2DOK-1}}{2}\right)\right]m\vec{a} & \text{if } 0 \le P_r \le 0.5 \\ \left[\left(\dfrac{1+\sqrt{2DOK-1}}{2}\right)+i\left(\dfrac{1-\sqrt{2DOK-1}}{2}\right)\right]m\vec{a} & \text{if } 0.5 \le P_r \le 1 \end{cases}$$

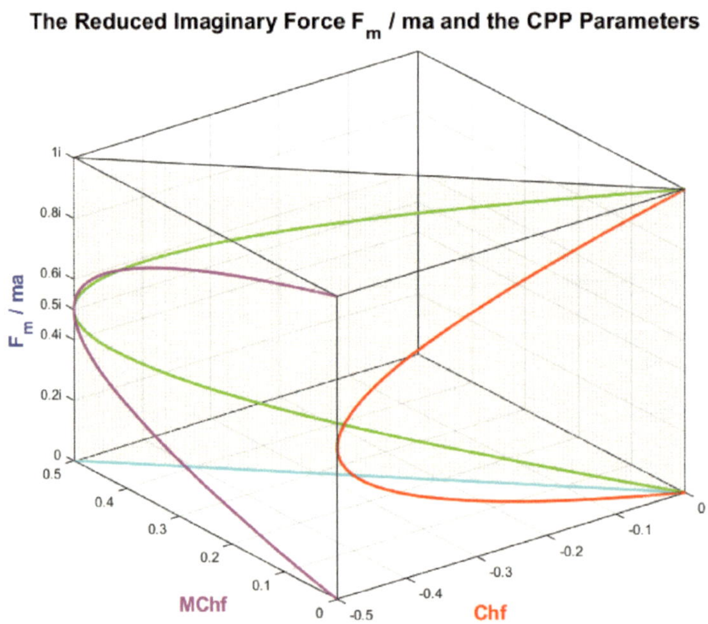

Figure 14.
The graphs of the reduced imaginary force F_m (MChf) / ma *in pink and of* F_m (Chf) / ma *in red and of* Chf + MChf = 0 *in cyan and of* F_m (Chf, MChf) / ma *in green in the* Chf + MChf = 0 *plane in light gray.*

And

$$\vec{F} = \begin{cases} \left[\left(\dfrac{1-\sqrt{1+2Chf}}{2}\right)+i\left(\dfrac{1+\sqrt{1+2Chf}}{2}\right)\right]m\vec{a} & \text{if } 0 \le P_r \le 0.5 \\ \left[\left(\dfrac{1+\sqrt{1+2Chf}}{2}\right)+i\left(\dfrac{1-\sqrt{1+2Chf}}{2}\right)\right]m\vec{a} & \text{if } 0.5 \le P_r \le 1 \end{cases}$$

And

$$\vec{F} = \begin{cases} \left[\left(\dfrac{1-\sqrt{1-2MChf}}{2}\right)+i\left(\dfrac{1+\sqrt{1-2MChf}}{2}\right)\right]m\vec{a} & \text{if } 0 \le P_r \le 0.5 \\ \left[\left(\dfrac{1+\sqrt{1-2MChf}}{2}\right)+i\left(\dfrac{1-\sqrt{1-2MChf}}{2}\right)\right]m\vec{a} & \text{if } 0.5 \le P_r \le 1 \end{cases}$$

We can deduce also that:

$$\vec{F} = \begin{cases} \left[\left(\dfrac{1-\sqrt{DOK+Chf}}{2}\right)+i\left(\dfrac{1+\sqrt{DOK+Chf}}{2}\right)\right]m\vec{a} & \text{if } 0 \le P_r \le 0.5 \\ \left[\left(\dfrac{1+\sqrt{DOK+Chf}}{2}\right)+i\left(\dfrac{1-\sqrt{DOK+Chf}}{2}\right)\right]m\vec{a} & \text{if } 0.5 \le P_r \le 1 \end{cases}$$

And

$$\vec{F} = \begin{cases} \left[\left(\dfrac{1-\sqrt{DOK-MChf}}{2}\right)+i\left(\dfrac{1+\sqrt{DOK-MChf}}{2}\right)\right]m\vec{a} & \text{if } 0 \le P_r \le 0.5 \\ \left[\left(\dfrac{1+\sqrt{DOK-MChf}}{2}\right)+i\left(\dfrac{1-\sqrt{DOK-MChf}}{2}\right)\right]m\vec{a} & \text{if } 0.5 \le P_r \le 1 \end{cases}$$

And

$$\vec{F} = \begin{cases} \left[\left(\dfrac{1-\sqrt{1+Chf-MChf}}{2}\right)+i\left(\dfrac{1+\sqrt{1+Chf-MChf}}{2}\right)\right]m\vec{a} & \text{if } 0 \le P_r \le 0.5 \\ \left[\left(\dfrac{1+\sqrt{1+Chf-MChf}}{2}\right)+i\left(\dfrac{1-\sqrt{1+Chf-MChf}}{2}\right)\right]m\vec{a} & \text{if } 0.5 \le P_r \le 1 \end{cases}$$

But according to *CPP*: $P_c^{\,2} = DOK - Chf = DOK + MChf = 1 = P_c$ hence the complex resultant force $\vec{F} = \vec{F}_r + \vec{F}_m$ in the set $\mathcal{C} = \mathcal{R} + \mathcal{M}$ as a function of all the *CPP* parameters is the following (**Figure 15**):

$$\vec{F} = \begin{cases} \left[\left(\dfrac{P_c-\sqrt{DOK-Chf-2MChf}}{2}\right)+i\left(\dfrac{P_c+\sqrt{DOK-Chf-2MChf}}{2}\right)\right]m\vec{a} & \text{if } 0 \le P_r \le 0.5 \\ \left[\left(\dfrac{P_c+\sqrt{DOK-Chf-2MChf}}{2}\right)+i\left(\dfrac{P_c-\sqrt{DOK-Chf-2MChf}}{2}\right)\right]m\vec{a} & \text{if } 0.5 \le P_r \le 1 \end{cases}$$

And since the deterministic force in $\mathcal{C} = \mathcal{R} + \mathcal{M}$ is $\vec{F}_c = m\vec{a}$ then:

$$\vec{F} = \begin{cases} \left[\left(\frac{P_c - \sqrt{DOK - Chf - 2MChf}}{2} \right) + i \left(\frac{P_c + \sqrt{DOK - Chf - 2MChf}}{2} \right) \right] \vec{F_c} & \text{if } 0 \le P_r \le 0.5 \\ \left[\left(\frac{P_c + \sqrt{DOK - Chf - 2MChf}}{2} \right) + i \left(\frac{P_c - \sqrt{DOK - Chf - 2MChf}}{2} \right) \right] \vec{F_c} & \text{if } 0.5 \le P_r \le 1 \end{cases}$$
$$= zm\vec{a}$$

In this cube (**Figure 15**), we can notice the simulation of the complex resultant reduced force $F / ma = z(X)$ in $\mathcal{C} = \mathcal{R} + \mathcal{M}$ as a function of the real reduced force $F_r / ma = P_r(X) = \text{Re}(z)$ in $\mathcal{R}$ and of its complementary imaginary reduced force $F_m / ma = P_m(X) = i \times \text{Im}(z)$ in $\mathcal{M}$, and this in terms of the random variable X for any probability and stochastic distribution. The red curve represents F_r / ma in the plane $P_m(X) = 0$ and the blue curve represents F_m / ma in the plane $P_r(X) = 0$. The green curve represents the complex resultant reduced force $F / ma = F_r / ma + F_m / ma = z(X) = P_r(X) + P_m(X) = \text{Re}(z) + i \times \text{Im}(z)$ in the plane $z(X) = P_r(X) + P_m(X)$ or $z(X)$ plane in cyan. The curve of F / ma starts at the point J ($P_r = 0, P_m = i, X = L_b$ = lower bound of X) at $z = i$ and ends at the point L ($P_r = 1, P_m = 0, X = U_b$ = upper bound of X) at $z = 1$. The thick line in cyan is $P_r(X = L_b) + P_m(X = L_b) = z(X = L_b)$ and it is the projection of the F / ma curve on the complex probability plane whose equation is

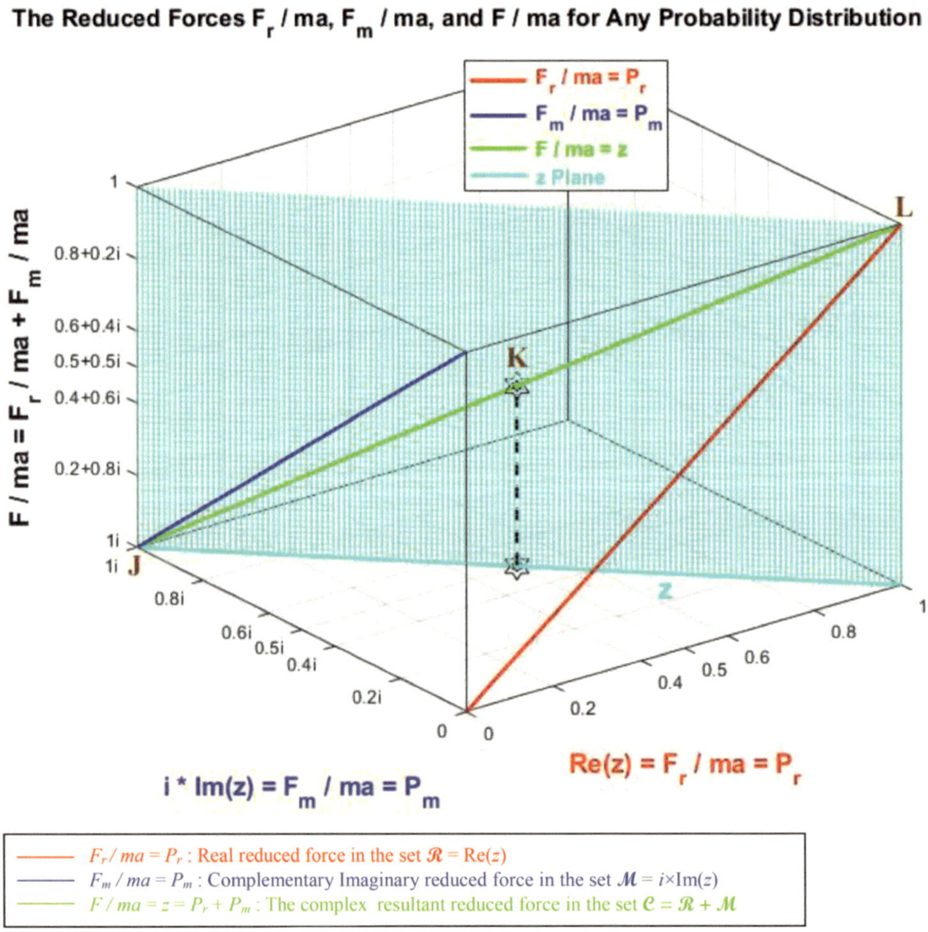

Figure 15.
The graphs of the reduced real force $F_r / ma = P_r = Re(z)$ *in red and of the reduced imaginary force* $F_m / ma = P_m = i \times Im(z)$ *in blue and of* $z = P_r + P_m$ *in cyan and of the reduced complex resultant force* $F / ma = F_r / ma + F_m / ma$ *in green in the z plane in light cyan.*

$z = i$. This projected thick line starts at the point J ($P_r = 0, P_m = i, X = L_b$) and ends at the point ($P_r = 1, P_m = 0, X = L_b$). Notice the importance of the point K corresponding to $z = 0.5 + 0.5i$ when $P_r = 0.5$ and $P_m = 0.5i$.

5.4.3.1 *The relationships between the norm of the resultant complex force and all the* CPP *parameters*

We have: $\vec{F} = \vec{F}_r + \vec{F}_m = (P_r + P_m)m\vec{a} = zm\vec{a}$ then the norm of the complex force $\vec{F}$ can be computed as follows: $\left|\vec{F}\right|^2 = |z|^2 \times m^2\left|\vec{a}\right|^2$.

But from *CPP* we have: $|z|^2 = DOK \Leftrightarrow \left|\vec{F}\right|^2 = DOK \times m^2\left|\vec{a}\right|^2 \Leftrightarrow \left|\vec{F}\right| = \sqrt{DOK} \times m\left|\vec{a}\right|$.

According to *CPP*: $P_c^2 = DOK - Chf = 1 \Leftrightarrow DOK = 1 + Chf \Leftrightarrow \left|\vec{F}\right|^2 = (1 + Chf) \times m^2\left|\vec{a}\right|^2 \Leftrightarrow \left|\vec{F}\right| = \sqrt{1 + Chf} \times m\left|\vec{a}\right|$.

Since also $P_c^2 = DOK + MChf = 1 \Leftrightarrow DOK = 1 - MChf \Leftrightarrow \left|\vec{F}\right|^2 = (1 - MChf) \times m^2\left|\vec{a}\right|^2 \Leftrightarrow \left|\vec{F}\right| = \sqrt{1 - MChf} \times m\left|\vec{a}\right|$.

Since we have: $P_c^2 = DOK - Chf = DOK + MChf = 1 = P_c \Leftrightarrow \left|\vec{F}\right|^2 = (P_c + Chf) \times m^2\left|\vec{a}\right|^2 \Leftrightarrow \left|\vec{F}\right| = \sqrt{P_c + Chf} \times m\left|\vec{a}\right|$

And $\Leftrightarrow \left|\vec{F}\right|^2 = (P_c - MChf) \times m^2\left|\vec{a}\right|^2 \Leftrightarrow \left|\vec{F}\right| = \sqrt{P_c - MChf} \times m\left|\vec{a}\right|$

5.4.4 *The relationships between the real deterministic force in* $\mathcal{C} = \mathcal{R} + \mathcal{M}$ *and all the* CPP *parameters*

Furthermore, since $\vec{F}_c = P_c m\vec{a}$ and since $P_c^2 = DOK - Chf = DOK + MChf = 1 = P_c$ therefore:

$$
\begin{aligned}
\vec{F}_c = P_c m\vec{a} &= \sqrt{DOK - Chf}.m\vec{a} \\
&= \sqrt{DOK + MChf}.m\vec{a} \\
&= \sqrt{1 + Chf + MChf}.m\vec{a} \\
&= P_c^2 m\vec{a} \\
&= 1 \times m\vec{a} = m\vec{a}
\end{aligned}
$$

Hence, we can conclude that no chaos, no ignorance, no disorder, no unpredictability, no chance, and no randomness exist in the probability universe $\mathcal{C} = \mathcal{R} + \mathcal{M}$, but complete and perfect and deterministic knowledge and experiment.

6. The resultant complex random vector *Z* of *CPP* and statistical physics

A powerful tool will be described in the current section which was developed in my personal previous research papers and which is founded on the concept of a complex random vector that is a vector combining the real and the imaginary probabilities of a random particle, defined in the three added axioms of *CPP* by the term $z_j = P_{rj} + P_{mj}$. Accordingly, we will define the vector Z as the resultant complex random vector which is the sum of all the complex random vectors z_j in the complex probability plane $\mathcal{C}$. This procedure is illustrated by considering first a general Bernoulli distribution, then we will discuss a discrete probability distribution with N equiprobable random vectors as a general case. In fact, if z represents one particle in a macrosystem from the uniform distribution U, then Z_U represents all the particles in the whole macrosystem from the uniform distribution U that means that Z_U represents the whole random distribution in the complex probability plane $\mathcal{C}$. So, in this context, it follows directly that a Bernoulli distribution can be understood as a simplified system with two random particles (section 6-1), whereas the general case is a random system with N random particles (section 6-2). Afterward, I will prove an important property at the foundation of statistical mechanics and physics using this new powerful concept (section 6-3) [45–61].

6.1 The resultant complex random vector Z of a general Bernoulli distribution (a distribution with two random particles)

First, let us consider the following general Bernoulli distribution and let us define its complex random vectors and their resultant (**Table 1**):

Where,

x_1 and x_2 are the outcomes of the first and second random vectors respectively.

P_{r1} and P_{r2} are the real probabilities of x_1 and x_2 respectively.

P_{m1} and P_{m2} are the imaginary probabilities of x_1 and x_2 respectively.

We have:

$$\sum_{j=1}^{2} P_{rj} = P_{r1} + P_{r2} = p + q = 1$$

and

$$\sum_{j=1}^{2} P_{mj} = P_{m1} + P_{m2} = iq + ip = i(1-p) + ip$$
$$= i - ip + ip = i = i(2-1) = i(N-1)$$

Where N is the number of random vectors or outcomes which is equal to 2 for a Bernoulli distribution.

The complex random vector corresponding to the random outcome x_1 is:

$$z_1 = P_{r1} + P_{m1} = p + i(1-p) = p + iq$$

Outcome	x_j	x_1	x_2
In $\mathcal{R}$	P_{rj}	$P_{r1} = p$	$P_{r2} = q$
In $\mathcal{M}$	P_{mj}	$P_{m1} = i(1-p) = iq$	$P_{m2} = i(1-q) = ip$
In $\mathcal{C} = \mathcal{R} + \mathcal{M}$	z_j	$z_1 = P_{r1} + P_{m1}$	$z_2 = P_{r2} + P_{m2}$

Table 1.
A general Bernoulli distribution in $\mathcal{R}$, $\mathcal{M}$, and $\mathcal{C}$.

The complex random vector corresponding to the random outcome x_2 is:

$$z_2 = P_{r2} + P_{m2} = q + i(1 - q) = q + ip$$

The resultant complex random vector is defined as follows:

$$\begin{aligned} Z &= \sum_{j=1}^{2} z_j = z_1 + z_2 = \sum_{j=1}^{2} P_{rj} + \sum_{j=1}^{2} P_{mj} \\ &= (p + iq) + (q + ip) = (p + q) + i(p + q) \\ &= 1 + i = 1 + i(2 - 1) \\ &\Rightarrow Z = 1 + i(N - 1) \end{aligned}$$

The probability P_{c1} in the complex plane $\mathcal{C} = \mathcal{R} + \mathcal{M}$ which corresponds to the complex random vector z_1 is computed as follows:

$$\begin{aligned} &|z_1|^2 = P_{r1}^2 + (P_{m1}/i)^2 = p^2 + q^2 \\ &Chf_1 = -2P_{r1}P_{m1}/i = -2pq \\ &\Rightarrow P_{c1}^2 = |z_1|^2 - Chf_1 \\ &\qquad = p^2 + q^2 + 2pq = (p + q)^2 = 1^2 = 1 \\ &\Rightarrow P_{c1} = 1 \end{aligned}$$

This is coherent with the three novel complementary axioms defined for the *CPP*.

Similarly, P_{c2} corresponding to z_2 is:

$$\begin{aligned} &|z_2|^2 = P_{r2}^2 + (P_{m2}/i)^2 = q^2 + p^2 \\ &Chf_2 = -2P_{r2}P_{m2}/i = -2qp \\ &\Rightarrow P_{c2}^2 = |z_2|^2 - Chf_2 \\ &\qquad = q^2 + p^2 + 2qp = (q + p)^2 = 1^2 = 1 \\ &\Rightarrow P_{c2} = 1 \end{aligned}$$

The probability P_c in the complex plane $\mathcal{C}$ which corresponds to the resultant complex random vector $Z = 1 + i$ is computed as follows:

$$|Z|^2 = \left(\sum_{j=1}^{2} P_{rj}\right)^2 + \left(\sum_{j=1}^{2} P_{mj}/i\right)^2 = 1^2 + 1^2 = 2$$

$$Chf = -2\sum_{j=1}^{2} P_{rj} \sum_{j=1}^{2} P_{mj}/i = -2(1)(1) = -2$$

$$\begin{aligned} &\text{Let } s^2 = |Z|^2 - Chf = 2 + 2 = 4 \Rightarrow s = 2 \\ &\Rightarrow P_c^2 = \frac{s^2}{N^2} = \frac{|Z|^2 - Chf}{N^2} = \frac{|Z|^2}{N^2} - \frac{Chf}{N^2} = \frac{4}{2^2} = \frac{4}{4} = 1 \\ &\Rightarrow P_c = \frac{s}{N} = \frac{2}{2} = 1 \end{aligned}$$

Where s is an intermediary quantity used in our computation of P_c.

P_c is the probability corresponding to the resultant complex random vector Z in the probability universe $\mathcal{C} = \mathcal{R} + \mathcal{M}$ and is also equal to 1. Actually, Z represents both z_1 and z_2 that means the whole distribution of random vectors of the general Bernoulli distribution in the complex plane $\boldsymbol{C}$ and its probability P_c is computed in the same way as P_{c1} and P_{c2}.

By analogy, for the case of one random vector z_j we have:

$$P_{cj}^2 = |z_j|^2 - Chf_j \quad \text{with} \quad (N = 1).$$

In general, for the vector Z we have:

$$P_c{}^2 = \frac{|Z|^2}{N^2} - \frac{Chf}{N^2}; \quad (N \geq 1)$$

Where the degree of our knowledge of the whole distribution is equal to $DOK_Z = \frac{|Z|^2}{N^2}$, its relative chaotic factor is $Chf_Z = \frac{Chf}{N^2}$, and its relative magnitude of the chaotic factor is $MChf_Z = |Chf_Z|$.

Notice, if $N = 1$ in the previous formula, then:

$$P_c{}^2 = \frac{|Z|^2}{N^2} - \frac{Chf}{N^2} = \frac{|Z|^2}{1^2} - \frac{Chf}{1^2} = |Z|^2 - Chf = |z_j|^2 - Chf_j = P_{cj}^2$$

which is coherent with the calculations already done.

To illustrate the concept of the resultant complex random vector Z, I will use the following graph (**Figure 16**).

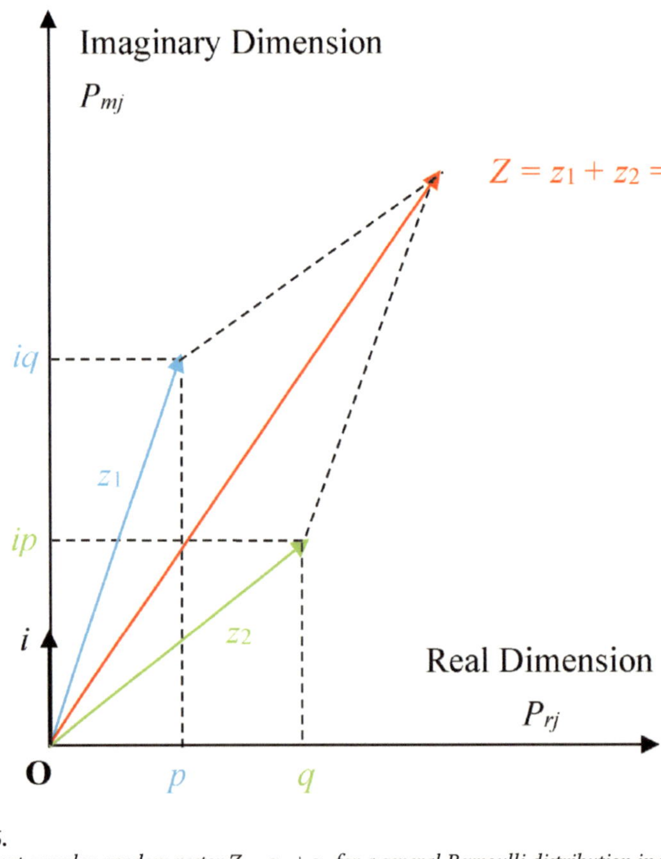

Figure 16.
The resultant complex random vector $Z = z_1 + z_2$ for a general Bernoulli distribution in the complex probability plane $\boldsymbol{C}$.

6.2 The general case: A discrete distribution with *N* Equiprobable random vectors (a uniform distribution *U* with *N* random particles)

As a general case, let us consider then this discrete probability distribution with *N* equiprobable random vectors which is a discrete uniform probability distribution *U* with *N* particles (**Table 2**):

We have here in $\mathcal{C} = \mathcal{R} + \mathcal{M}$:

$$z_j = P_{rj} + P_{mj}, \quad \forall j: \ 1 \le j \le N,$$

$$\text{and } z_1 = z_2 = \ldots = z_N = \frac{1}{N} + \frac{i(N-1)}{N}$$

$$\Rightarrow Z_U = \sum_{j=1}^{N} z_j = z_1 + z_2 + \ldots + z_N = N z_j = N\left(\frac{1}{N} + \frac{i(N-1)}{N}\right) = 1 + i(N-1)$$

Moreover, we can notice that: $|z_1| = |z_2| = \cdots = |z_N|$, hence,

$$|Z_U| = |z_1 + z_2 + \ldots + z_N| = N|z_1| = N|z_2| = \ldots = N|z_N|$$

$$\Rightarrow |Z_U|^2 = N^2 |z_j|^2 = N^2\left(\frac{1}{N^2} + \frac{(N-1)^2}{N^2}\right) = 1 + (N-1)^2, \text{ where } 1 \le j \le N;$$

And

$$Chf = N^2 \times Chf_j = -2 \times P_{rj} \times (P_{mj}/i) \times N^2 = -2N^2 \times \left(\frac{1}{N}\right)\left(\frac{N-1}{N}\right)$$
$$= -2(1)(N-1) = -2(N-1)$$

$$\Rightarrow s^2 = |Z_U|^2 - Chf = 1 + (N-1)^2 + 2(N-1) = [1 + (N-1)]^2 = N^2$$

$$\Rightarrow Pc^2|_{Z_U} = \frac{s^2}{N^2} = \frac{N^2}{N^2} = 1$$

$$= \frac{|Z_U|^2}{N^2} - \frac{Chf}{N^2} = \frac{1 + (N-1)^2}{N^2} - \frac{-2(N-1)}{N^2} = \frac{1 + (N-1)^2 + 2(N-1)}{N^2} = \frac{[1 + (N-1)]^2}{N^2} = \frac{N^2}{N^2} = 1$$

$$\Rightarrow Pc|_{Z_U} = 1$$

Where s is an intermediary quantity used in our computation of $Pc|_{Z_U}$.

Therefore, the degree of our knowledge corresponding to the resultant complex vector Z_U representing the whole uniform distribution is:

$$DOK_{Z_U} = \frac{|Z_U|^2}{N^2} = \frac{1 + (N-1)^2}{N^2},$$

and its relative chaotic factor is:

$$Chf_{Z_U} = \frac{Chf}{N^2} = -\frac{2(N-1)}{N^2},$$

Similarly, its relative magnitude of the chaotic factor is:

Outcome	x_j	x_1	x_2	...	x_N
In $\mathcal{R}$	P_{rj}	$P_{r1} = \frac{1}{N}$	$P_{r2} = \frac{1}{N}$	...	$P_{rN} = \frac{1}{N}$
In $\mathcal{M}$	P_{mj}	$P_{m1} = i(1-\frac{1}{N})$	$P_{m2} = i(1-\frac{1}{N})$	...	$P_{mN} = i(1-\frac{1}{N})$
In $\mathcal{C} = \mathcal{R} + \mathcal{M}$	z_j	$z_1 = P_{r1} + P_{m1}$	$z_2 = P_{r2} + P_{m2}$	...	$z_N = P_{rN} + P_{mN}$

Table 2.
A discrete uniform distribution with N *equiprobable random vectors in $\mathcal{R}$, $\mathcal{M}$, and $\mathcal{C}$.*

$$MChf_{Z_U} = \left|Chf_{Z_U}\right| = \left|\frac{Chf}{N^2}\right| = \left|-\frac{2(N-1)}{N^2}\right| = \frac{2(N-1)}{N^2}.$$

Thus, we can verify that we have always:

$$P_c^2|_{Z_U} = \frac{|Z_U|^2}{N^2} - \frac{Chf}{N^2} = DOK_{Z_U} - Chf_{Z_U} = DOK_{Z_U} + MChf_{Z_U} = 1 \Leftrightarrow P_c|_{Z_U} = 1$$

What is important here is that we can notice the following fact. Take for example:

$$N = 2 \Rightarrow DOK_{Z_U} = \frac{1+(2-1)^2}{2^2} = 0.5 \text{ and } Chf_{Z_U} = \frac{-2(2-1)}{2^2} = -0.5$$

$$N = 4 \Rightarrow DOK_{Z_U} = \frac{1+(4-1)^2}{4^2} = 0.625 \geq 0.5 \text{ and } Chf_{Z_U} = \frac{-2(4-1)}{4^2} = -0.375 \geq -0.5$$

$$N = 5 \Rightarrow DOK_{Z_U} = \frac{1+(5-1)^2}{5^2} = 0.68 \geq 0.625 \text{ and } Chf_{Z_U} = \frac{-2(5-1)}{5^2} = -0.32 \geq -0.375$$

$$N = 10 \Rightarrow DOK_{Z_U} = \frac{1+(10-1)^2}{10^2} = 0.82 \geq 0.68 \text{ and } Chf_{Z_U} = \frac{-2(10-1)}{10^2} = -0.18 \geq -0.32$$

$$N = 100 \Rightarrow DOK_{Z_U} = \frac{1+(100-1)^2}{100^2} = 0.9802 \geq 0.82 \text{ and } Chf_{Z_U} = \frac{-2(100-1)}{100^2} = -0.0198 \geq -0.18$$

$$N = 1000 \Rightarrow DOK_{Z_U} = \frac{1+(1000-1)^2}{1000^2} = 0.998002 \geq 0.9802 \text{ and}$$

$$Chf_{Z_U} = \frac{-2(1000-1)}{1000^2} = -0.001998 \geq -0.0198$$

$$N = 1,000,000 \Rightarrow DOK_{Z_U} = \frac{1+\left(10^6-1\right)^2}{10^{12}} = 0.999998 \geq 0.998002 \text{ and}$$

$$Chf_{Z_U} = \frac{-2\left(10^6-1\right)}{10^{12}} = -0.000001999998 \geq -0.001998$$

We can deduce mathematically using calculus that:

$$\lim_{N\to+\infty} \frac{|Z_U|^2}{N^2} = \lim_{N\to+\infty} DOK_{Z_U} = \lim_{N\to+\infty} \frac{1+(N-1)^2}{N^2} = 1,$$

$$\text{and } \lim_{N\to+\infty}\frac{Chf}{N^2} = \lim_{N\to+\infty} Chf_{Z_U} = \lim_{N\to+\infty} -\frac{2(N-1)}{N^2} = 0.$$

From the above, we can also deduce this conclusion:

As much as N increases, as much as the degree of our knowledge in $\mathcal{R}$ corresponding to the resultant complex vector is perfect and absolute, that means, it is equal to one, and as much as the chaotic factor that prevents us from foretelling exactly and totally the outcome of the stochastic phenomenon in $\mathcal{R}$ approaches zero. Mathematically we state that: If N tends to infinity then the degree of our knowledge in $\mathcal{R}$ tends to one and the chaotic factor always in $\mathcal{R}$ tends to zero.

6.3 Statistical mechanics using *Z* and *CPP*

We have:

$P_r|_{Z_U} = \sum_{j=1}^{N} P_{rj}/N = N \times \frac{P_{rj}}{N} = P_{rj} = \frac{1}{N}$= the mean of the real probability of all the N complex random vectors z_j represented by Z_U, and.

$P_m|_{Z_U} = \sum_{j=1}^{N} P_{mj}/N = N \times \frac{P_{mj}}{N} = P_{mj} = i\left(1 - \frac{1}{N}\right)$= the mean of the imaginary probability of all the N complex random vectors z_j represented by Z_U, then:

$Z_U = Nz_j = N(P_r|_{Z_U} + P_m|_{Z_U}) = N\left[\frac{1}{N} + i\left(1 - \frac{1}{N}\right)\right] = 1 + i(N-1)$, as computed in section 6-2.

Where $\frac{Z_U}{N} = P_r|_{Z_U} + P_m|_{Z_U} = \frac{\sum_{j=1}^{N} z_j}{N} = \frac{Nz_j}{N} = z_j = P_{rj} + P_{mj} = \frac{1}{N} + i\left(1 - \frac{1}{N}\right), \ \forall j : 1 \le j \le N$

= the mean of all the N complex random vectors z_j represented by Z_U.

Therefore, $P_c|_{Z_U} = P_r|_{Z_U} + \frac{P_m|_{Z_U}}{i} = \frac{1}{N} + \left(1 - \frac{1}{N}\right) = 1 = P_{cj}, \ \forall j : 1 \le j \le N$, just as predicted by *CPP*.

Additionally, we have:

$\vec{F}_{rj} = P_{rj} m \vec{a}_j, \forall j : 1 \le j \le N$, that means for every particle j in the macrosystem of N particles, and

$$\begin{aligned}
\vec{F}_r|_{Z_U} &= \sum_{j=1}^{N} \vec{F}_{rj} = P_{r1} m\vec{a}_1 + P_{r2} m\vec{a}_2 + \ldots + P_{rj} m\vec{a}_j + \ldots + P_{rN} m\vec{a}_N \\
&= \frac{1}{N} m\vec{a}_1 + \frac{1}{N} m\vec{a}_2 + \ldots + \frac{1}{N} m\vec{a}_j + \ldots + \frac{1}{N} m\vec{a}_N \\
&= \frac{1}{N}\left(m\vec{a}_1 + m\vec{a}_2 + \ldots + m\vec{a}_j + \ldots + m\vec{a}_N\right) \\
&= P_r|_{Z_U}\left(m\vec{a}_1 + m\vec{a}_2 + \ldots + m\vec{a}_j + \ldots + m\vec{a}_N\right) \\
&= P_r|_{Z_U} m \sum_{j=1}^{N} \vec{a}_j = P_r|_{Z_U} m\vec{a} = \frac{m\vec{a}}{N}
\end{aligned}$$

= the mean real random force acting on the whole macrosystem in $\mathcal{R}$.

Moreover,

$\vec{F}_{mj} = P_{mj} m \vec{a}_j, \forall j : 1 \le j \le N$, that means for every particle j in the macrosystem of N particles, and

$$\vec{F}_m|_{Z_U} = \sum_{j=1}^{N} \vec{F}_{mj} = P_{m1}m\vec{a}_1 + P_{m2}m\vec{a}_2 + \ldots + P_{mj}m\vec{a}_j + \ldots + P_{mN}m\vec{a}_N$$

$$= i\left(1-\frac{1}{N}\right)m\vec{a}_1 + i\left(1-\frac{1}{N}\right)m\vec{a}_2 + \ldots + i\left(1-\frac{1}{N}\right)m\vec{a}_j + \ldots + i\left(1-\frac{1}{N}\right)m\vec{a}_N$$

$$= i\left(1-\frac{1}{N}\right)\left(m\vec{a}_1 + m\vec{a}_2 + \ldots + m\vec{a}_j + \ldots + m\vec{a}_N\right)$$

$$= P_m|_{Z_U}\left(m\vec{a}_1 + m\vec{a}_2 + \ldots + m\vec{a}_j + \ldots + m\vec{a}_N\right)$$

$$= P_m|_{Z_U} m \sum_{j=1}^{N} \vec{a}_j = P_m|_{Z_U} m\vec{a} = i\left(1-\frac{1}{N}\right)m\vec{a}$$

= the mean imaginary random force acting on the whole macrosystem in $\mathcal{M}$. Furthermore,

$$\vec{F}|_{Z_U} = \vec{F}_r|_{Z_U} + \vec{F}_m|_{Z_U} = \sum_{j=1}^{N} \vec{F}_{rj} + \sum_{j=1}^{N} \vec{F}_{mj} = P_r|_{Z_U} m\vec{a} + P_m|_{Z_U} m\vec{a}$$

$$= (P_r|_{Z_U} + P_m|_{Z_U})m\vec{a} = \frac{Z_U}{N} m\vec{a} = \left[\frac{1}{N} + i\left(1-\frac{1}{N}\right)\right]m\vec{a}$$

= the mean resultant complex random force acting on the whole macrosystem in $\mathcal{C} = \mathcal{R} + \mathcal{M}$.

Also, we have:

$\vec{F}_{cj} = P_{cj}m\vec{a}_j = 1 \times m\vec{a}_j = m\vec{a}_j, \forall j : 1 \leq j \leq N$, that means for every particle j in the macrosystem of N particles, just as predicted by *CPP*.

And $\vec{F}_c|_{Z_U} = P_c\big|_{Z_U} m\vec{a} = 1 \times m\vec{a} = m\vec{a}$ = the deterministic force acting on the whole macrosystem in $\mathcal{C} = \mathcal{R} + \mathcal{M}$, as predicted by *CPP* also.

Correspondingly, we can deduce the following result:

$$\text{If } DOK_{Z_U} = \frac{|Z_U|^2}{N^2} = (P_r|_{Z_U})^2 + \left(\frac{P_m|_{Z_U}}{i}\right)^2 = P_r^2|_{Z_U} + (1 - P_r|_{Z_U})^2 = 1$$

$$\Leftrightarrow \begin{cases} P_r|_{Z_U} = \frac{1}{N} = 0 \\ \text{or} \\ P_r|_{Z_U} = \frac{1}{N} = 1 \end{cases} \Leftrightarrow \begin{cases} N \to +\infty \\ \text{or} \\ N = 1 \end{cases} \Leftrightarrow \begin{cases} \vec{F}_r|_{Z_U} = P_r|_{Z_U} \times m\vec{a} = 0 \times m\vec{a} = \vec{0} \\ \text{or} \\ \vec{F}_r|_{Z_U} = P_r|_{Z_U} \times m\vec{a} = 1 \times m\vec{a} = m\vec{a} \end{cases}$$

$$\Leftrightarrow \begin{cases} P_m|_{Z_U} = i(1 - P_r|_{Z_U}) = i(1-0) = i \\ \text{or} \\ P_m|_{Z_U} = i(1 - P_r|_{Z_U}) = i(1-1) = 0 \end{cases} \Leftrightarrow \begin{cases} \vec{F}_m|_{Z_U} = P_m|_{Z_U} \times m\vec{a} = im\vec{a} \\ \text{or} \\ \vec{F}_m|_{Z_U} = P_m|_{Z_U} \times m\vec{a} = 0 \times m\vec{a} = \vec{0} \end{cases}$$

Therefore, this means that in the first case the mean real force acting on the macrosystem in the real set $\mathcal{R}$ is equal to $\vec{0}$, or that in the second case the experiment on the macrosystem is totally deterministic always in the real probability set $\mathcal{R}$.

$$\Leftrightarrow \begin{cases} \vec{F}|_{Z_U} = \vec{F}_r|_{Z_U} + \vec{F}_m|_{Z_U} = \vec{0} + im\vec{a} = im\vec{a} \\ \text{or} \\ \vec{F}|_{Z_U} = \vec{F}_r|_{Z_U} + \vec{F}_m|_{Z_U} = m\vec{a} + \vec{0} = m\vec{a} \end{cases}$$

$$\Leftrightarrow \left|\vec{F}|_{Z_U}\right| = \begin{cases} \left|im\vec{a}\right| = m\left|\vec{a}\right| \\ \text{or} \\ \left|m\vec{a}\right| = m\left|\vec{a}\right| \end{cases} \Leftrightarrow \left|\vec{F}|_{Z_U}\right| = m\left|\vec{a}\right| = \left|\vec{F}_c|_{Z_U}\right| \text{ in both cases.}$$

That means that the mean norm of the resultant force acting on the whole macrosystem is totally deterministic in both cases in the probability set $\mathcal{C} = \mathcal{R} + \mathcal{M}$ and is always equal accordingly to $m\left|\vec{a}\right|$.

Similarly, we can deduce also the following similar result:

$$\text{If } DOK_{Z_U} = 1 \Leftrightarrow Chf_{Z_U} = 2iP_r|_{Z_U} \times P_m|_{Z_U} = -2P_r|_{Z_U} \times (1 - P_r|_{Z_U}) = 0$$

$$\Leftrightarrow \begin{cases} P_r|_{Z_U} = \dfrac{1}{N} = 0 \\ \text{or} \\ P_r|_{Z_U} = \dfrac{1}{N} = 1 \end{cases} \Leftrightarrow \begin{cases} N \to +\infty \\ \text{or} \\ N = 1 \end{cases} \Leftrightarrow \begin{cases} \vec{F}_r|_{Z_U} = P_r|_{Z_U} \times m\vec{a} = 0 \times m\vec{a} = \vec{0} \\ \text{or} \\ \vec{F}_r|_{Z_U} = P_r|_{Z_U} \times m\vec{a} = 1 \times m\vec{a} = m\vec{a} \end{cases}$$

$$\Leftrightarrow \begin{cases} P_m|_{Z_U} = i(1 - P_r|_{Z_U}) = i(1-0) = i \\ \text{or} \\ P_m|_{Z_U} = i(1 - P_r|_{Z_U}) = i(1-1) = 0 \end{cases} \Leftrightarrow \begin{cases} \vec{F}_m|_{Z_U} = P_m|_{Z_U} \times m\vec{a} = im\vec{a} \\ \text{or} \\ \vec{F}_m|_{Z_U} = P_m|_{Z_U} \times m\vec{a} = 0 \times m\vec{a} = \vec{0} \end{cases}$$

Therefore, this means that in the first case the mean real force acting on the macrosystem in the real set $\mathcal{R}$ is equal to $\vec{0}$, or that in the second case the experiment on the macrosystem is totally deterministic always in the real probability set $\mathcal{R}$.

$$\Leftrightarrow \begin{cases} \vec{F}|_{Z_U} = \vec{F}_r|_{Z_U} + \vec{F}_m|_{Z_U} = \vec{0} + im\vec{a} = im\vec{a} \\ \text{or} \\ \vec{F}|_{Z_U} = \vec{F}_r|_{Z_U} + \vec{F}_m|_{Z_U} = m\vec{a} + \vec{0} = m\vec{a} \end{cases}$$

$$\Leftrightarrow \left|\vec{F}|_{Z_U}\right| = \begin{cases} \left|im\vec{a}\right| = m\left|\vec{a}\right| \\ \text{or} \\ \left|m\vec{a}\right| = m\left|\vec{a}\right| \end{cases} \Leftrightarrow \left|\vec{F}|_{Z_U}\right| = m\left|\vec{a}\right| = \left|\vec{F}_c|_{Z_U}\right| \text{ in both cases.}$$

That means that the mean norm of the resultant force acting on the whole macrosystem is totally deterministic in both cases in the probability set $\mathcal{C} = \mathcal{R} + \mathcal{M}$ and is always equal accordingly to $m\left|\vec{a}\right|$. Consequently, we reach the same conclusion if we consider Chf_{Z_U} as above when we have considered DOK_{Z_U}.

In addition, for $N = 1 \Rightarrow \frac{|Z_U|^2}{N^2} = DOK_{Z_U} = \frac{1+(N-1)^2}{N^2} = \frac{1+(1-1)^2}{1^2} = 1$

$$\text{and } \frac{Chf}{N^2} = Chf_{Z_U} = -\frac{2(N-1)}{N^2} = -\frac{2(1-1)}{1^2} = 0$$

This means that we have a random experiment with only one outcome or vector, hence, $P_r|_{Z_U} = \frac{1}{N} = \frac{1}{1} = 1$, that means we have a sure event in $\mathcal{R}$. Consequently, we have accordingly the degree of our knowledge is equal to one (perfect macrosystem knowledge) and the chaotic factor is equal to zero (no chaos) since the experiment is certain and totally deterministic in $\mathcal{R}$, which is absolutely logical.

6.4 Analysis and interpretation of all the results

The law of large numbers states that:

"As N increases, then the probability that the value of sample mean to be close to population mean approaches 1".

We can deduce now the following conclusion related to the law of large numbers:

We can see, as we have proved, that as much as N increases, as much as the degree of knowledge of the resultant complex vector $DOK_{Z_U} = \frac{|Z_U|^2}{N^2}$ tends to 1 and its relative chaotic factor $Chf_{Z_U} = \frac{Chf}{N^2}$ tends to 0. Assume now that the random variables x_j's correspond to the atoms or particles or molecules moving randomly in a gas or a liquid. So, if we study a gas or a liquid with billions of such particles, then N is big enough (e.g. Avogadro's number $\approx 6.02214 \times 10^{23}$ / mole in the International System of Units) to allow that its corresponding temperature, pressure, energy etc. ... tend to the mean of these quantities corresponding to the whole system. This because the chaotic factor of the whole macrosystem (gas, liquid, etc.), that is, of the resultant complex random vector Z_U representing all the random particles or vectors, tends to 0; thus, the behavior and characteristics of the whole system in $\mathcal{R}$ is predictable with great precision since the degree of our knowledge of

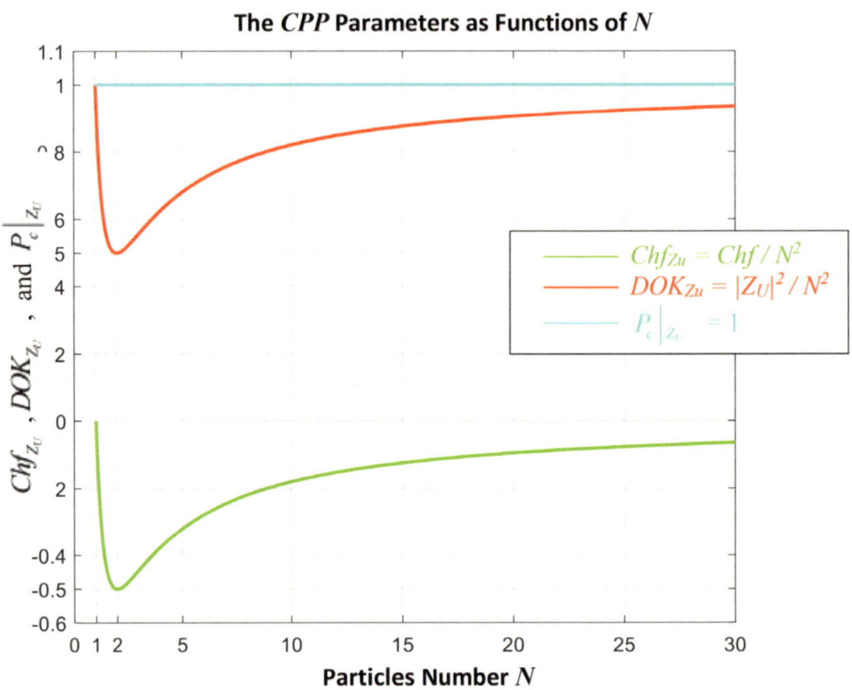

Figure 17.
Chf_{Z_U}, DOK_{Z_U}, and $P_c|_{Z_U}$, as functions of the particles number N in 2D.

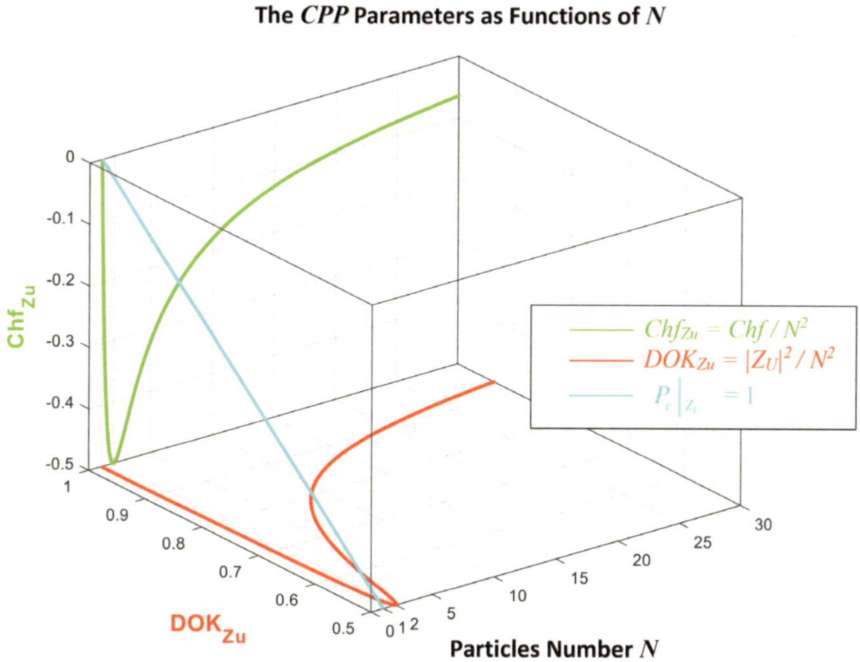

Figure 18.
Chf_{Z_U}, DOK_{Z_U}, and $P_c|_{Z_U}$, as functions of the particles number N *in 3D.*

the whole macrosystem tends to 1. Subsequently, we can deduce from the above that since for $DOK_{Z_U} = 1$ or for $Chf_{Z_U} = 0$ the mean norm of the resultant force acting on the macrosystem that consists of $N >> 1$ individual particles is totally known and deterministic in $\mathcal{R}$ then all the properties of the macrosystem are totally and completely known and determined like the macrosystem energy which should be equal to the mean of the individual particles energies, or the macrosystem pressure which should be equal to the mean of the individual particles pressures or the macrosystem temperature which should be equal to the mean of the individual particles temperatures, etc.

Hence, what we have done here is that we have proved the law of large numbers (already discussed in the published papers [46, 50, 57, 61]) as well as an important property of statistical mechanics using *CPP*. In fact, as it is very well known in the classical probability theory and statistics, the law of large numbers is tightly related and linked to statistical mechanics. Here *CPP* comes and proves both of them in a novel and original way. This looks very interesting and fruitful and shows the validity and the benefits of extending Kolmogorov's axioms to the complex probability set $\mathcal{C} = \mathcal{R} + \mathcal{M}$. The following figures (**Figures 17** and **18**) show the convergence of Chf_{Z_U} to 0 and of DOK_{Z_U} to 1 as functions of the particles or atoms or molecules number N.

7. Flowchart of the complex probability and Newton's mechanics prognostic model

The following flowchart summarizes all the procedures of the proposed complex probability prognostic model where X is between the lower bound L_b and the upper bound U_b:

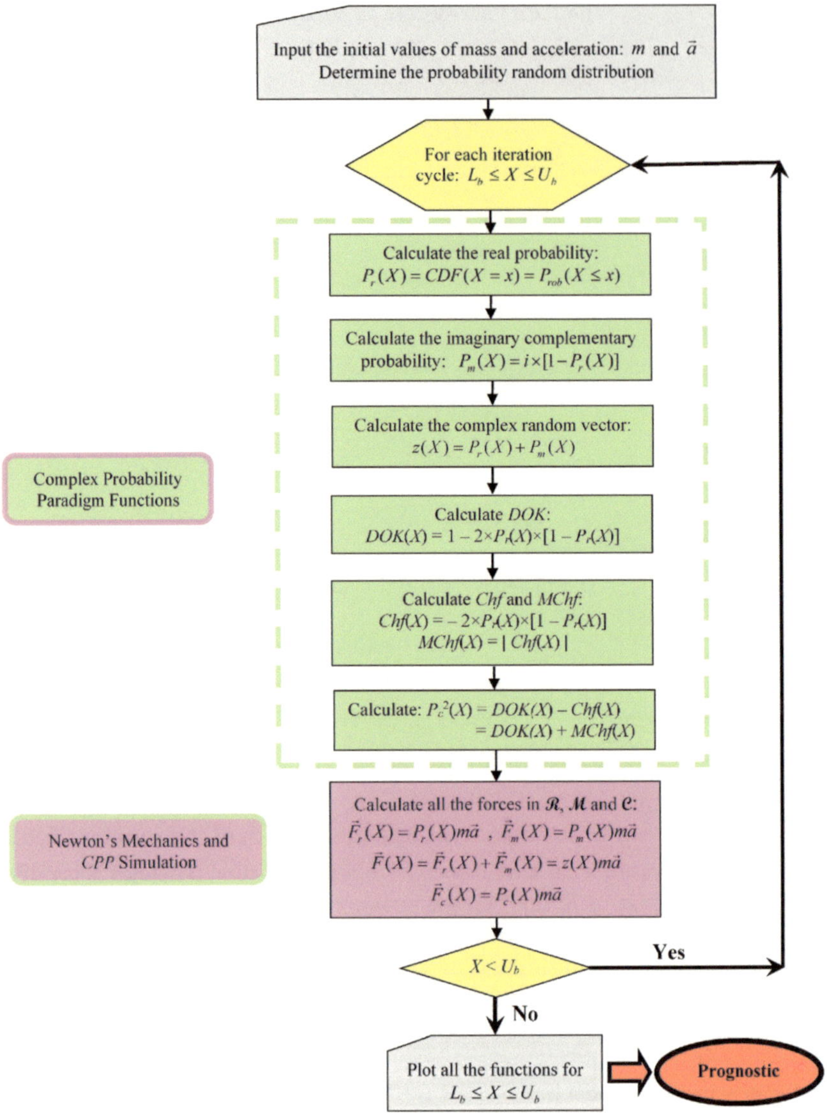

8. The new paradigm applied to various discrete and continuous stochastic distributions

In this section, the simulation of the novel *CPP* model for various discrete and continuous random distributions will be done. Note that all the numerical values found in the paradigm functions analysis for all the simulations were computed using the 64-Bit MATLAB version 2020 software. It is important to mention here that a few important and well-known probability distributions were considered although the original *CPP* model can be applied to any stochastic distribution beside the studied random cases below. This will lead to similar results and conclusions. Hence, the new paradigm is successful with any discrete or continuous random case.

8.1 Simulation of discrete probability distributions

8.1.1 The discrete uniform probability distribution

The probability density function (*PDF*) of this discrete stochastic distribution is:

$$f(X = x_k; N) = \begin{cases} 0 & \text{for } X = x_0 = L_b, k = 0 \\ \dfrac{1}{N} & \text{for } X = x_1, x_2, \ldots, x_k, \ldots, (x_N = U_b), \forall k : 1 \le k \le N \end{cases}$$

Note that in the simulation we have considered: $L_b = -21$ and $U_b = 21$ and $N = 60$ and $\forall k : 1 \le k \le (N = 60)$ we have : $\Delta x_k = x_k - x_{k-1} = 0.7$.

The cumulative distribution function (CDF) is:

$$CDF(x) = P_{rob}(X \le x) = \sum_{j=0}^{k} f(x_j; N) = f(x_0; N) + \sum_{j=1}^{k} f(x_j; N) = 0 + \sum_{j=1}^{k} \frac{1}{N} = \frac{k}{N}$$
$$= \frac{k}{60}, \forall k : 0 \le k \le (N = 60)$$

Note that:

If $k = 0 \Leftrightarrow CDF(x) = P_{rob}(X \le x) = f(X = x_0 = L_b; N) = 0$.

If $k = N \Leftrightarrow X = x_N = U_b$

$$\Leftrightarrow CDF(x) = P_{rob}(X \le x) = f(x_0; N) + \sum_{j=1}^{k=N} f(x_j; N) = 0 + \sum_{j=1}^{k=N} \frac{1}{N} = \frac{N}{N} = \frac{60}{60} = 1$$

The mean or average or expectation is:

$$\mu = \frac{\sum_{j=0}^{N} x_j}{N+1} = 0$$

The variance is:

$$\sigma^2 = \frac{\sum_{j=0}^{N} (x_j - \mu)^2}{N+1} = 151.9000$$

The standard deviation is:

$$\sigma = \sqrt{\frac{\sum_{j=0}^{N} (x_j - \mu)^2}{N+1}} = \sqrt{151.9000} = 12.3247718$$

The median $Md = 0 = \mu$ since it is a symmetric distribution.

Since the distribution is uniform then it has no mode.

The real probability $P_r(x)$ and force are:

$$P_r(x) = CDF(x) = \sum_{j=0}^{k} f(x_j; N) = \frac{k}{N} = \frac{k}{60}, \forall k : 0 \le k \le (N = 60)$$
$$\Leftrightarrow \vec{F}_r(x) = P_r(x) m\vec{a} = \left(\frac{k}{N}\right) m\vec{a} = \left(\frac{k}{60}\right) m\vec{a}$$

The imaginary complementary probability $P_m(x)$ and force are:

$$P_m(x) = i[1 - P_r(x)] = i[1 - CDF(x)] = i\left[1 - \sum_{j=0}^{k} f(x_j; N)\right]$$
$$= i \sum_{j=k+1}^{N} f(x_j; N) = i\left(1 - \frac{k}{N}\right) = i\left(1 - \frac{k}{60}\right), \quad \forall k : 0 \le k \le (N = 60)$$

$$\Leftrightarrow \vec{F}_m(x) = P_m(x)m\vec{a} = i\left(1-\frac{k}{N}\right)m\vec{a} = i\left(1-\frac{k}{60}\right)m\vec{a}$$

The real complementary probability $P_m(x)/i$ and force are:

$$P_m(x)/i = 1 - P_r(x) = 1 - CDF(x) = 1 - \sum_{j=0}^{k} f(x_j; N) = \sum_{j=k+1}^{N} f(x_j; N) = 1 - \frac{k}{N}$$
$$= 1 - \frac{k}{60}, \forall k : 0 \leq k \leq (N = 60)$$
$$\Leftrightarrow \vec{F}_m(x)/i = \frac{P_m(x)}{i}m\vec{a} = \left(1-\frac{k}{N}\right)m\vec{a} = \left(1-\frac{k}{60}\right)m\vec{a}$$

The complex probability or random vector and force are:

$$z(x) = P_r(x) + P_m(x) = \frac{k}{N} + i\left(1-\frac{k}{N}\right) = \frac{k}{60} + i\left(1-\frac{k}{60}\right), \forall k : 0 \leq k \leq (N = 60)$$

$$\Leftrightarrow \vec{F}(x) = \vec{F}_r(x) + \vec{F}_m(x) = P_r(x)m\vec{a} + P_m(x)m\vec{a} = [P_r(x) + P_m(x)]m\vec{a} = zm\vec{a}$$
$$= \left(\frac{k}{N}\right)m\vec{a} + i\left(1-\frac{k}{N}\right)m\vec{a} = \left[\left(\frac{k}{N}\right) + i\left(1-\frac{k}{N}\right)\right]m\vec{a}$$
$$= \left[\left(\frac{k}{60}\right) + i\left(1-\frac{k}{60}\right)\right]m\vec{a}$$

The Degree of Our Knowledge:

$$DOK(x) = |z(x)|^2 = P_r^2(x) + [P_m(x)/i]^2 = \left(\frac{k}{N}\right)^2 + \left(1-\frac{k}{N}\right)^2$$
$$= 1 + 2iP_r(x)P_m(x) = 1 - 2P_r(x)[1 - P_r(x)] = 1 - 2P_r(x) + 2P_r^2(x)$$
$$= 1 - 2\left(\frac{k}{N}\right) + 2\left(\frac{k}{N}\right)^2$$
$$= 1 - 2\left(\frac{k}{60}\right) + 2\left(\frac{k}{60}\right)^2, \quad \forall k : 0 \leq k \leq (N = 60)$$

$DOK(x)$ is equal to 1 when $P_r(x) = P_r(L_b = -21) = 0$ and when $P_r(x) = P_r(U_b = 21) = 1$.

The Chaotic Factor:

$$Chf(x) = 2iP_r(x)P_m(x) = -2P_r(x)[1 - P_r(x)] = -2P_r(x) + 2P_r^2(x)$$
$$= -2\left(\frac{k}{N}\right) + 2\left(\frac{k}{N}\right)^2$$
$$= -2\left(\frac{k}{60}\right) + 2\left(\frac{k}{60}\right)^2, \quad \forall k : 0 \leq k \leq (N = 60)$$

$Chf(x)$ is null when $P_r(x) = P_r(L_b = -21) = 0$ and when $P_r(x) = P_r(U_b = 21) = 1$.

The Magnitude of the Chaotic Factor $MChf$:

$$MChf(x) = |Chf(x)| = -2iP_r(x)P_m(x) = 2P_r(x)[1 - P_r(x)] = 2P_r(x) - 2P_r^2(x)$$

$$= 2\left(\frac{k}{N}\right) - 2\left(\frac{k}{N}\right)^2$$
$$= 2\left(\frac{k}{60}\right) - 2\left(\frac{k}{60}\right)^2, \quad \forall k : 0 \le k \le (N = 60)$$

$MChf(x)$ is null when $P_r(x) = P_r(L_b = -21) = 0$ and when $P_r(x) = P_r(U_b = 21) = 1$.

At any value of x: $\forall x : (L_b = -21) \le x \le (U_b = 21)$ and $\forall k : 0 \le k \le (N = 60)$, the probability expressed in the complex probability set $\mathcal{C} = \mathcal{R} + \mathcal{M}$ is the following:

$$\begin{aligned} P_c^2(x) &= [P_r(x) + P_m(x)/i]^2 = |z(x)|^2 - 2iP_r(x)P_m(x) \\ &= DOK(x) - Chf(x) \\ &= DOK(x) + MChf(x) \\ &= 1 \end{aligned}$$

then,

$$P_c^2(x) = [P_r(x) + P_m(x)/i]^2 = \{P_r(x) + [1 - P_r(x)]\}^2 = 1^2 = 1 \Leftrightarrow P_c(x) = 1 \text{ always}$$

$\Leftrightarrow \vec{F}_c(x) = P_c(x)m\vec{a} = 1 \times m\vec{a} = m\vec{a}$ always also.

Hence, the prediction of all the probabilities and forces of the stochastic experiment in the universe $\mathcal{C} = \mathcal{R} + \mathcal{M}$ is permanently certain and perfectly deterministic (**Figure 19**).

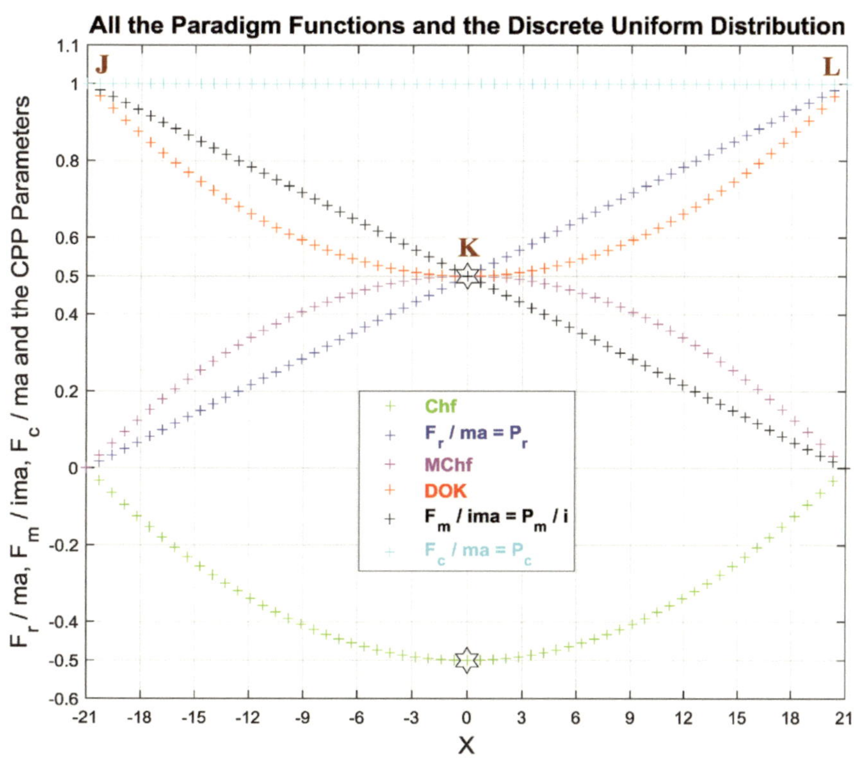

Figure 19.
The graphs of F_r / ma, F_m / ima, *and* F_c / ma *and of all the* CPP *parameters as functions of the random variable* X *for this discrete uniform probability distribution.*

8.1.1.1 The complex probability cubes

In the first cube (**Figure 20**), the simulation of *DOK* and *Chf* as functions of each other and of the random variable *X* for the discrete uniform probability distribution can be seen. The dotted line in cyan is the projection of the plane $P_c^2(X) = DOK(X) - Chf(X) = 1 = P_c(X) = F_c / ma$ on the plane $X = L_b$ = lower bound of $X = -21$. This dotted line starts at the point J ($DOK = 1$, $Chf = 0$) when $X = L_b = -21$, reaches the point ($DOK = 0.5$, $Chf = -0.5$) when $X = 0$, and returns at the end to J ($DOK = 1$, $Chf = 0$) when $X = U_b$ = upper bound of $X = 21$. The other curves are the graphs of $DOK(X)$ (red) and $Chf(X)$ (green, blue, pink) in different simulation planes. Notice that they all have a minimum at the point K ($DOK = 0.5$, $Chf = -0.5$, $X = 0$). The point L corresponds to ($DOK = 1$, $Chf = 0$, $X = U_b = 21$). The three points J, K, L are the same as in **Figure 19**.

In the second cube (**Figure 21**), we can notice the simulation of the real reduced force $F_r / ma = P_r(X)$ in $\mathcal{R}$ and its complementary real reduced force $F_m / ima = P_m(X)/i$ in $\mathcal{R}$ also in terms of the random variable *X* for the discrete uniform probability distribution. The dotted line in cyan is the projection of the plane $P_c^2(X) = P_r(X) + P_m(X)/i = 1 = P_c(X) = F_c / ma$ on the plane $X = L_b$ = lower

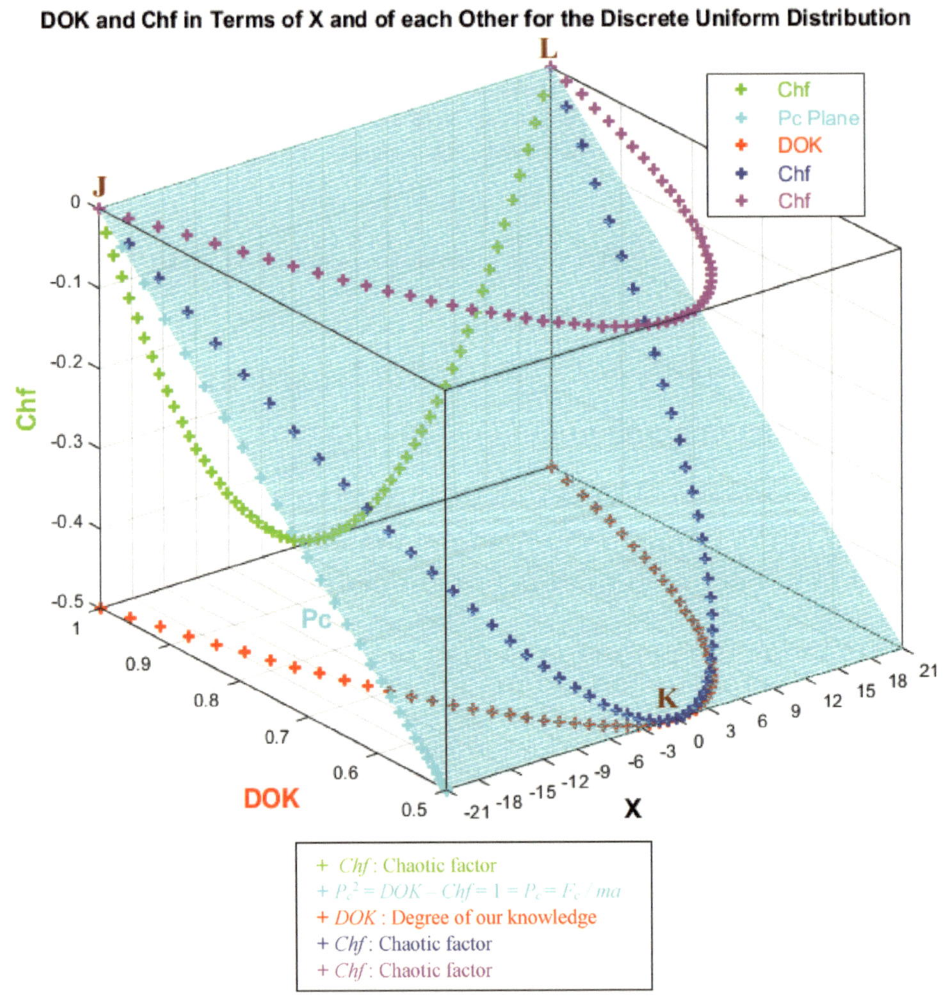

Figure 20.
The graphs of DOK *and* Chf *and the deterministic reduced force* F_c / ma = P_c *in terms of* X *and of each other for this discrete uniform probability distribution.*

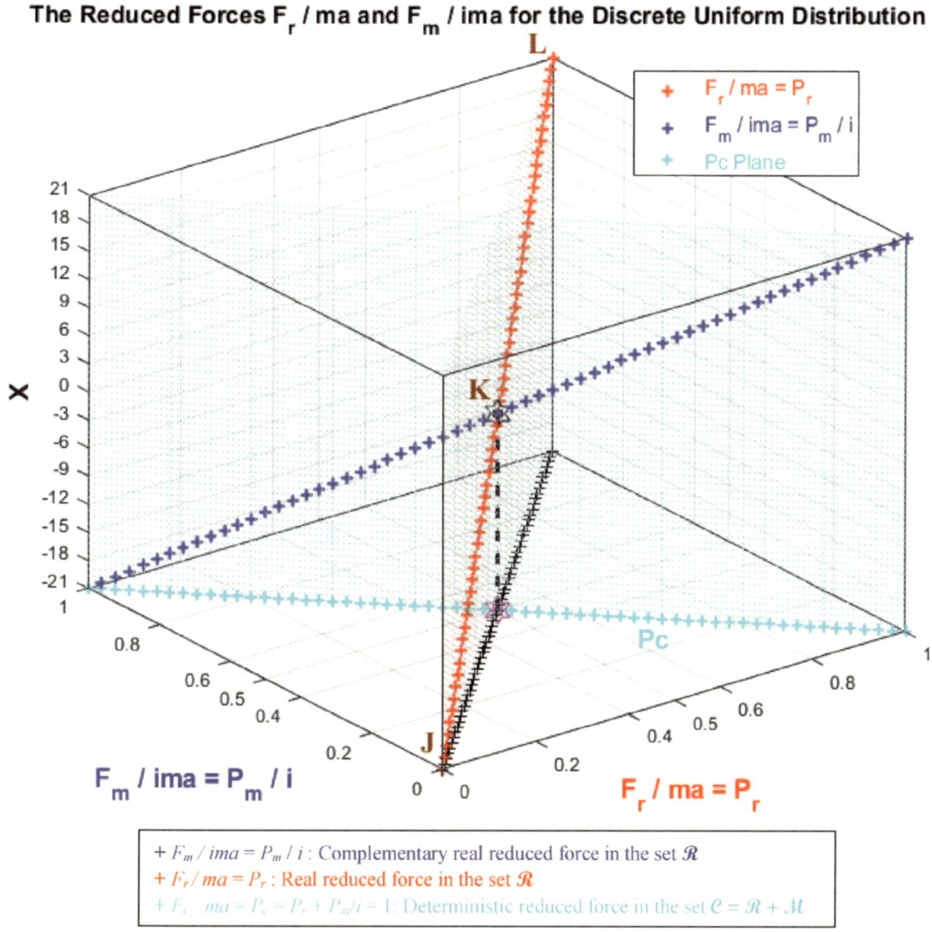

Figure 21.
The graphs of F_r / ma = P_r *and* F_m / ima = P_m / i *and* F_c / ma = P_c *in terms of* X *and of each other for this discrete uniform probability distribution.*

bound of $X = -21$. This dotted line starts at the point ($P_r = 0$, $P_m/i = 1$) and ends at the point ($P_r = 1$, $P_m/i = 0$). The red curve represents $F_r / ma = P_r(X)$ in the plane $P_r(X) = P_m(X)/i$ in light gray. This curve starts at the point J ($P_r = 0$, $P_m/i = 1$, $X = L_b$ = lower bound of $X = -21$), reaches the point K ($P_r = 0.5$, $P_m/i = 0.5$, $X = 0$), and gets at the end to L ($P_r = 1$, $P_m/i = 0$, $X = U_b$ = upper bound of $X = 21$). The blue curve represents $F_m / ima = P_m(X)/i$ in the plane in cyan $P_r(X) + P_m(X)/i = 1 = P_c(X) = F_c / ma$. Notice the importance of the point K which is the intersection of the red and blue curves at $X = 0$ and when $P_r(X) = P_m(X)/i = 0.5$. The three points J, K, L are the same as in **Figure 19**.

In the third cube (**Figure 22**), we can notice the simulation of the complex resultant reduced force $F / ma = z(X)$ in $\mathcal{C} = \mathcal{R} + \mathcal{M}$ as a function of the real reduced force $F_r / ma = P_r(X) = \text{Re}(z)$ in $\mathcal{R}$ and of its complementary imaginary reduced force $F_m / ma = P_m(X) = i \times \text{Im}(z)$ in $\mathcal{M}$, and this in terms of the random variable X for the discrete uniform probability distribution. The red curve represents F_r / ma in the plane $P_m(X) = 0$ and the blue curve represents F_m / ma in the plane $P_r(X) = 0$. The green curve represents the complex resultant reduced force $F / ma = F_r / ma + F_m / ma = z(X) = P_r(X) + P_m(X) = \text{Re}(z) + i \times \text{Im}(z)$ in the plane $P_r(X) = iP_m(X) + 1$ or $z(X)$ plane in cyan. The curve of F / ma starts at the point J ($P_r = 0$, $P_m = i$, $X = L_b$ = lower bound of $X = -21$) and ends at the point L ($P_r = 1$,

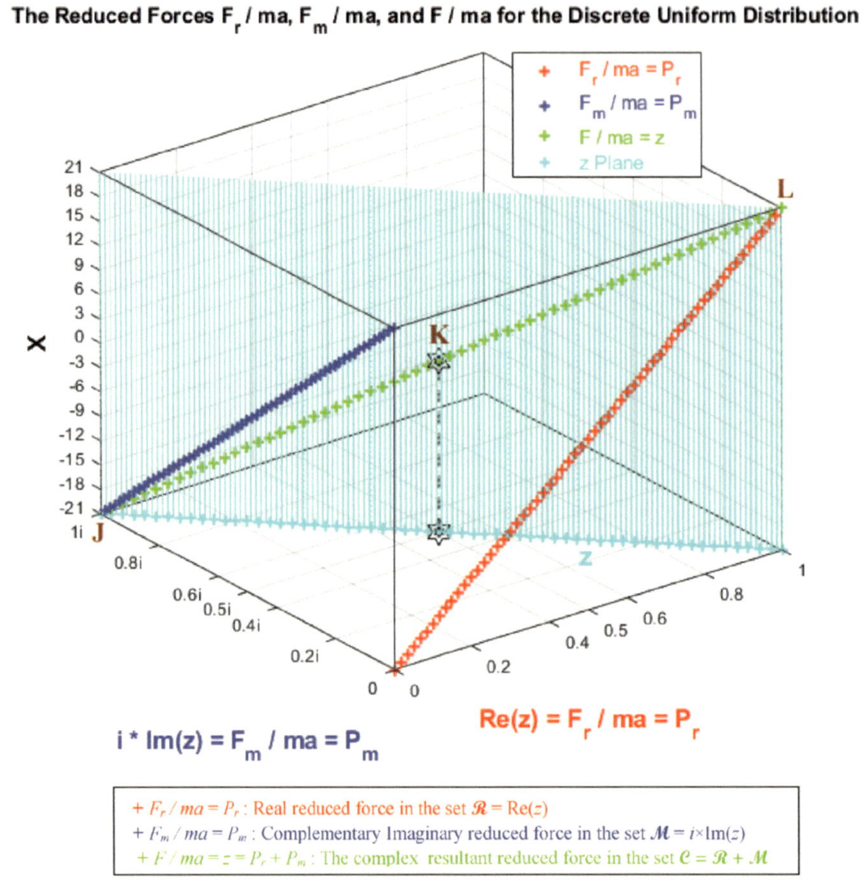

Figure 22.
The graphs of the reduced forces F_r / ma = P_r *and* F_m / ma = P_m *and* F / ma = z *in terms of* X *for this discrete uniform probability distribution.*

P_m = 0, X = U_b = upper bound of X = 21). The dotted line in cyan is $P_r(X = L_b = -21)$ = $iP_m(X = L_b = -21)$ + 1 and it is the projection of the F / ma curve on the complex probability plane whose equation is $X = L_b = -21$. This projected dotted line starts at the point J ($P_r = 0$, $P_m = i$, $X = L_b = -21$) and ends at the point ($P_r = 1$, $P_m = 0$, $X = L_b = -21$). Notice the importance of the point K corresponding to $X = 0$ and $z = 0.5 + 0.5i$ when $P_r = 0.5$ and $P_m = 0.5i$. The three points J, K, L are the same as in **Figure 19**.

8.1.2 The binomial probability distribution

The probability density function (*PDF*) of this discrete stochastic distribution is:

$$f(x) = {}_N C_x p^x q^{N-x} = \binom{N}{x} p^x q^{N-x}, \quad \text{for } (L_b = 0) \leq x \leq (U_b = N)$$

I have taken the domain for the binomial random variable to be:
$x \in [L_b = 0, U_b = N = 12]$ and $\forall k : 1 \leq k \leq 12$ we have $\Delta x_k = x_k - x_{k-1} = 1$, then: $x = 0, 1, 2, \ldots, 12$.

Taking in our simulation $N = 12$ and $p + q = 1, p = q = 0.5$ then:

The mean of this binomial discrete random distribution is: $\mu = Np = 12 \times 0.5 = 6$.

The standard deviation is: $\sigma = \sqrt{Npq} = \sqrt{12 \times 0.5 \times 0.5} = \sqrt{3} = 1.73205\ldots$.
The median is $Md = \mu = 6$.
The mode for this symmetric distribution is = 6 = Md = μ.
The cumulative distribution function (CDF) is:

$$CDF(x) = P_{rob}(X \le x) = \sum_{k=0}^{x} f(k;N) = \sum_{k=0}^{x} {}_N C_k p^k q^{N-k} = \sum_{k=0}^{x} {}_{12}C_k p^k q^{12-k},$$
$$\forall x : 0 \le x \le (N = 12)$$

Note that:
If $x = 0 \Leftrightarrow X = L_b \Leftrightarrow CDF(x) = P_{rob}(X \le 0) = f(X = L_b; N) = {}_N C_0 p^0 q^{N-0} = q^N = 0.5^{12} \cong 0$.
If $x = N = 12 \Leftrightarrow X = U_b \Leftrightarrow CDF(x) = P_{rob}(X \le x) = \sum_{k=0}^{N} {}_N C_k p^k q^{N-k} = (p+q)^N = 1^N = 1^{12} = 1$ by the binomial theorem.
The real probability $P_r(x)$ and force are:

$$P_r(x) = CDF(x) = \sum_{k=0}^{x} f(k;N) = \sum_{k=0}^{x} {}_N C_k p^k q^{N-k} = \sum_{k=0}^{x} {}_{12}C_k p^k q^{12-k},$$
$$\forall x : 0 \le x \le (N = 12)$$

$$\Leftrightarrow \vec{F}_r(x) = P_r(x) m\vec{a} = \left(\sum_{k=0}^{x} {}_{12}C_k p^k q^{12-k} \right) m\vec{a}$$

The imaginary complementary probability $P_m(x)$ and force are:

$$P_m(x) = i[1 - P_r(x)] = i[1 - CDF(x)] = i\left[1 - \sum_{k=0}^{x} f(k;N)\right]$$

$$= i\left(1 - \sum_{k=0}^{x} {}_N C_k p^k q^{N-k}\right) = i \sum_{k=x+1}^{N} {}_N C_k p^k q^{N-k} = i \sum_{k=x+1}^{12} {}_{12}C_k p^k q^{12-k},$$
$$\forall x : 0 \le x \le (N = 12)$$

$$\Leftrightarrow \vec{F}_m(x) = P_m(x) m\vec{a} = i\left(\sum_{k=x+1}^{12} {}_{12}C_k p^k q^{12-k} \right) m\vec{a}$$

The real complementary probability $P_m(x)/i$ and force are:

$$P_m(x)/i = 1 - P_r(x) = 1 - CDF(x) = 1 - \sum_{k=0}^{x} f(k;N) = \sum_{k=x+1}^{N} {}_N C_k p^k q^{N-k}$$

$$= \sum_{k=x+1}^{12} {}_{12}C_k p^k q^{12-k}, \quad \forall x : 0 \le x \le (N = 12)$$

$$\Leftrightarrow \vec{F}_m(x)/i = \frac{P_m(x)}{i} m\vec{a} = \left(\sum_{k=x+1}^{12} {}_{12}C_k p^k q^{12-k} \right) m\vec{a}$$

The complex probability or random vector and force are:

$$z(x) = P_r(x) + P_m(x) = \sum_{k=0}^{x} {}_N C_k p^k q^{N-k} + i\left(\sum_{k=x+1}^{N} {}_N C_k p^k q^{N-k}\right)$$

$$= \sum_{k=0}^{x} {}_{12} C_k p^k q^{12-k} + i\left(\sum_{k=x+1}^{12} {}_{12} C_k p^k q^{12-k}\right), \quad \forall x : 0 \le x \le (N = 12)$$

$$\Leftrightarrow \vec{F}(x) = \vec{F}_r(x) + \vec{F}_m(x) = P_r(x) m\vec{a} + P_m(x) m\vec{a} = [P_r(x) + P_m(x)] m\vec{a} = z m\vec{a}$$

$$= \left(\sum_{k=0}^{x} {}_N C_k p^k q^{N-k}\right) m\vec{a} + i\left(\sum_{k=x+1}^{N} {}_N C_k p^k q^{N-k}\right) m\vec{a}$$

$$= \left[\left(\sum_{k=0}^{x} {}_N C_k p^k q^{N-k}\right) + i\left(\sum_{k=x+1}^{N} {}_N C_k p^k q^{N-k}\right)\right] m\vec{a}$$

$$= \left[\left(\sum_{k=0}^{x} {}_{12} C_k p^k q^{12-k}\right) + i\left(\sum_{k=x+1}^{12} {}_{12} C_k p^k q^{12-k}\right)\right] m\vec{a}, \quad \forall x : 0 \le x \le (N = 12)$$

The Degree of Our Knowledge:

$$DOK(x) = |z(x)|^2 = P_r^2(x) + [P_m(x)/i]^2 = \left(\sum_{k=0}^{x} {}_N C_k p^k q^{N-k}\right)^2 + \left(1 - \sum_{k=0}^{x} {}_N C_k p^k q^{N-k}\right)^2$$

$$= \left(\sum_{k=0}^{x} {}_N C_k p^k q^{N-k}\right)^2 + \left(\sum_{k=x+1}^{N} {}_N C_k p^k q^{N-k}\right)^2 = \left(\sum_{k=0}^{x} {}_{12} C_k p^k q^{12-k}\right)^2 + \left(\sum_{k=x+1}^{12} {}_{12} C_k p^k q^{12-k}\right)^2$$

$$= 1 + 2iP_r(x)P_m(x) = 1 - 2P_r(x)[1 - P_r(x)] = 1 - 2P_r(x) + 2P_r^2(x)$$

$$= 1 - 2\left(\sum_{k=0}^{x} {}_N C_k p^k q^{N-k}\right) + 2\left(\sum_{k=0}^{x} {}_N C_k p^k q^{N-k}\right)^2$$

$$= 1 - 2\left(\sum_{k=0}^{x} {}_{12} C_k p^k q^{12-k}\right) + 2\left(\sum_{k=0}^{x} {}_{12} C_k p^k q^{12-k}\right)^2, \quad \forall x : 0 \le x \le (N = 12)$$

$DOK(x)$ is equal to 1 when $P_r(x) = P_r(L_b = 0) = 0$ and when $P_r(x) = P_r(U_b = 12) = 1$.

The Chaotic Factor:

$$Chf(x) = 2iP_r(x)P_m(x) = -2P_r(x)[1 - P_r(x)] = -2P_r(x) + 2P_r^2(x)$$

$$= -2\left(\sum_{k=0}^{x} {}_N C_k p^k q^{N-k}\right) + 2\left(\sum_{k=0}^{x} {}_N C_k p^k q^{N-k}\right)^2$$

$$= -2\left(\sum_{k=0}^{x} {}_{12} C_k p^k q^{12-k}\right) + 2\left(\sum_{k=0}^{x} {}_{12} C_k p^k q^{12-k}\right)^2, \forall x : 0 \le x \le (N = 12)$$

$Chf(x)$ is null when $P_r(x) = P_r(L_b = 0) = 0$ and when $P_r(x) = P_r(U_b = 12) = 1$.

The Magnitude of the Chaotic Factor *MChf*:

$$MChf(x) = |Chf(x)| = -2iP_r(x)P_m(x) = 2P_r(x)[1 - P_r(x)] = 2P_r(x) - 2P_r^2(x)$$

$$= 2\left(\sum_{k=0}^{x} {}_N C_k p^k q^{N-k}\right) - 2\left(\sum_{k=0}^{x} {}_N C_k p^k q^{N-k}\right)^2$$

$$= 2\left(\sum_{k=0}^{x} {}_{12} C_k p^k q^{12-k}\right) - 2\left(\sum_{k=0}^{x} {}_{12} C_k p^k q^{12-k}\right)^2, \quad \forall x : 0 \le x \le (N = 12)$$

$MChf(x)$ is null when $P_r(x) = P_r(L_b = 0) = 0$ and when $P_r(x) = P_r(U_b = 12) = 1$.

At any value of x: $\forall x : (L_b = 0) \le x \le (U_b = N = 12)$, the probability expressed in the complex probability set $\mathcal{C} = \mathcal{R} + \mathcal{M}$ is the following:

$$P_c^2(x) = [P_r(x) + P_m(x)/i]^2 = |z(x)|^2 - 2iP_r(x)P_m(x)$$

$$= DOK(x) - Chf(x)$$

$$= DOK(x) + MChf(x)$$

$$= 1$$

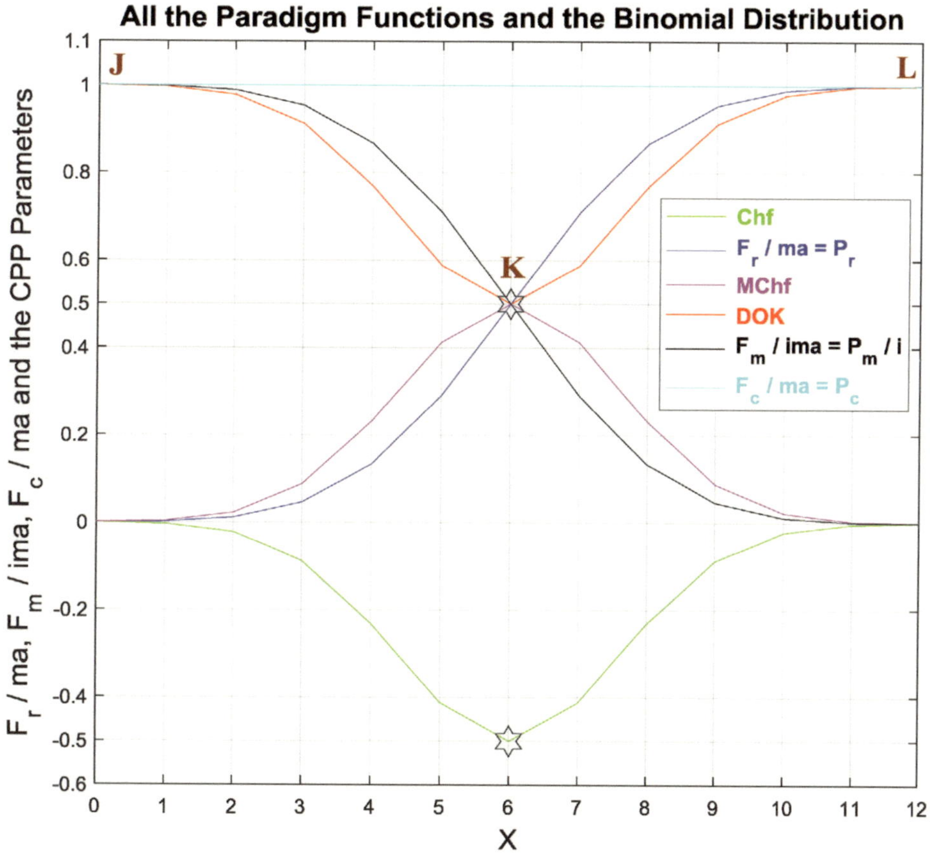

Figure 23.
The graphs of F_r / ma, F_m / ima, *and* F_c / ma *and of all the* CPP *parameters as functions of the random variable* X *for this discrete binomial probability distribution.*

then,

$$P_c^{\,2}(x) = [P_r(x) + P_m(x)/i]^2 = \{P_r(x) + [1 - P_r(x)]\}^2 = 1^2 = 1 \Leftrightarrow P_c(x) = 1 \text{ always}$$

$$\Leftrightarrow \vec{F}_c(x) = P_c(x) m\vec{a} = 1 \times m\vec{a} = m\vec{a} \text{ always also.}$$

Hence, the prediction of all the probabilities and forces of the stochastic experiment in the universe $\mathcal{C} = \mathcal{R} + \mathcal{M}$ is permanently certain and perfectly deterministic (**Figure 23**).

8.1.2.1 The complex probability cubes

In the first cube (**Figure 24**), the simulation of *DOK* and *Chf* as functions of each other and of the random variable *X* for the binomial probability distribution can be seen. The thick line in cyan is the projection of the plane $P_c^2(X) = DOK(X) - Chf(X) = 1 = P_c(X) = F_c / ma$ on the plane $X = L_b$ = lower bound of $X = 0$. This thick line starts at the point J ($DOK = 1$, $Chf = 0$) when $X = L_b = 0$, reaches the point ($DOK = 0.5$, $Chf = -0.5$) when $X = 6$, and returns at the end to J ($DOK = 1$, $Chf = 0$)

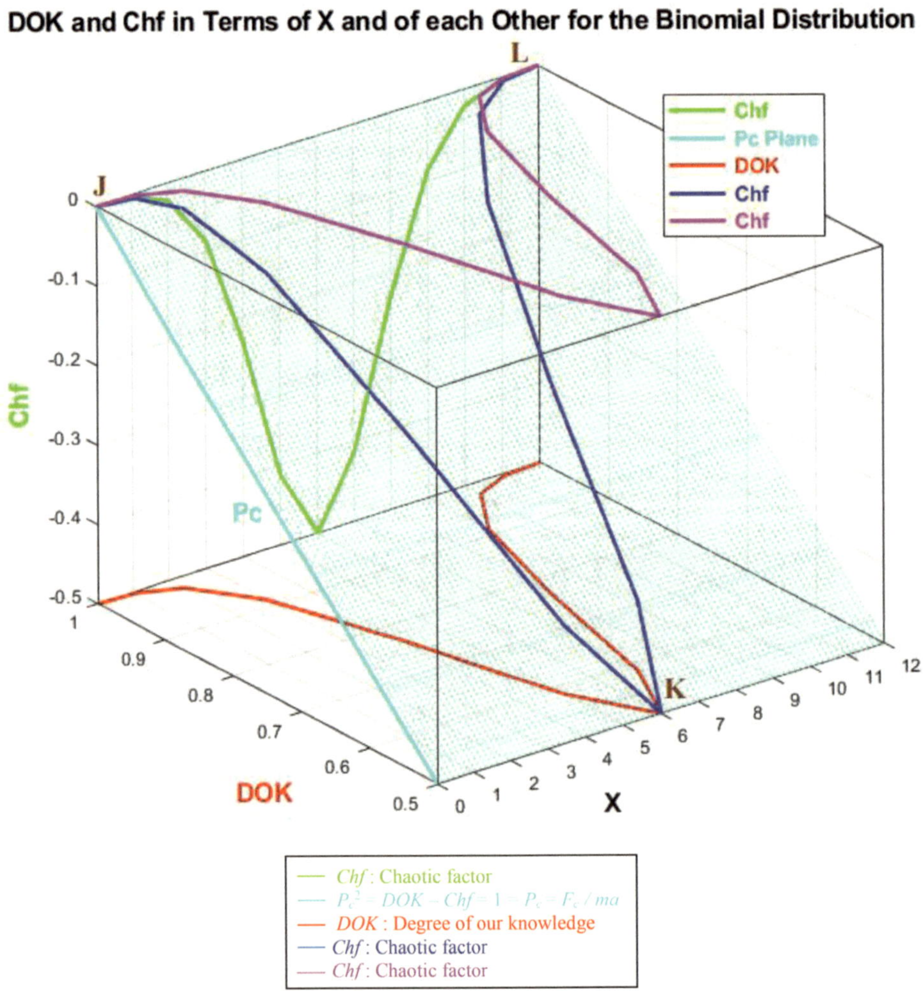

Figure 24.
The graphs of DOK *and* Chf *and the deterministic reduced force* F_c / ma = P_c *in terms of* X *and of each other for this binomial probability distribution.*

when $X = U_b$ = upper bound of $X = 12$. The other curves are the graphs of $DOK(X)$ (red) and $Chf(X)$ (green, blue, pink) in different simulation planes. Notice that they all have a minimum at the point K ($DOK = 0.5$, $Chf = -0.5$, $X = 6$). The point L corresponds to ($DOK = 1$, $Chf = 0$, $X = U_b = 12$). The three points J, K, L are the same as in **Figure 23**.

In the second cube (**Figure 25**), we can notice the simulation of the real reduced force $F_r / ma = P_r(X)$ in $\mathcal{R}$ and its complementary real reduced force $F_m / ima = P_m(X)/i$ in $\mathcal{R}$ also in terms of the random variable X for the binomial probability distribution. The thick line in cyan is the projection of the plane $P_c^2(X) = P_r(X) + P_m(X)/i = 1 = P_c(X) = F_c / ma$ on the plane $X = L_b$ = lower bound of $X = 0$. This thick line starts at the point ($P_r = 0$, $P_m/i = 1$) and ends at the point ($P_r = 1$, $P_m/i = 0$). The red curve represents $F_r / ma = P_r(X)$ in the plane $P_r(X) = P_m(X)/i$ in light gray. This curve starts at the point J ($P_r = 0$, $P_m/i = 1$, $X = L_b$ = lower bound of $X = 0$), reaches the point K ($P_r = 0.5$, $P_m/i = 0.5$, $X = 6$), and gets at the end to L ($P_r = 1$, $P_m/i = 0$, $X = U_b$ = upper bound of $X = 12$). The blue curve represents $F_m / ima = P_m(X)/i$ in the plane in cyan $P_r(X) + P_m(X)/i = 1 = P_c(X) = F_c / ma$. Notice the importance of the point K which is the intersection of the red and blue curves at $X = 6$ and when $P_r(X) = P_m(X)/i = 0.5$. The three points J, K, L are the same as in **Figure 23**.

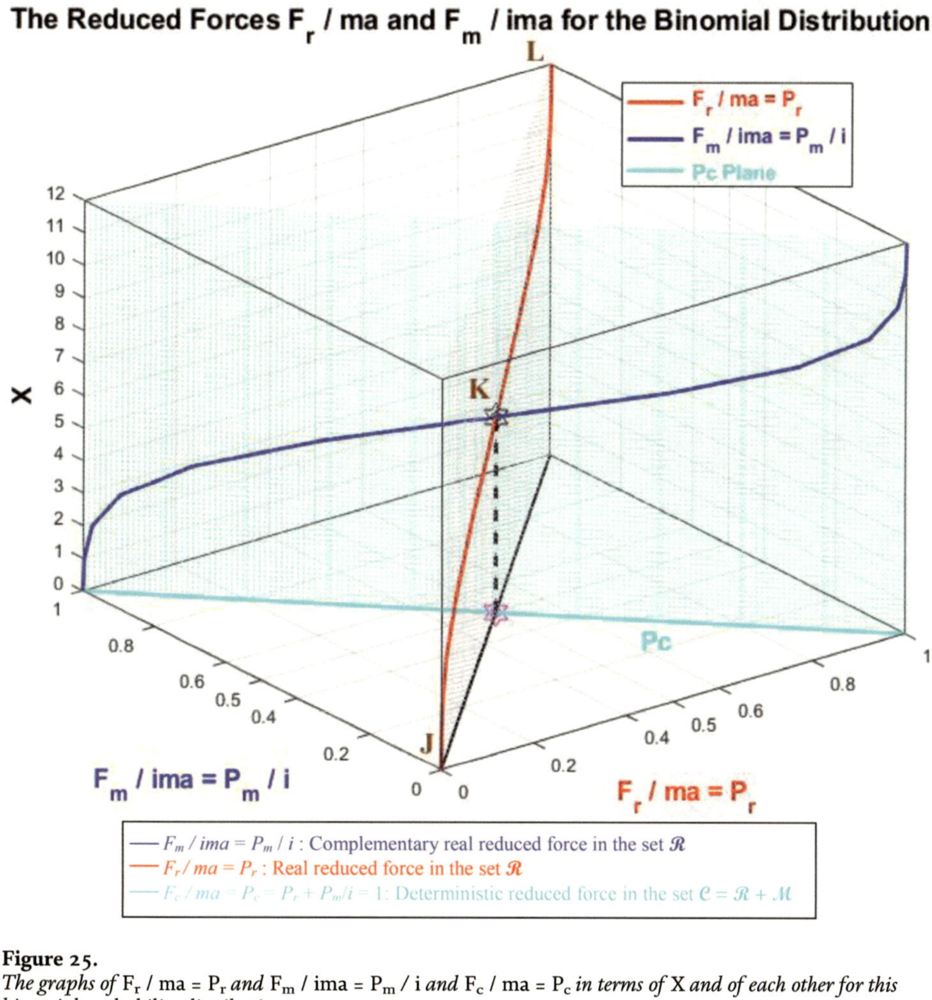

Figure 25.
The graphs of $F_r / ma = P_r$ *and* $F_m / ima = P_m / i$ *and* $F_c / ma = P_c$ *in terms of* X *and of each other for this binomial probability distribution.*

In the third cube (**Figure 26**), we can notice the simulation of the complex resultant reduced force $F / ma = z(X)$ in $\mathcal{C} = \mathcal{R} + \mathcal{M}$ as a function of the real reduced force $F_r / ma = P_r(X) = \text{Re}(z)$ in $\mathcal{R}$ and of its complementary imaginary reduced force $F_m / ma = P_m(X) = i \times \text{Im}(z)$ in $\mathcal{M}$, and this in terms of the random variable X for the binomial probability distribution. The red curve represents F_r / ma in the plane $P_m(X) = 0$ and the blue curve represents F_m / ma in the plane $P_r(X) = 0$. The green curve represents the complex resultant reduced force $F / ma = F_r / ma + F_m / ma = z(X) = P_r(X) + P_m(X) = \text{Re}(z) + i \times \text{Im}(z)$ in the plane $P_r(X) = iP_m(X) + 1$ or $z(X)$ plane in cyan. The curve of F / ma starts at the point J ($P_r = 0, P_m = i, X = L_b$ = lower bound of $X = 0$) and ends at the point L ($P_r = 1, P_m = 0, X = U_b$ = upper bound of $X = 12$). The thick line in cyan is $P_r(X = L_b = 0) = iP_m(X = L_b = 0) + 1$ and it is the projection of the F / ma curve on the complex probability plane whose equation is $X = L_b = 0$. This projected thick line starts at the point J ($P_r = 0, P_m = i, X = L_b = 0$) and ends at the point ($P_r = 1, P_m = 0, X = L_b = 0$). Notice the importance of the point K corresponding to $X = 6$ and $z = 0.5 + 0.5i$ when $P_r = 0.5$ and $P_m = 0.5i$. The three points J, K, L are the same as in **Figure 23**.

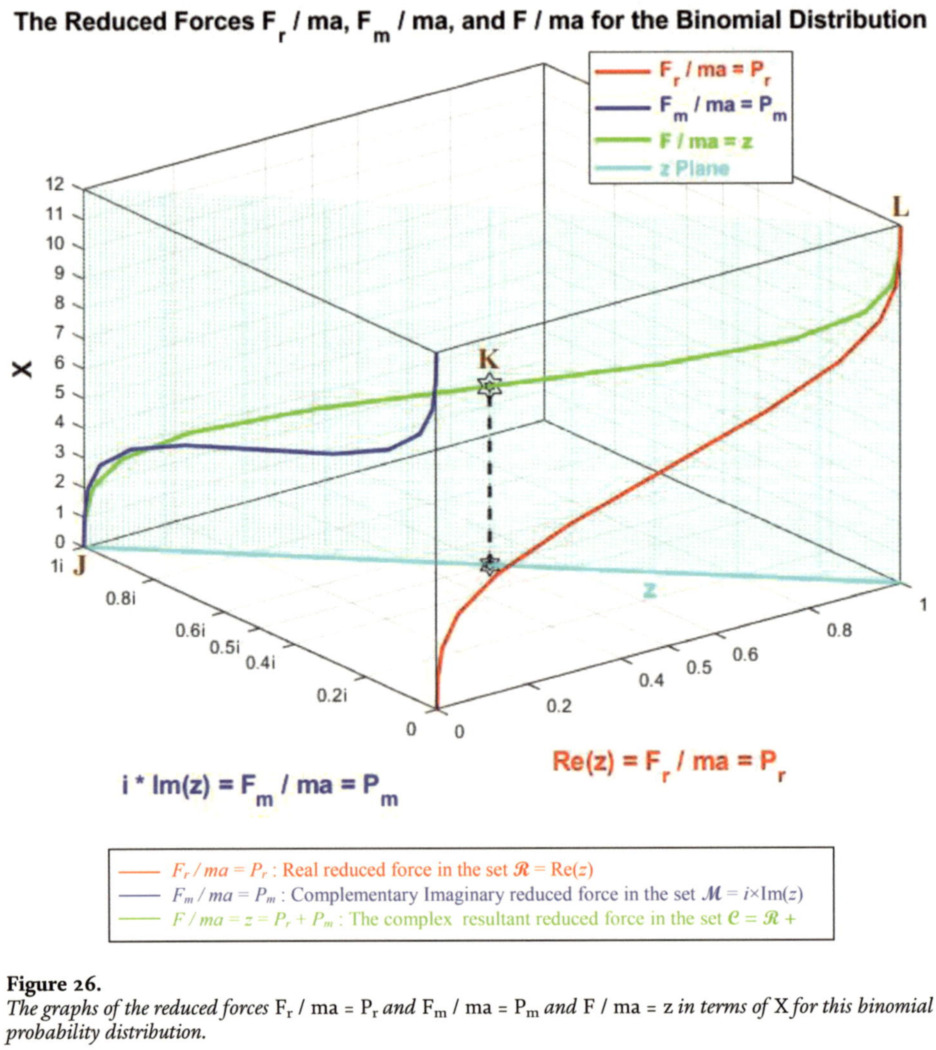

Figure 26.
The graphs of the reduced forces F_r / ma = P_r *and* F_m / ma = P_m *and* F / ma = z *in terms of* X *for this binomial probability distribution.*

8.1.3 The Poisson probability distribution

The probability density function (*PDF*) of this discrete stochastic distribution is:

$$f(x;\lambda) = \frac{e^{-\lambda}\lambda^x}{x!} \text{ where } 0 \le x < \infty.$$

For the Poisson discrete random variable: $x \in [L_b = 0, \infty)$ and $\forall k$: $k \ge 1$ we have $\Delta x_k = x_k - x_{k-1} = 1$, then $x = 0, 1, 2, \dots, \infty$.

I have taken in the simulation the domain for the Poisson random variable to be equal to: $x \in [L_b = 0, U_b = 16]$, then: $x = 0, 1, 2, \dots, 16$.

The mean of this Poisson discrete random distribution is: $\mu = \lambda = 6.7$.

The standard deviation is: $\sigma = \sqrt{\lambda} = \sqrt{6.7} = 2.588435821\dots$.

The median Md is $= 6$.

The mode is = $= \lfloor \lambda \rfloor = \lfloor 6.7 \rfloor = 6$.

Since Md = mode $< \mu$ then this distribution is skewed to the right or positively skewed.

The cumulative distribution function (*CDF*) is:

$$CDF(x) = P_{rob}(X \le x) = \sum_{k=0}^{x} f(k;\lambda) = \sum_{k=0}^{x} \frac{e^{-\lambda}\lambda^k}{k!} = \sum_{k=0}^{x} \frac{e^{-6.7}6.7^k}{k!}, \quad \forall x : 0 \le x \le 16$$

Note that:

If $x = 0 \Leftrightarrow CDF(x) = P_{rob}(X \le 0) = f(X = L_b;\lambda) = e^{-\lambda} = e^{-6.7} \cong 0$.

If $x = U_b \Leftrightarrow X >> 1 \Leftrightarrow X \to +\infty \Leftrightarrow CDF(x) = P_{rob}(X \le x) \to \sum_{k=0}^{+\infty} \frac{e^{-\lambda}\lambda^k}{k!} =$ $e^{-\lambda} \sum_{k=0}^{+\infty} \frac{\lambda^k}{k!} = e^{-\lambda} \times e^{\lambda} = 1$ by the properties of infinite series from calculus.

The real probability $P_r(x)$ and force are:

$$P_r(x) = CDF(x) = \sum_{k=0}^{x} f(k;\lambda) = \sum_{k=0}^{x} \frac{e^{-\lambda}\lambda^k}{k!} = \sum_{k=0}^{x} \frac{e^{-6.7}6.7^k}{k!}, \quad \forall x : 0 \le x \le 16$$

$$\Leftrightarrow \vec{F}_r(x) = P_r(x) m\vec{a} = \left(\sum_{k=0}^{x} \frac{e^{-\lambda}\lambda^k}{k!} \right) m\vec{a} = \left(\sum_{k=0}^{x} \frac{e^{-6.7}6.7^k}{k!} \right) m\vec{a}$$

The imaginary complementary probability $P_m(x)$ and force are:

$$P_m(x) = i[1 - P_r(x)] = i[1 - CDF(x)] = i\left[1 - \sum_{k=0}^{x} f(k;\lambda)\right]$$

$$= i\left(1 - \sum_{k=0}^{x} \frac{e^{-\lambda}\lambda^k}{k!}\right) = i\left(\sum_{k=x+1}^{+\infty} \frac{e^{-\lambda}\lambda^k}{k!}\right) = i\left(\sum_{k=x+1}^{16} \frac{e^{-6.7}6.7^k}{k!}\right), \forall x : 0 \le x \le 16$$

$$\Leftrightarrow \vec{F}_m(x) = P_m(x) m\vec{a} = i\left(\sum_{k=x+1}^{+\infty} \frac{e^{-\lambda}\lambda^k}{k!}\right) m\vec{a} = i\left(\sum_{k=x+1}^{16} \frac{e^{-6.7}6.7^k}{k!}\right) m\vec{a}$$

The real complementary probability $P_m(x)/i$ and force are:

$$P_m(x)/i = 1 - P_r(x) = 1 - CDF(x) = 1 - \sum_{k=0}^{x} \frac{e^{-\lambda}\lambda^k}{k!}$$

$$= \sum_{k=x+1}^{+\infty} \frac{e^{-\lambda}\lambda^k}{k!} = \sum_{k=x+1}^{16} \frac{e^{-6.7}6.7^k}{k!}, \forall x : 0 \leq x \leq 16$$

$$\Leftrightarrow \vec{F}_m(x)/i = \frac{P_m(x)}{i}m\vec{a} = \left(\sum_{k=x+1}^{+\infty} \frac{e^{-\lambda}\lambda^k}{k!}\right)m\vec{a} = \left(\sum_{k=x+1}^{16} \frac{e^{-6.7}6.7^k}{k!}\right)m\vec{a}$$

The complex probability or random vector and force are:

$$z(x) = P_r(x) + P_m(x) = \sum_{k=0}^{x} \frac{e^{-\lambda}\lambda^k}{k!} + i\left(\sum_{k=x+1}^{+\infty} \frac{e^{-\lambda}\lambda^k}{k!}\right)$$
$$= \sum_{k=0}^{x} \frac{e^{-6.7}6.7^k}{k!} + i\left(\sum_{k=x+1}^{16} \frac{e^{-6.7}6.7^k}{k!}\right),$$

$$\Leftrightarrow \vec{F}(x) = \vec{F}_r(x) + \vec{F}_m(x) = P_r(x)m\vec{a} + P_m(x)m\vec{a} = [P_r(x) + P_m(x)]m\vec{a} = zm\vec{a}$$

$$= \left(\sum_{k=0}^{x} \frac{e^{-\lambda}\lambda^k}{k!}\right)m\vec{a} + i\left(\sum_{k=x+1}^{+\infty} \frac{e^{-\lambda}\lambda^k}{k!}\right)m\vec{a}$$

$$= \left[\left(\sum_{k=0}^{x} \frac{e^{-\lambda}\lambda^k}{k!}\right) + i\left(\sum_{k=x+1}^{+\infty} \frac{e^{-\lambda}\lambda^k}{k!}\right)\right]m\vec{a}$$

$$= \left[\left(\sum_{k=0}^{x} \frac{e^{-6.7}6.7^k}{k!}\right) + i\left(\sum_{k=x+1}^{16} \frac{e^{-6.7}6.7^k}{k!}\right)\right]m\vec{a}, \quad \forall x : 0 \leq x \leq 16$$

The Degree of Our Knowledge:

$$DOK(x) = |z(x)|^2 = P_r^2(x) + [P_m(x)/i]^2 = \left(\sum_{k=0}^{x} \frac{e^{-\lambda}\lambda^k}{k!}\right)^2 + \left(1 - \sum_{k=0}^{x} \frac{e^{-\lambda}\lambda^k}{k!}\right)^2$$

$$= \left(\sum_{k=0}^{x} \frac{e^{-\lambda}\lambda^k}{k!}\right)^2 + \left(\sum_{k=x+1}^{+\infty} \frac{e^{-\lambda}\lambda^k}{k!}\right)^2 = \left(\sum_{k=0}^{x} \frac{e^{-6.7}6.7^k}{k!}\right)^2 + \left(\sum_{k=x+1}^{16} \frac{e^{-6.7}6.7^k}{k!}\right)^2$$

$$= 1 + 2iP_r(x)P_m(x) = 1 - 2P_r(x)[1 - P_r(x)] = 1 - 2P_r(x) + 2P_r^2(x)$$

$$= 1 - 2\left(\sum_{k=0}^{x} \frac{e^{-\lambda}\lambda^k}{k!}\right) + 2\left(\sum_{k=0}^{x} \frac{e^{-\lambda}\lambda^k}{k!}\right)^2$$

$$= 1 - 2\left(\sum_{k=0}^{x} \frac{e^{-6.7}6.7^k}{k!}\right) + 2\left(\sum_{k=0}^{x} \frac{e^{-6.7}6.7^k}{k!}\right)^2, \quad \forall x : 0 \leq x \leq 16$$

$DOK(x)$ is equal to 1 when $P_r(x) = P_r(L_b = 0) = 0$ and when $P_r(x) = P_r(U_b = 16) = 1$.

The Chaotic Factor:

$$Chf(x) = 2iP_r(x)P_m(x) = -2P_r(x)[1 - P_r(x)] = -2P_r(x) + 2P_r^2(x)$$

$$= -2\left(\sum_{k=0}^{x} \frac{e^{-\lambda}\lambda^k}{k!}\right) + 2\left(\sum_{k=0}^{x} \frac{e^{-\lambda}\lambda^k}{k!}\right)^2$$

$$= -2\left(\sum_{k=0}^{x}\frac{e^{-6.7}6.7^k}{k!}\right) + 2\left(\sum_{k=0}^{x}\frac{e^{-6.7}6.7^k}{k!}\right)^2, \quad \forall x : 0 \le x \le 16$$

$Chf(x)$ is null when $P_r(x) = P_r(L_b = 0) = 0$ and when$P_r(x) = P_r(U_b = 16) = 1$.
The Magnitude of the Chaotic Factor $MChf$:

$$MChf(x) = |Chf(x)| = -2iP_r(x)P_m(x) = 2P_r(x)[1 - P_r(x)] = 2P_r(x) - 2P_r^2(x)$$

$$= 2\left(\sum_{k=0}^{x}\frac{e^{-\lambda}\lambda^k}{k!}\right) - 2\left(\sum_{k=0}^{x}\frac{e^{-\lambda}\lambda^k}{k!}\right)^2$$

$$= 2\left(\sum_{k=0}^{x}\frac{e^{-6.7}6.7^k}{k!}\right) - 2\left(\sum_{k=0}^{x}\frac{e^{-6.7}6.7^k}{k!}\right)^2, \quad \forall x : 0 \le x \le 16$$

$MChf(x)$ is null when $P_r(x) = P_r(L_b = 0) = 0$ and when $P_r(x) = P_r(U_b = 16) = 1$.
At any value of x: $\forall x : (L_b = 0) \le x \le (U_b = 16)$, the probability expressed in the complex probability set $\mathcal{C} = \mathcal{R} + \mathcal{M}$ is the following:

$$\begin{aligned} P_c^2(x) &= [P_r(x) + P_m(x)/i]^2 = |z(x)|^2 - 2iP_r(x)P_m(x) \\ &= DOK(x) - Chf(x) \\ &= DOK(x) + MChf(x) \\ &= 1 \end{aligned}$$

then,

$$P_c^2(x) = [P_r(x) + P_m(x)/i]^2 = \{P_r(x) + [1 - P_r(x)]\}^2 = 1^2 = 1 \Leftrightarrow P_c(x) = 1 \text{ always}$$

$\Leftrightarrow \vec{F}_c(x) = P_c(x)m\vec{a} = 1 \times m\vec{a} = m\vec{a}$ always also.
Hence, the prediction of all the probabilities and forces of the stochastic experiment in the universe $\mathcal{C} = \mathcal{R} + \mathcal{M}$ is permanently certain and perfectly deterministic (**Figure 27**).

8.1.3.1 The complex probability cubes

In the first cube (**Figure 28**), the simulation of DOK and Chf as functions of each other and of the random variable X for the Poisson probability distribution can be seen. The thick line in cyan is the projection of the plane $P_c^2(X) = DOK(X) - Chf(X) = 1 = P_c(X) = F_c / ma$ on the plane $X = L_b$ = lower bound of $X = 0$. This thick line starts at the point J ($DOK = 1$, $Chf = 0$) when $X = L_b = 0$, reaches the point ($DOK = 0.5$, $Chf = -0.5$) when $X = 6$, and returns at the end to J ($DOK = 1$, $Chf = 0$) when $X = U_b$ = upper bound of $X = 16$. The other curves are the graphs of $DOK(X)$ (red) and $Chf(X)$ (green, blue, pink) in different simulation planes. Notice that they all have a minimum at the point K ($DOK = 0.5$, $Chf = -0.5$, $X = 6$). The point L corresponds to ($DOK = 1$, $Chf = 0$, $X = U_b = 16$). The three points J, K, L are the same as in **Figure 27**.
In the second cube (**Figure 29**), we can notice the simulation of the real reduced force $F_r / ma = P_r(X)$ in $\mathcal{R}$ and its complementary real reduced force $F_m / ima = P_m(X)/i$ in $\mathcal{R}$ also in terms of the random variable X for the Poisson probability distribution. The thick line in cyan is the projection of the plane $P_c^2(X) = P_r(X) + P_m(X)/i = 1 = P_c(X) = F_c / ma$ on the plane $X = L_b$ = lower bound of $X = 0$. This thick line starts at the point ($P_r = 0$, $P_m/i = 1$) and ends at the point

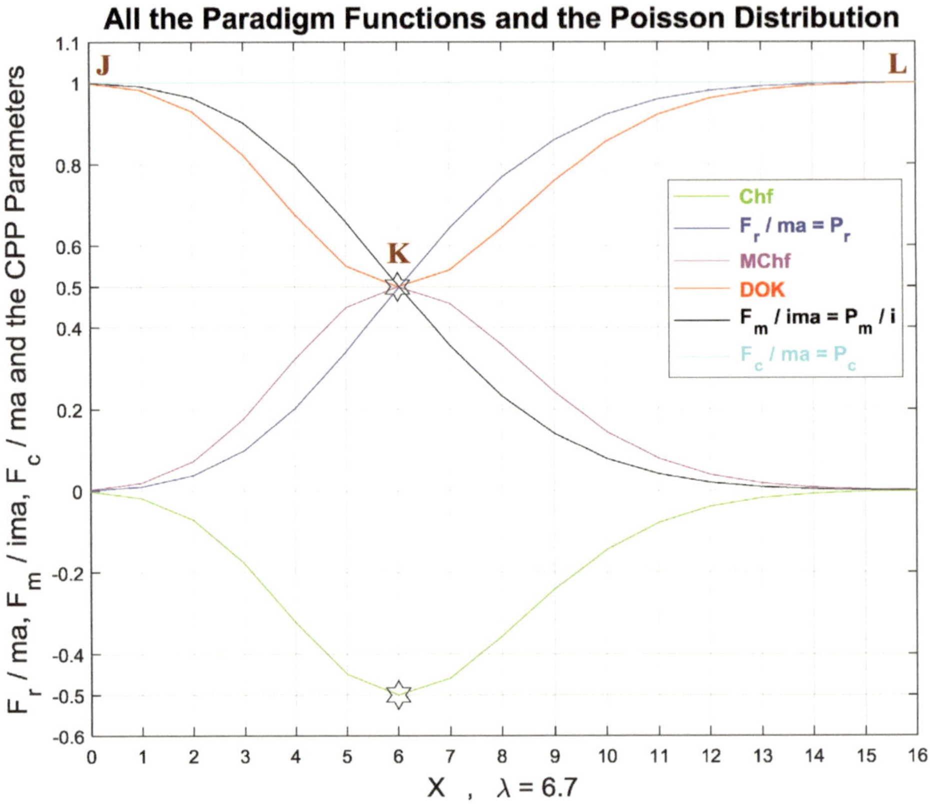

Figure 27.
The graphs of F_r / ma, F_m / ima, *and* F_c / ma *and of all the* CPP *parameters as functions of the random variable* X *for this discrete Poisson probability distribution.*

($P_r = 1$, $P_m/i = 0$). The red curve represents $F_r / ma = P_r(X)$ in the plane $P_r(X) = P_m(X)/i$ in light gray. This curve starts at the point J ($P_r = 0$, $P_m/i = 1$, $X = L_b$ = lower bound of $X = 0$), reaches the point K ($P_r = 0.5$, $P_m/i = 0.5$, $X = 6$), and gets at the end to L ($P_r = 1$, $P_m/i = 0$, $X = U_b$ = upper bound of $X = 16$). The blue curve represents $F_m / ima = P_m(X)/i$ in the plane in cyan $P_r(X) + P_m(X)/i = 1 = P_c(X) = F_c / ma$. Notice the importance of the point K which is the intersection of the red and blue curves at $X = 6$ and when $P_r(X) = P_m(X)/i = 0.5$. The three points J, K, L are the same as in **Figure 27**.

In the third cube (**Figure 30**), we can notice the simulation of the complex resultant reduced force $F / ma = z(X)$ in $\mathcal{C} = \mathcal{R} + \mathcal{M}$ as a function of the real reduced force $F_r / ma = P_r(X) = \text{Re}(z)$ in $\mathcal{R}$ and of its complementary imaginary reduced force $F_m / ma = P_m(X) = i \times \text{Im}(z)$ in $\mathcal{M}$, and this in terms of the random variable X for the Poisson probability distribution. The red curve represents F_r / ma in the plane $P_m(X) = 0$ and the blue curve represents F_m / ma in the plane $P_r(X) = 0$. The green curve represents the complex resultant reduced force $F / ma = F_r / ma + F_m / ma = z(X) = P_r(X) + P_m(X) = \text{Re}(z) + i \times \text{Im}(z)$ in the plane $P_r(X) = iP_m(X) + 1$ or $z(X)$ plane in cyan. The curve of F / ma starts at the point J ($P_r = 0$, $P_m = i$, $X = L_b$ = lower bound of $X = 0$) and ends at the point L ($P_r = 1$, $P_m = 0$, $X = U_b$ = upper bound of $X = 16$). The thick line in cyan is $P_r(X = L_b = 0) = iP_m(X = L_b = 0) + 1$ and it is the projection of the F / ma curve on the complex probability plane whose equation is $X = L_b = 0$. This projected thick line starts at the point J ($P_r = 0$, $P_m = i$, $X = L_b = 0$) and ends at the point ($P_r = 1$, $P_m = 0$, $X = L_b = 0$). Notice the importance of the point K corresponding to $X = 6$ and $z = 0.5 + 0.5i$ when $P_r = 0.5$ and $P_m = 0.5i$. The three points J, K, L are the same as in **Figure 27**.

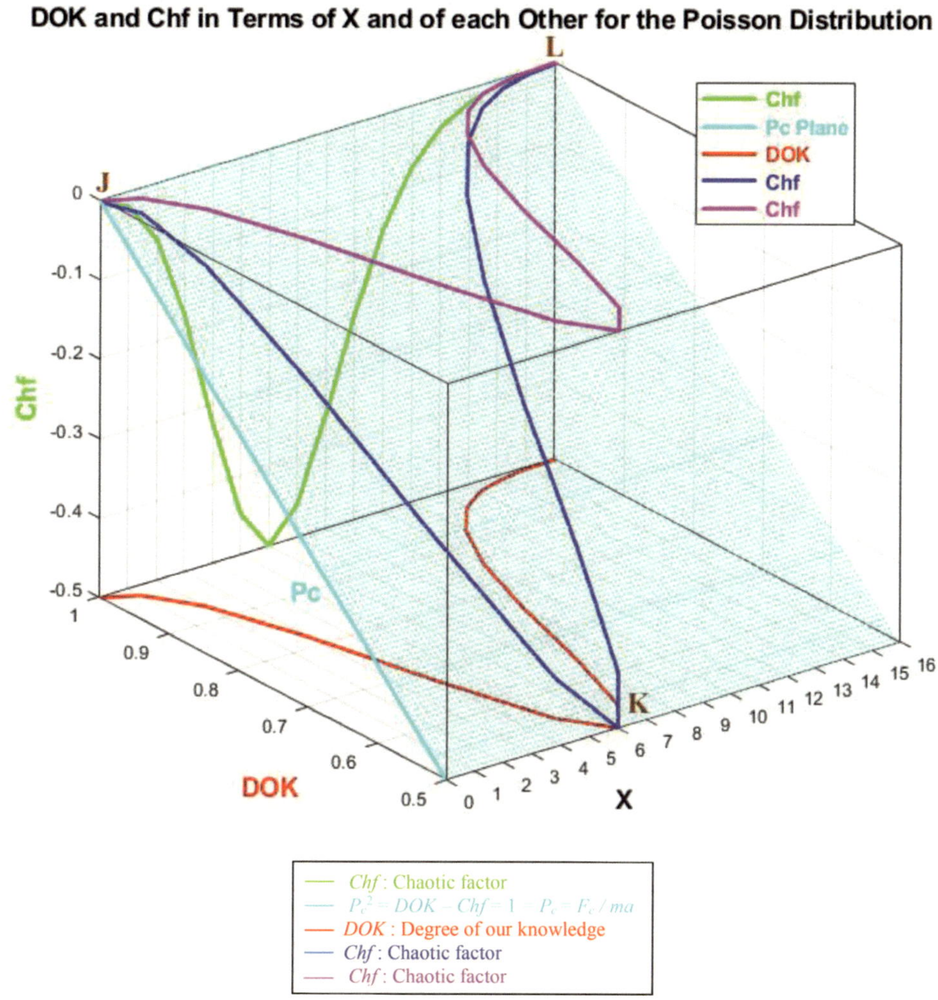

Figure 28.
The graphs of DOK *and* Chf *and the deterministic reduced force* F_c / ma = P_c *in terms of* X *and of each other for this Poisson probability distribution.*

8.2 Simulation of continuous probability distributions

8.2.1 The continuous uniform probability distribution

The probability density function (*PDF*) of this continuous stochastic distribution is:

$$f(x) = \frac{d\,[CDF(x)]}{dx} = \begin{cases} \dfrac{1}{U_b - L_b} & \text{if} \quad L_b \le x \le U_b \\ 0 & \text{elsewhere} \end{cases}$$

and the cumulative distribution function (*CDF*) is:

$$CDF(x) = P_{rob}(X \le x) = \int_{-\infty}^{x} f(t)dt = \int_{L_b}^{x} f(t)dt = \begin{cases} \dfrac{x - L_b}{U_b - L_b} & \text{if } L_b \le x \le U_b \\ 0 & \text{elsewhere} \end{cases}$$

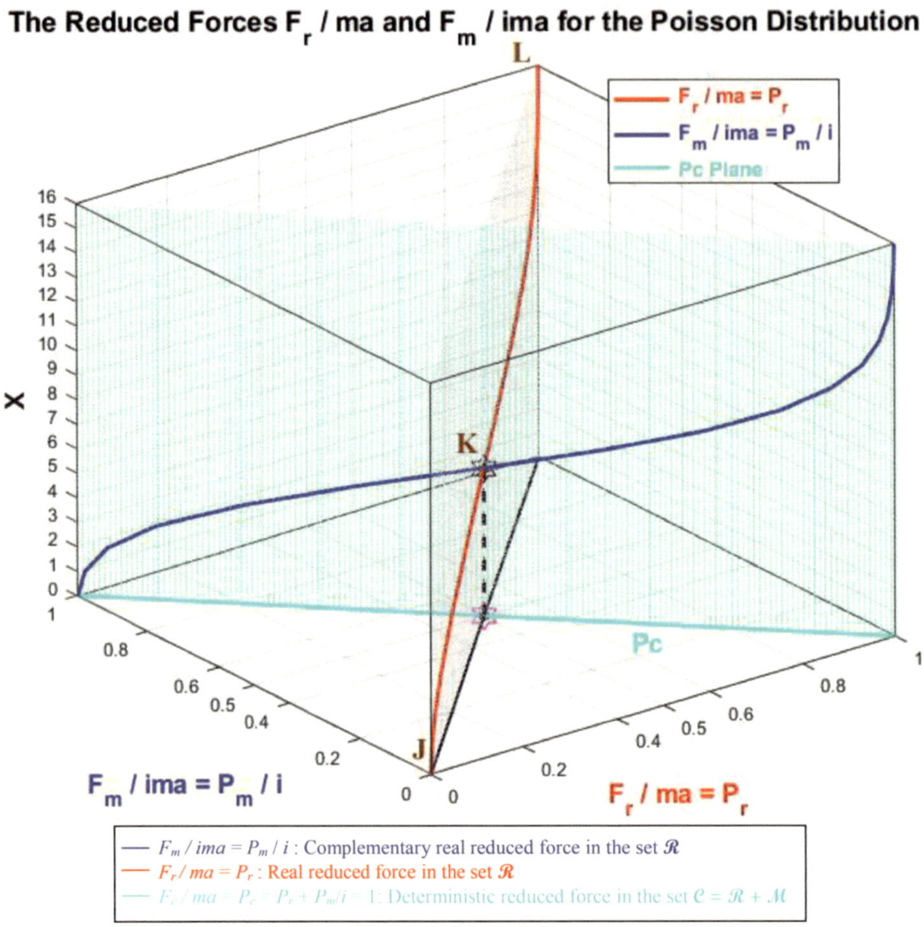

Figure 29.
The graphs of F_r / ma = P_r *and* F_m / ima = P_m / i *and* F_c / ma = P_c *in terms of* X *and of each other for this Poisson probability distribution.*

I have taken the domain for the continuous uniform random variable to be equal to: $x \in [L_b = -3, U_b = 3]$ and $dx = 0.01$.

$$\text{Then } CDF(x) = \begin{cases} \dfrac{x+3}{6} & \text{if } (L_b = -3) \le x \le (U_b = 3) \\ 0 & \text{elsewhere} \end{cases}$$

Note that:

If $x = L_b = -3 \Leftrightarrow CDF(x) = P_{rob}(X \le -3) = \frac{-3+3}{6} = 0$.

If $x = U_b = +3 \Leftrightarrow CDF(x) = P_{rob}(X \le +3) = \frac{3+3}{6} = 1$.

The mean of this continuous uniform random distribution is: $\mu = \frac{L_b + U_b}{2} = \frac{-3+3}{2} = 0$.

The variance is: $\sigma^2 = \frac{(L_b - U_b)^2}{12} = \frac{(-3-3)^2}{12} = \frac{36}{12} = 3$.

The standard deviation is: $\sigma = \frac{|L_b - U_b|}{\sqrt{12}} = \frac{|-3-3|}{\sqrt{12}} = \frac{6}{\sqrt{12}} = \sqrt{3} = 1.732050808\ldots$.

The median is $Md = 0 = \mu$ since the distribution is symmetric.

Since the distribution is uniform then it has no mode.

The real probability $P_r(x)$ and force are:

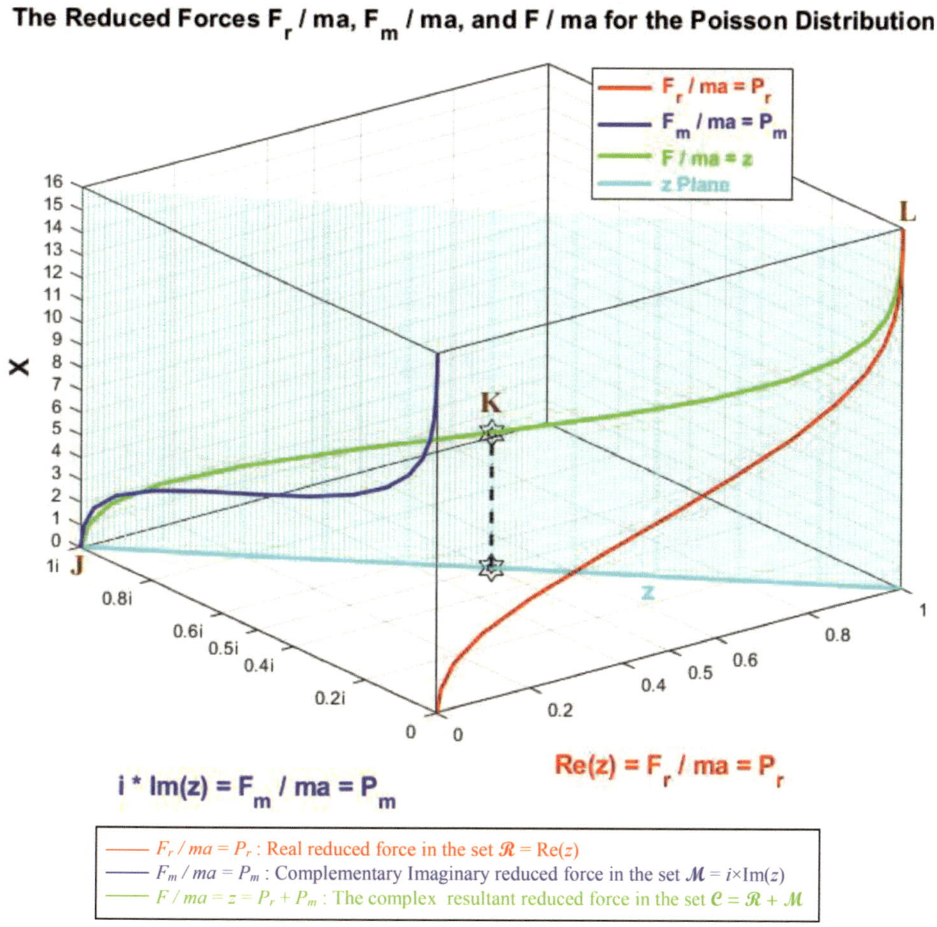

Figure 30.
The graphs of the reduced forces F_r / ma = P_r *and* F_m / ma = P_m *and* F / ma = z *in terms of* X *for this Poisson probability distribution.*

$$P_r(x) = CDF(x) = \frac{x+3}{6}, \quad \forall x: -3 \le x \le 3$$

$$\Leftrightarrow \vec{F}_r(x) = P_r(x)m\vec{a} = \left(\frac{x+3}{6}\right)m\vec{a}$$

The imaginary complementary probability $P_m(x)$ and force are:

$$P_m(x) = i[1 - P_r(x)] = i[1 - CDF(x)] = i\left[1 - \int_{-\infty}^{x} f(t)dt\right] = i\left[1 - \int_{-3}^{x} f(t)dt\right]$$

$$= i\left[\int_{x}^{+\infty} f(t)dt\right] = i\left[\int_{x}^{3} f(t)dt\right] = i\left(1 - \frac{x+3}{6}\right) = i\left(\frac{3-x}{6}\right), \forall x: -3 \le x \le 3$$

$$\Leftrightarrow \vec{F}_m(x) = P_m(x)m\vec{a} = i\left(\frac{3-x}{6}\right)m\vec{a}$$

The real complementary probability $P_m(x)/i$ and force are:

$$P_m(x)/i = 1 - P_r(x) = 1 - CDF(x) = 1 - \int_{-\infty}^{x} f(t)dt = \int_{x}^{+\infty} f(t)dt = \int_{x}^{3} f(t)dt$$
$$= \frac{3-x}{6}, \forall x : -3 \leq x \leq 3$$
$$\Leftrightarrow \vec{F}_m(x)/i = \frac{P_m(x)}{i} m\vec{a} = \left(\frac{3-x}{6}\right) m\vec{a}$$

The complex probability or random vector and force are:

$$z(x) = P_r(x) + P_m(x) = \left(\frac{x+3}{6}\right) + i\left(\frac{3-x}{6}\right), \forall x : -3 \leq x \leq 3$$
$$\Leftrightarrow \vec{F}(x) = \vec{F}_r(x) + \vec{F}_m(x) = P_r(x)m\vec{a} + P_m(x)m\vec{a} = [P_r(x) + P_m(x)]m\vec{a} = zm\vec{a}$$
$$= \left(\frac{x+3}{6}\right)m\vec{a} + i\left(\frac{3-x}{6}\right)m\vec{a}$$
$$= \left[\left(\frac{x+3}{6}\right) + i\left(\frac{3-x}{6}\right)\right]m\vec{a}$$

The Degree of Our Knowledge:

$$DOK(x) = |z(x)|^2 = P_r^2(x) + [P_m(x)/i]^2 = \left(\frac{x+3}{6}\right)^2 + \left(1 - \frac{x+3}{6}\right)^2$$
$$= \left(\frac{x+3}{6}\right)^2 + \left(\frac{3-x}{6}\right)^2$$
$$= 1 + 2iP_r(x)P_m(x) = 1 - 2P_r(x)[1 - P_r(x)] = 1 - 2P_r(x) + 2P_r^2(x)$$
$$= 1 - 2\left(\frac{x+3}{6}\right) + 2\left(\frac{x+3}{6}\right)^2, \quad \forall x : -3 \leq x \leq 3$$

$DOK(x)$ is equal to 1 when $P_r(x) = P_r(L_b = -3) = 0$ and when $P_r(x) = P_r(U_b = 3) = 1$.

The Chaotic Factor:

$$Chf(x) = 2iP_r(x)P_m(x) = -2P_r(x)[1 - P_r(x)] = -2P_r(x) + 2P_r^2(x)$$
$$= -2\left(\frac{x+3}{6}\right) + 2\left(\frac{x+3}{6}\right)^2, \quad \forall x : -3 \leq x \leq 3$$

$Chf(x)$ is null when $P_r(x) = P_r(L_b = -3) = 0$ and when $P_r(x) = P_r(U_b = 3) = 1$.

The Magnitude of the Chaotic Factor $MChf$:

$$MChf(x) = |Chf(x)| = -2iP_r(x)P_m(x) = 2P_r(x)[1 - P_r(x)] = 2P_r(x) - 2P_r^2(x)$$
$$= 2\left(\frac{x+3}{6}\right) - 2\left(\frac{x+3}{6}\right)^2, \quad \forall x : -3 \leq x \leq 3$$

$MChf(x)$ is null when $P_r(x) = P_r(L_b = -3) = 0$ and when $P_r(x) = P_r(U_b = 3) = 1$.

At any value of x: $\forall x : (L_b = -3) \leq x \leq (U_b = 3)$, the probability expressed in the complex probability set $\mathcal{C} = \mathcal{R} + \mathcal{M}$ is the following:

$$\begin{aligned} P_c^{\,2}(x) &= [P_r(x) + P_m(x)/i]^2 = |z(x)|^2 - 2iP_r(x)P_m(x) \\ &= DOK(x) - Chf(x) \\ &= DOK(x) + MChf(x) \\ &= 1 \end{aligned}$$

then,

$$P_c^{\,2}(x) = [P_r(x) + P_m(x)/i]^2 = \{P_r(x) + [1 - P_r(x)]\}^2 = 1^2 = 1 \Leftrightarrow P_c(x) = 1 \text{ always}$$

$\Leftrightarrow \vec{F}_c(x) = P_c(x)m\vec{a} = 1 \times m\vec{a} = m\vec{a}$ always also.

Hence, the prediction of all the probabilities and forces of the stochastic experiment in the universe $\mathcal{C} = \mathcal{R} + \mathcal{M}$ is permanently certain and perfectly deterministic (**Figure 31**).

8.2.1.1 The complex probability cubes

In the first cube (**Figure 32**), the simulation of *DOK* and *Chf* as functions of each other and of the random variable X for the continuous uniform probability distribution can be seen. The thick line in cyan is the projection of the plane $P_c^{\,2}(X) = DOK(X) - Chf(X) = 1 = P_c(X) = F_c / ma$ on the plane $X = L_b$ = lower bound of $X = -3$. This thick line starts at the point J ($DOK = 1$, $Chf = 0$) when $X = L_b = -3$,

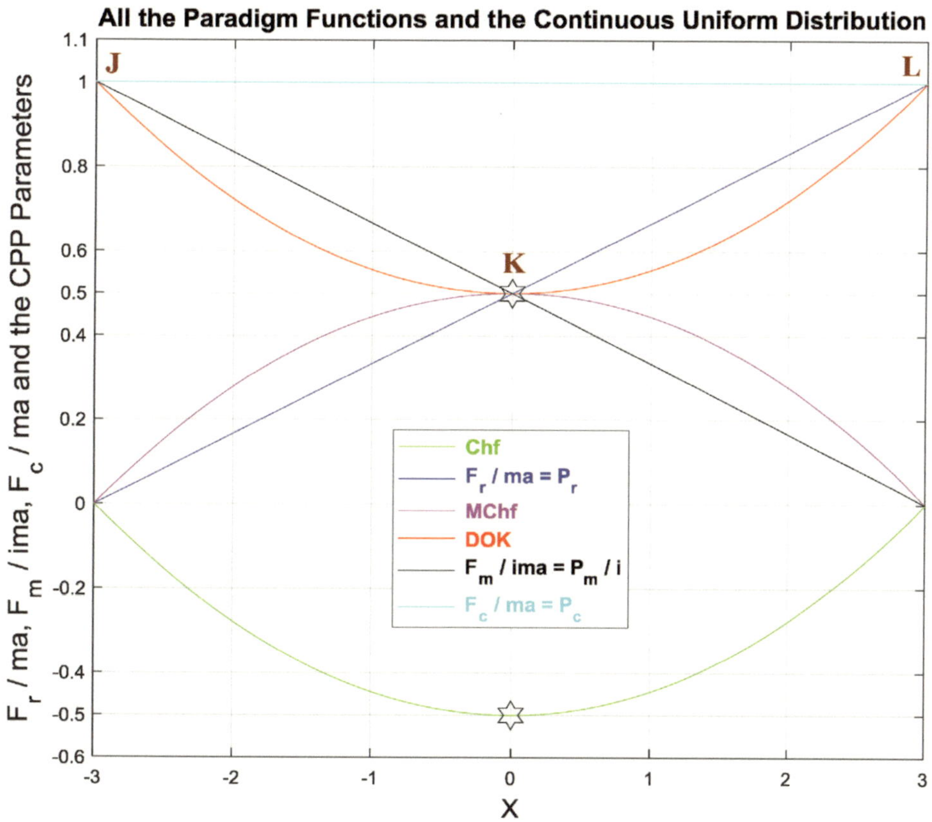

Figure 31.
The graphs of F_r / ma, F_m / ima, *and* F_c / ma *and of all the* CPP *parameters as functions of the random variable* X *for this continuous uniform probability distribution.*

reaches the point ($DOK = 0.5$, $Chf = -0.5$) when $X = 0$, and returns at the end to J ($DOK = 1$, $Chf = 0$) when $X = U_b$ = upper bound of $X = 3$. The other curves are the graphs of $DOK(X)$ (red) and $Chf(X)$ (green, blue, pink) in different simulation planes. Notice that they all have a minimum at the point K ($DOK = 0.5$, $Chf = -0.5$, $X = 0$). The point L corresponds to ($DOK = 1$, $Chf = 0$, $X = U_b = 3$). The three points J, K, L are the same as in **Figure 31**.

In the second cube (**Figure 33**), we can notice the simulation of the real reduced force $F_r / ma = P_r(X)$ in $\mathcal{R}$ and its complementary real reduced force $F_m / ima = P_m(X)/i$ in $\mathcal{R}$ also in terms of the random variable X for the continuous uniform probability distribution. The thick line in cyan is the projection of the plane $P_c^2(X) = P_r(X) + P_m(X)/i = 1 = P_c(X) = F_c / ma$ on the plane $X = L_b$ = lower bound of $X = -3$. This thick line starts at the point ($P_r = 0$, $P_m/i = 1$) and ends at the point ($P_r = 1$, $P_m/i = 0$). The red curve represents $F_r / ma = P_r(X)$ in the plane $P_r(X) = P_m(X)/i$ in light gray. This curve starts at the point J ($P_r = 0$, $P_m/i = 1$, $X = L_b$ = lower bound of $X = -3$), reaches the point K ($P_r = 0.5$, $P_m/i = 0.5$, $X = 0$), and gets at the end to L ($P_r = 1$, $P_m/i = 0$, $X = U_b$ = upper bound of $X = 3$). The blue curve represents $F_m / ima = P_m(X)/i$ in the plane in cyan $P_r(X) + P_m(X)/i = 1 = P_c(X) = F_c / ma$. Notice the importance of the point K which is the intersection of the red and blue

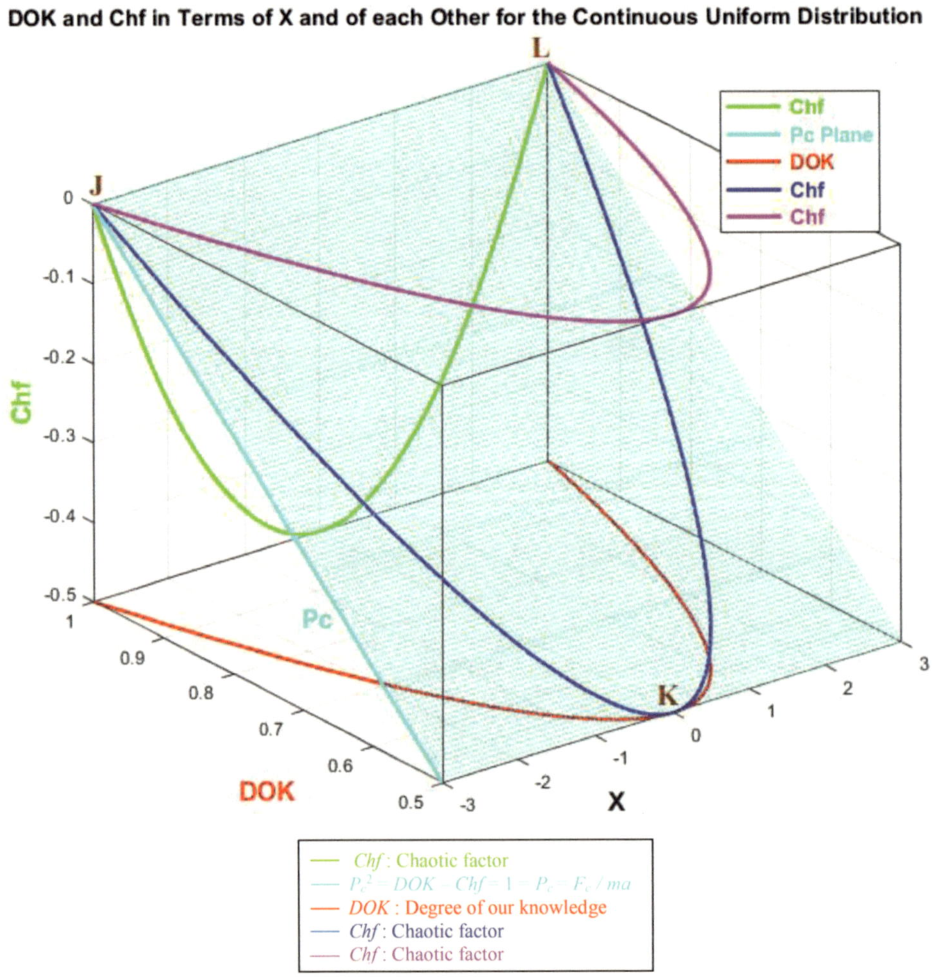

Figure 32.
The graphs of DOK *and* Chf *and the deterministic reduced force* $F_c / ma = P_c$ *in terms of* X *and of each other for this continuous uniform probability distribution.*

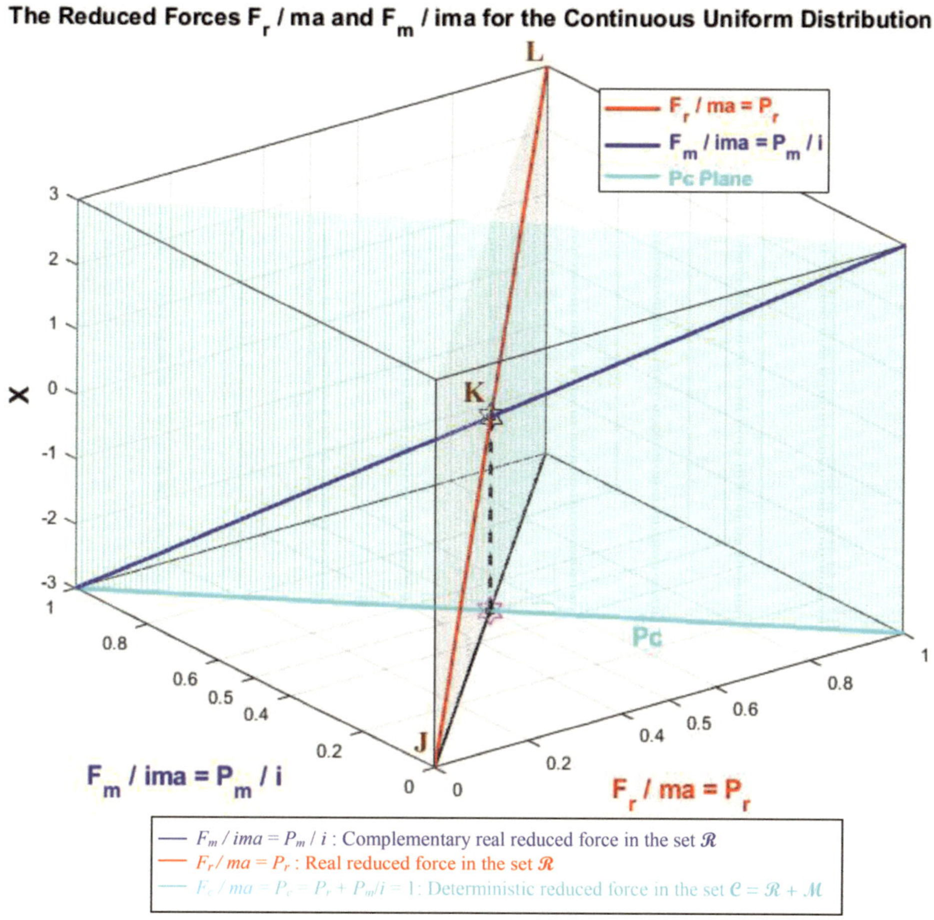

Figure 33.
The graphs of F_r / ma = P_r *and* F_m / ima = P_m / i *and* F_c / ma = P_c *in terms of* X *and of each other for this continuous uniform probability distribution.*

curves at $X = 0$ and when $P_r(X) = P_m(X)/i = 0.5$. The three points J, K, L are the same as in **Figure 31**.

In the third cube (**Figure 34**), we can notice the simulation of the complex resultant reduced force $F / ma = z(X)$ in $\mathcal{C} = \mathcal{R} + \mathcal{M}$ as a function of the real reduced force $F_r / ma = P_r(X) = \text{Re}(z)$ in $\mathcal{R}$ and of its complementary imaginary reduced force $F_m / ma = P_m(X) = i \times \text{Im}(z)$ in $\mathcal{M}$, and this in terms of the random variable X for the continuous uniform probability distribution. The red curve represents F_r / ma in the plane $P_m(X) = 0$ and the blue curve represents F_m / ma in the plane $P_r(X) = 0$. The green curve represents the complex resultant reduced force $F / ma = F_r / ma + F_m / ma = z(X) = P_r(X) + P_m(X) = \text{Re}(z) + i \times \text{Im}(z)$ in the plane $P_r(X) = iP_m(X) + 1$ or $z(X)$ plane in cyan. The curve of F / ma starts at the point J ($P_r = 0, P_m = i, X = L_b =$ lower bound of $X = -3$) and ends at the point L ($P_r = 1$, $P_m = 0, X = U_b =$ upper bound of $X = 3$). The thick line in cyan is $P_r(X = L_b = -3) = iP_m(X = L_b = -3) + 1$ and it is the projection of the F / ma curve on the complex probability plane whose equation is $X = L_b = -3$. This projected thick line starts at the point J ($P_r = 0, P_m = i, X = L_b = -3$) and ends at the point ($P_r = 1$, $P_m = 0, X = L_b = -3$). Notice the importance of the point K corresponding to $X = 0$ and $z = 0.5 + 0.5i$ when $P_r = 0.5$ and $P_m = 0.5i$. The three points J, K, L are the same as in **Figure 31**.

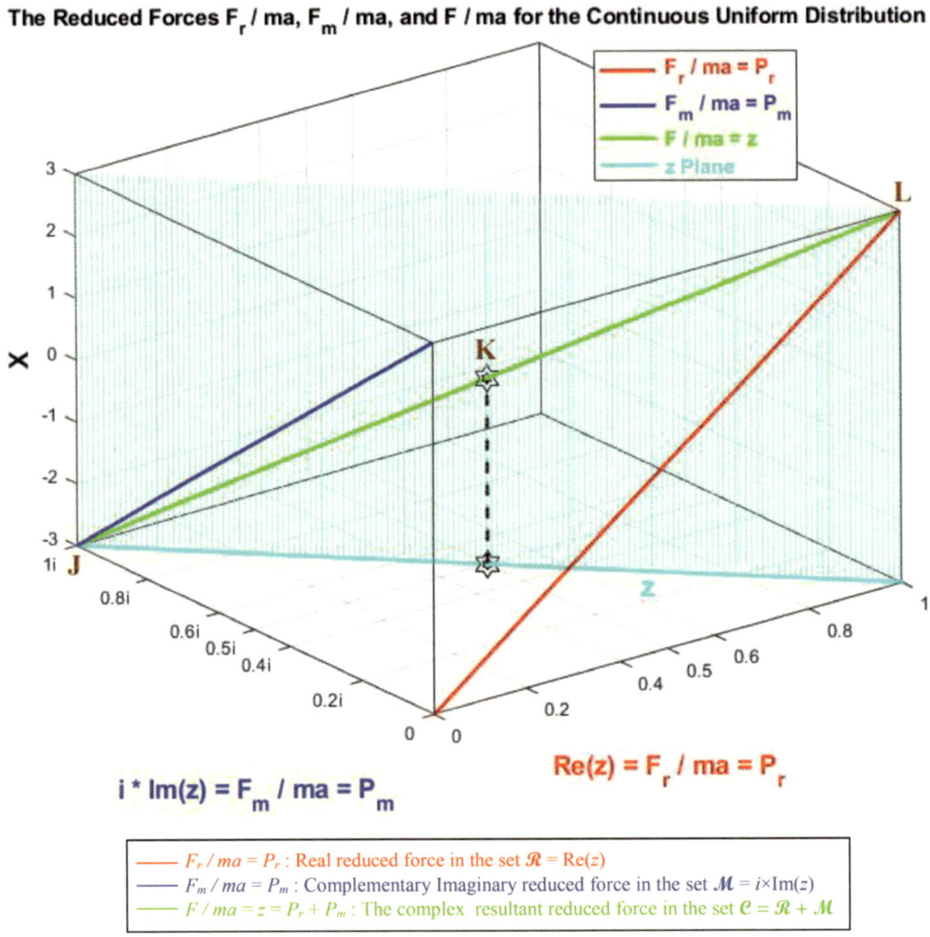

Figure 34.
The graphs of the reduced forces F_r / ma = P_r *and* F_m / ma = P_m *and* F / ma = z *in terms of* X *for this continuous uniform probability distribution.*

8.2.2 The standard Gaussian normal probability distribution

The probability density function (*PDF*) of this continuous stochastic distribution is:

$$f(x) = \frac{d\,[CDF(x)]}{dx} = \frac{1}{\sqrt{2\pi}} \exp\left(-\frac{x^2}{2}\right), \text{for } -\infty < x < \infty$$

and the cumulative distribution function (*CDF*) is:

$$CDF(x) = P_{rob}(X \le x) = \int_{-\infty}^{x} f(t)dt = \int_{-\infty}^{x} \frac{1}{\sqrt{2\pi}} \exp\left(-\frac{t^2}{2}\right) dt$$

The domain for this standard Gaussian normal variable is considered in the simulations to be equal to: $x \in [L_b = -4, U_b = 4]$ and I have taken $dx = 0.01$.

In the simulations, the mean of this standard normal random distribution is $\mu = 0$.

The variance is $\sigma^2 = 1$.

The standard deviation is $\sigma = 1$.
The median is $Md = 0$.
The mode for this symmetric distribution is = 0 = Md = μ.
The real probability $P_r(x)$ and force are:

$$P_r(x) = CDF(x) = \int_{-\infty}^{x} \frac{1}{\sqrt{2\pi}} \exp\left(-\frac{t^2}{2}\right) dt = \int_{-4}^{x} \frac{1}{\sqrt{2\pi}} \exp\left(-\frac{t^2}{2}\right) dt, \forall x : -4 \leq x \leq 4$$

$$\Leftrightarrow \vec{F}_r(x) = P_r(x) m\vec{a} = \left[\int_{-\infty}^{x} \frac{1}{\sqrt{2\pi}} \exp\left(-\frac{t^2}{2}\right) dt\right] m\vec{a}$$
$$= \left[\int_{-4}^{x} \frac{1}{\sqrt{2\pi}} \exp\left(-\frac{t^2}{2}\right) dt\right] m\vec{a}$$

The imaginary complementary probability $P_m(x)$ and force are:

$$P_m(x) = i[1 - P_r(x)] = i[1 - CDF(x)] = i\left[1 - \int_{-\infty}^{x} f(t)dt\right]$$

$$= i\left[\int_{x}^{+\infty} f(t)dt\right] = i\left[\int_{x}^{+\infty} \frac{1}{\sqrt{2\pi}} \exp\left(-\frac{t^2}{2}\right) dt\right] = i\left[\int_{x}^{4} \frac{1}{\sqrt{2\pi}} \exp\left(-\frac{t^2}{2}\right) dt\right], \forall x : -4 \leq x \leq 4$$

$$\Leftrightarrow \vec{F}_m(x) = P_m(x) m\vec{a} = i\left[\int_{x}^{+\infty} \frac{1}{\sqrt{2\pi}} \exp\left(-\frac{t^2}{2}\right) dt\right] m\vec{a}$$
$$= i\left[\int_{x}^{4} \frac{1}{\sqrt{2\pi}} \exp\left(-\frac{t^2}{2}\right) dt\right] m\vec{a}$$

The real complementary probability $P_m(x)/i$ and force are:

$$P_m(x)/i = 1 - P_r(x) = 1 - CDF(x) = 1 - \int_{-\infty}^{x} f(t)dt = \int_{x}^{+\infty} f(t)dt$$
$$= \int_{x}^{4} \frac{1}{\sqrt{2\pi}} \exp\left(-\frac{t^2}{2}\right) dt, \forall x : -4 \leq x \leq 4$$

$$\Leftrightarrow \vec{F}_m(x)/i = \frac{P_m(x)}{i} m\vec{a} = \left[\int_{x}^{4} \frac{1}{\sqrt{2\pi}} \exp\left(-\frac{t^2}{2}\right) dt\right] m\vec{a}$$

The complex probability or random vector and force are:

$$z(x) = P_r(x) + P_m(x) = \left[\int_{-4}^{x} \frac{1}{\sqrt{2\pi}} \exp\left(-\frac{t^2}{2}\right) dt\right] + i\left[\int_{x}^{4} \frac{1}{\sqrt{2\pi}} \exp\left(-\frac{t^2}{2}\right) dt\right], \forall x : -4 \leq x \leq 4$$

$$\Leftrightarrow \vec{F}(x) = \vec{F}_r(x) + \vec{F}_m(x) = P_r(x)m\vec{a} + P_m(x)m\vec{a} = [P_r(x) + P_m(x)]m\vec{a} = zm\vec{a}$$

$$= \left[\int_{-4}^{x} \frac{1}{\sqrt{2\pi}} \exp\left(-\frac{t^2}{2}\right) dt\right] m\vec{a} + i\left[\int_{x}^{4} \frac{1}{\sqrt{2\pi}} \exp\left(-\frac{t^2}{2}\right) dt\right] m\vec{a}$$

$$= \left\{\left[\int_{-4}^{x} \frac{1}{\sqrt{2\pi}} \exp\left(-\frac{t^2}{2}\right) dt\right] + i\left[\int_{x}^{4} \frac{1}{\sqrt{2\pi}} \exp\left(-\frac{t^2}{2}\right) dt\right]\right\} m\vec{a}$$

The Degree of Our Knowledge:

$$DOK(x) = |z(x)|^2 = P_r^2(x) + [P_m(x)/i]^2 = \left[\int_{-4}^{x} \frac{1}{\sqrt{2\pi}} \exp\left(-\frac{t^2}{2}\right) dt\right]^2$$

$$+\left(1 - \left[\int_{-4}^{x} \frac{1}{\sqrt{2\pi}} \exp\left(-\frac{t^2}{2}\right) dt\right]\right)^2$$

$$= \left[\int_{-4}^{x} \frac{1}{\sqrt{2\pi}} \exp\left(-\frac{t^2}{2}\right) dt\right]^2 + \left[\int_{x}^{4} \frac{1}{\sqrt{2\pi}} \exp\left(-\frac{t^2}{2}\right) dt\right]^2$$

$$= 1 + 2iP_r(x)P_m(x) = 1 - 2P_r(x)[1 - P_r(x)] = 1 - 2P_r(x) + 2P_r^2(x)$$

$$= 1 - 2\left[\int_{-4}^{x} \frac{1}{\sqrt{2\pi}} \exp\left(-\frac{t^2}{2}\right) dt\right] + 2\left[\int_{-4}^{x} \frac{1}{\sqrt{2\pi}} \exp\left(-\frac{t^2}{2}\right) dt\right]^2, \quad \forall x : -4 \le x \le 4$$

$DOK(x)$ is equal to 1 when $P_r(x) = P_r(L_b = -4) = 0$ and when $P_r(x) = P_r(U_b = 4) = 1$.

The Chaotic Factor:

$$Chf(x) = 2iP_r(x)P_m(x) = -2P_r(x)[1 - P_r(x)] = -2P_r(x) + 2P_r^2(x)$$

$$= -2\left[\int_{-4}^{x} \frac{1}{\sqrt{2\pi}} \exp\left(-\frac{t^2}{2}\right) dt\right] + 2\left[\int_{-4}^{x} \frac{1}{\sqrt{2\pi}} \exp\left(-\frac{t^2}{2}\right) dt\right]^2, \quad \forall x : -4 \le x \le 4$$

$Chf(x)$ is null when $P_r(x) = P_r(L_b = -4) = 0$ and when$P_r(x) = P_r(U_b = 4) = 1$.

The Magnitude of the Chaotic Factor *MChf*:

$$MChf(x) = |Chf(x)| = -2iP_r(x)P_m(x) = 2P_r(x)[1 - P_r(x)] = 2P_r(x) - 2P_r^2(x)$$

$$= 2\left[\int_{-4}^{x} \frac{1}{\sqrt{2\pi}} \exp\left(-\frac{t^2}{2}\right) dt\right] - 2\left[\int_{-4}^{x} \frac{1}{\sqrt{2\pi}} \exp\left(-\frac{t^2}{2}\right) dt\right]^2, \quad \forall x : -4 \le x \le 4$$

$MChf(x)$ is null when $P_r(x) = P_r(L_b = -4) = 0$ and when $P_r(x) = P_r(U_b = 4) = 1$.

At any value of x: $\forall x : (L_b = -4) \le x \le (U_b = 4)$, the probability expressed in the complex probability set $\mathcal{C} = \mathcal{R} + \mathcal{M}$ is the following:

$$\begin{aligned} P_c^2(x) &= [P_r(x) + P_m(x)/i]^2 = |z(x)|^2 - 2iP_r(x)P_m(x) \\ &= DOK(x) - Chf(x) \\ &= DOK(x) + MChf(x) \\ &= 1 \end{aligned}$$

then,

$$P_c^2(x) = [P_r(x) + P_m(x)/i]^2 = \{P_r(x) + [1 - P_r(x)]\}^2 = 1^2 = 1 \Leftrightarrow P_c(x) = 1 \text{ always}$$

$\Leftrightarrow \vec{F}_c(x) = P_c(x)m\vec{a} = 1 \times m\vec{a} = m\vec{a}$ always also.

Hence, the prediction of all the probabilities and forces of the stochastic experiment in the universe $\mathcal{C} = \mathcal{R} + \mathcal{M}$ is permanently certain and perfectly deterministic (**Figure 35**).

8.2.2.1 The complex probability cubes

In the first cube (**Figure 36**), the simulation of *DOK* and *Chf* as functions of each other and of the random variable *X* for the standard Gaussian normal probability distribution can be seen. The thick line in cyan is the projection of the plane $P_c^2(X) = DOK(X) - Chf(X) = 1 = P_c(X) = F_c / ma$ on the plane $X = L_b$ = lower bound of $X = -4$. This thick line starts at the point J ($DOK = 1$, $Chf = 0$) when $X = L_b = -4$,

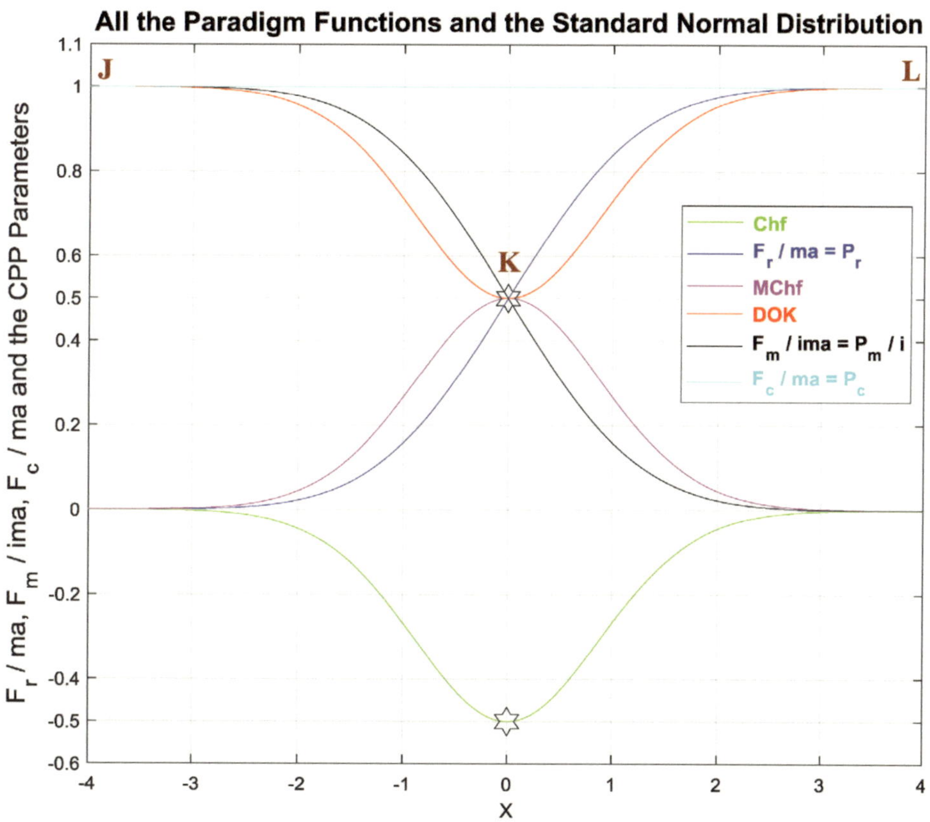

Figure 35.
The graphs of F_r / ma, F_m / ima, *and* F_c / ma *and of all the* CPP *parameters as functions of the random variable* X *for the continuous standard Gaussian normal distribution.*

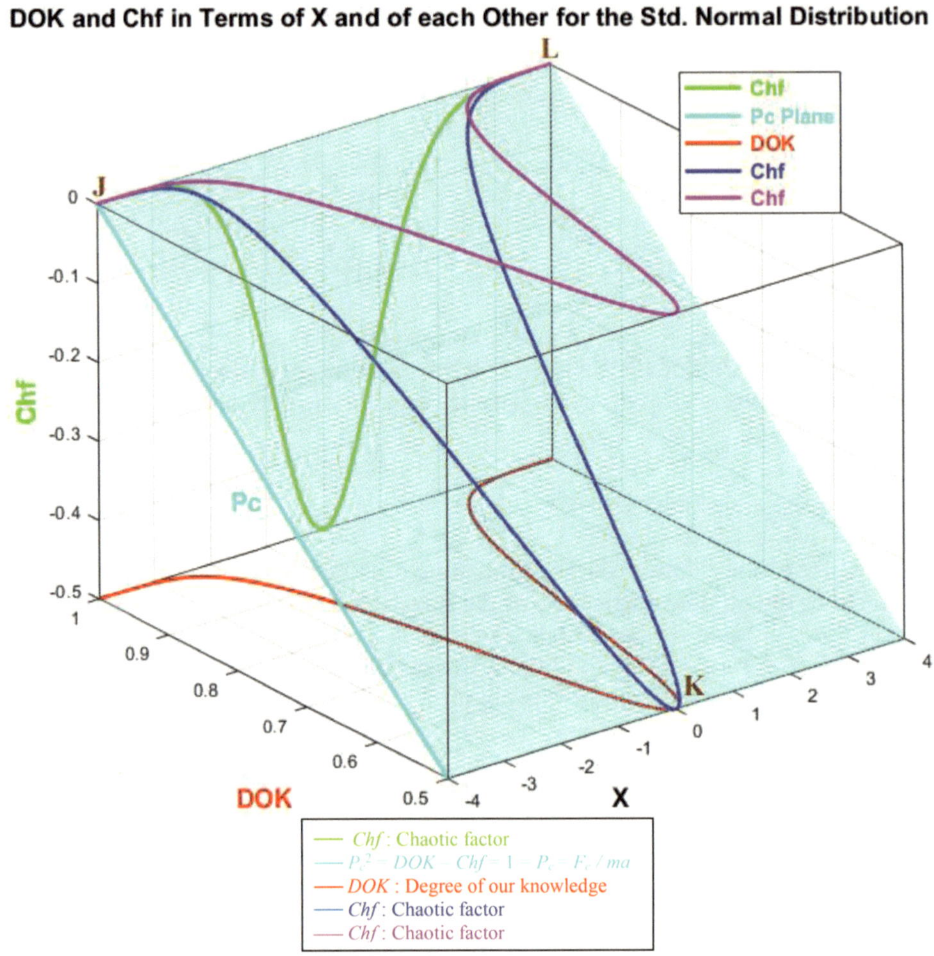

Figure 36.
The graphs of DOK *and* Chf *and the deterministic reduced force* F_c / ma = P_c *in terms of* X *and of each other for the standard Gaussian normal probability distribution.*

reaches the point ($DOK = 0.5$, $Chf = -0.5$) when $X = 0$, and returns at the end to J ($DOK = 1$, $Chf = 0$) when $X = U_b$ = upper bound of $X = 4$. The other curves are the graphs of $DOK(X)$ (red) and $Chf(X)$ (green, blue, pink) in different simulation planes. Notice that they all have a minimum at the point K ($DOK = 0.5$, $Chf = -0.5$, $X = 0$). The point L corresponds to ($DOK = 1$, $Chf = 0$, $X = U_b = 4$). The three points J, K, L are the same as in **Figure 35**.

In the second cube (**Figure 37**), we can notice the simulation of the real reduced force $F_r / ma = P_r(X)$ in $\mathcal{R}$ and its complementary real reduced force $F_m / ima = P_m(X)/i$ in $\mathcal{R}$ also in terms of the random variable X for the standard Gaussian normal probability distribution. The thick line in cyan is the projection of the plane $P_c^2(X) = P_r(X) + P_m(X)/i = 1 = P_c(X) = F_c / ma$ on the plane $X = L_b$ = lower bound of $X = -4$. This thick line starts at the point ($P_r = 0$, $P_m/i = 1$) and ends at the point ($P_r = 1$, $P_m/i = 0$). The red curve represents $F_r / ma = P_r(X)$ in the plane $P_r(X) = P_m(X)/i$ in light gray. This curve starts at the point J ($P_r = 0$, $P_m/i = 1$, $X = L_b$ = lower bound of $X = -4$), reaches the point K ($P_r = 0.5$, $P_m/i = 0.5$, $X = 0$), and gets at the end to L ($P_r = 1$, $P_m/i = 0$, $X = U_b$ = upper bound of $X = 4$). The blue curve represents $F_m / ima = P_m(X)/i$ in the plane in cyan $P_r(X) + P_m(X)/i = 1 = P_c(X) = F_c / ma$. Notice the importance of the point K which is the intersection of the

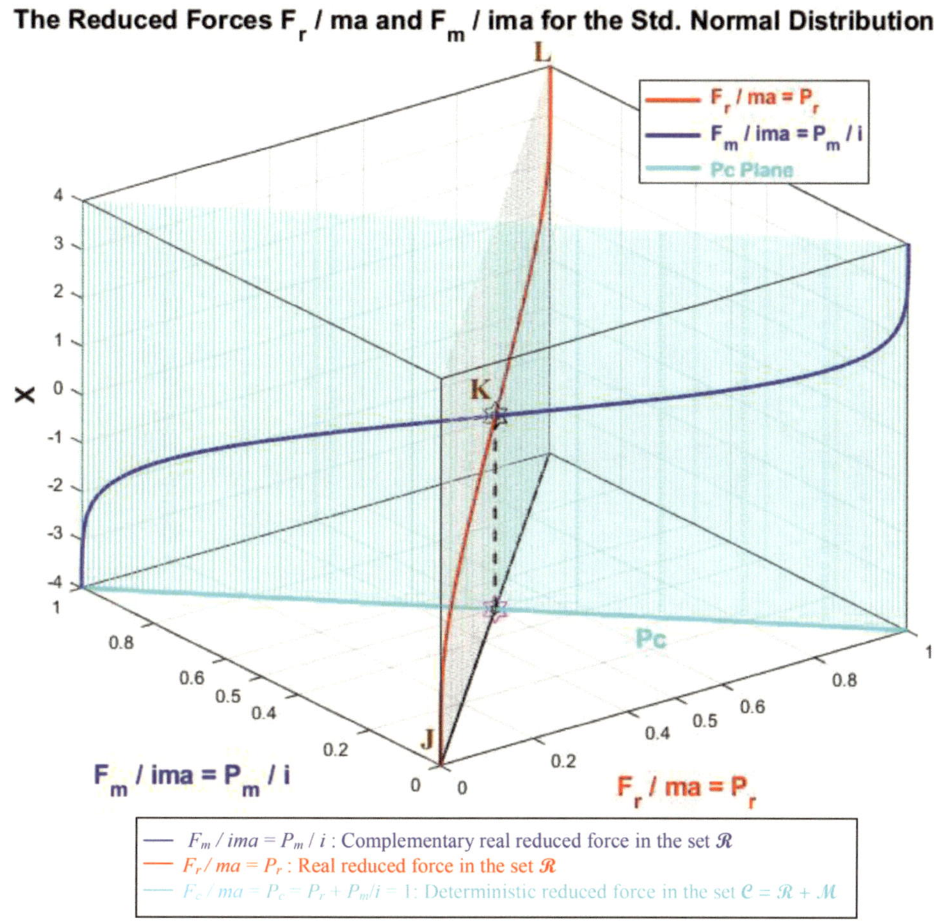

Figure 37.
The graphs of F_r / ma = P_r *and* F_m / ima = P_m / i *and* F_c / ma = P_c *in terms of* X *and of each other for the standard Gaussian normal probability distribution.*

red and blue curves at $X = 0$ and when $P_r(X) = P_m(X)/i = 0.5$. The three points J, K, L are the same as in **Figure 35**.

In the third cube (**Figure 38**), we can notice the simulation of the complex resultant reduced force $F / ma = z(X)$ in $\mathcal{C} = \mathcal{R} + \mathcal{M}$ as a function of the real reduced force $F_r / ma = P_r(X) = \text{Re}(z)$ in $\mathcal{R}$ and of its complementary imaginary reduced force $F_m / ma = P_m(X) = i \times \text{Im}(z)$ in $\mathcal{M}$, and this in terms of the random variable X for the standard Gaussian normal probability distribution. The red curve represents F_r / ma in the plane $P_m(X) = 0$ and the blue curve represents F_m / ma in the plane $P_r(X) = 0$. The green curve represents the complex resultant reduced force $F / ma = F_r / ma + F_m / ma = z(X) = P_r(X) + P_m(X) = \text{Re}(z) + i \times \text{Im}(z)$ in the plane $P_r(X) = iP_m(X) + 1$ or $z(X)$ plane in cyan. The curve of F / ma starts at the point J ($P_r = 0, P_m = i, X = L_b =$ lower bound of $X = -4$) and ends at the point L ($P_r = 1$, $P_m = 0, X = U_b =$ upper bound of $X = 4$). The thick line in cyan is $P_r(X = L_b = -4) = iP_m(X = L_b = -4) + 1$ and it is the projection of the F / ma curve on the complex probability plane whose equation is $X = L_b = -4$. This projected thick line starts at the point J ($P_r = 0, P_m = i, X = L_b = -4$) and ends at the point ($P_r = 1$, $P_m = 0, X = L_b = -4$). Notice the importance of the point K corresponding to $X = 0$ and $z = 0.5 + 0.5i$ when $P_r = 0.5$ and $P_m = 0.5i$. The three points J, K, L are the same as in **Figure 35**.

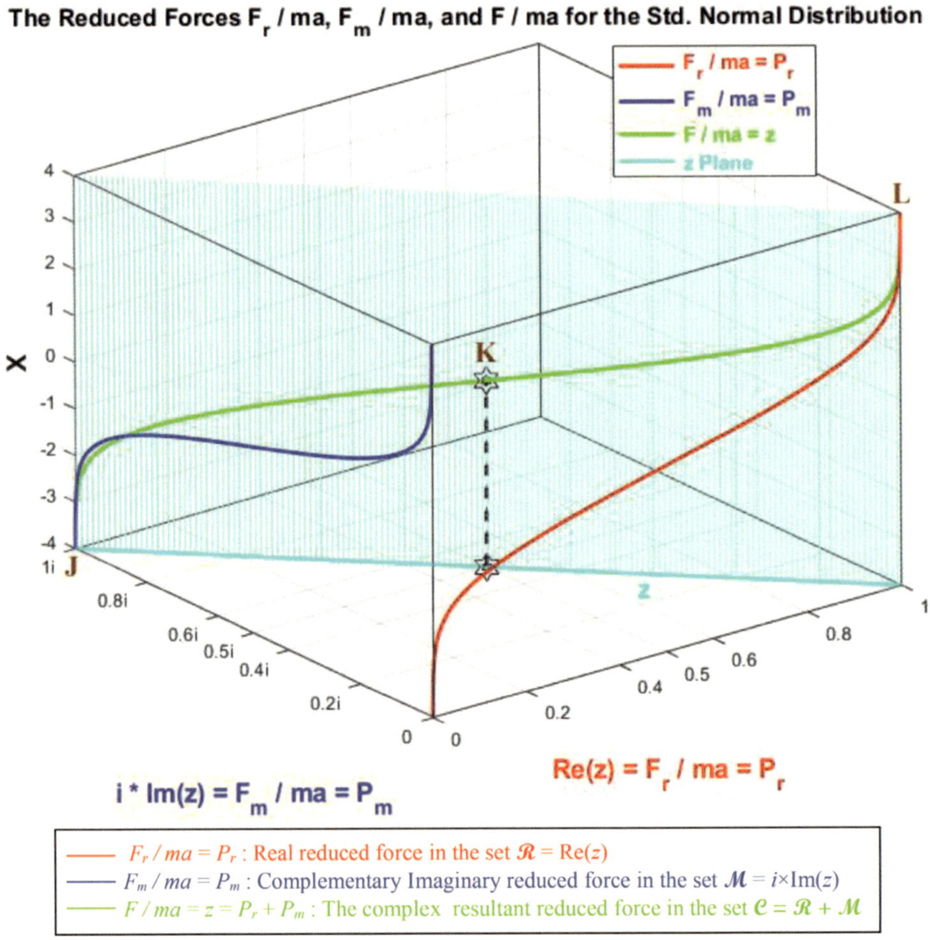

Figure 38.
The graphs of the reduced forces F_r / ma = P_r *and* F_m / ma = P_m *and* F / ma = z *in terms of* X *for the standard Gaussian normal probability distribution.*

9. Conclusion and perspectives

In the current research work, the original extended model of eight axioms (*EKA*) of A. N. Kolmogorov was connected and applied to Isaac Newton's classical mechanics theory. Thus, a tight link between classical mechanics and the novel paradigm was achieved. Consequently, the model of "Complex Probability" was more developed beyond the scope of my seventeen previous research works on this topic.

Additionally, as it was proved and verified in the novel model, before the beginning of the random phenomenon simulation and at its end we have the chaotic factor (*Chf* and *MChf*) is zero and the degree of our knowledge (*DOK*) is one since the stochastic fluctuations and effects have either not started yet or they have terminated and finished their task on the probabilistic phenomenon. During the execution of the nondeterministic phenomenon and experiment we also have: $0.5 \leq DOK < 1$, $-0.5 \leq Chf < 0$, and $0 < MChf \leq 0.5$. We can see that during this entire process we have incessantly and continually $P_c^2 = DOK - Chf = DOK + MChf = 1 = P_c$, that means that the simulation which behaved randomly and stochastically in the set $\mathcal{R}$ is now certain and deterministic in the probability set $\mathcal{C} = \mathcal{R} + \mathcal{M}$, and this after adding to the random experiment executed in $\mathcal{R}$ the contributions of the set $\mathcal{M}$ and hence after eliminating and subtracting the chaotic factor from the degree of our knowledge. Furthermore, the probabilities of the real, imaginary, complex, and deterministic forces acting on a body and that correspond to each value of the random variable X have been determined in the three probabilities sets which are $\mathcal{R}$, $\mathcal{M}$, and $\mathcal{C}$ by P_r, P_m, z and P_c respectively. Consequently, at each value of X, the novel classical mechanics and *CPP* parameters F_r, F_m, F, F_c, P_r, P_m, P_m/i, *DOK*, *Chf*, *MChf*, P_c, and z are surely and perfectly predicted in the complex probabilities set $\mathcal{C}$ with P_c maintained equal to one permanently and repeatedly. Also, as it was shown and proved in the equations above that if the real probability P_r is equal to one then we will return directly to the classical deterministic Newtonian mechanics theory which is a special deterministic case of the stochastic complex probability paradigm general case.

In addition, referring to all these obtained graphs and executed simulations throughout the whole research work, we are able to quantify and to visualize both the system chaos and stochastic effects and influences (expressed and materialized by *Chf* and *MChf*) and the certain knowledge (expressed and materialized by *DOK* and P_c) of the new paradigm. This is without any doubt very fruitful, wonderful, and fascinating and proves and reveals once again the advantages of extending A. N. Kolmogorov's five axioms of probability and hence the novelty and benefits of this inventive and original model in the fields of prognostics and applied mathematics that can be called truly: "The Complex Probability Paradigm".

Moreover, it is important to mention here that one very well-known and important random distribution was considered in the current work which is the discrete and uniform random distribution that was used to prove an important and essential result at the foundation of statistical mechanics and physics, knowing that the novel *CPP* paradigm can be implemented to any probability distribution that exists in literature as it was shown in the simulation section. This will lead without any doubt to analogous and similar conclusions and results and will confirm certainly the success of my innovative and original model.

As a future and prospective research and challenges, we aim to more develop the novel prognostic paradigm conceived and to implement it to a large set of random and nondeterministic events like for other probabilistic phenomena as in stochastic processes and in the classical theory of probability. Additionally, we will apply *CPP* to the random walk problems which have huge and very interesting consequences when implemented to chemistry, to physics, to economics, to applied and pure mathematics.

Conflicts of interest

The author declares that there are no conflicts of interest regarding the publication of this paper.

Data availability

The data used to support the findings of this study are available from the author upon request.

Nomenclature

$\mathcal{R}$	real set of events
$\mathcal{M}$	imaginary set of events
$\mathcal{C}$	complex set of events

i	the imaginary number where $i = \sqrt{-1}$ or $i^2 = -1$
EKA	Extended Kolmogorov's Axioms
CPP	Complex Probability Paradigm
P_{rob}	probability of any event
P_r	probability in the real set $\mathcal{R}$ = probability of the real random force in $\mathcal{R}$
P_m	probability in the imaginary set $\mathcal{M}$ corresponding to the real probability in $\mathcal{R}$ = probability of the imaginary random force in $\mathcal{M}$
P_c	probability of an event in $\mathcal{R}$ with its associated complementary event in $\mathcal{M}$ = probability of the real deterministic force in the complex probability set $\mathcal{C}$
$\vec{F}_r$	the real stochastic force in $\mathcal{R}$
$\vec{F}_m$	the imaginary stochastic force in $\mathcal{M}$
$\vec{F}$	the resultant complex stochastic force in $\mathcal{C}$
$\vec{F}_c$	the real deterministic force in $\mathcal{C}$
z	complex probability number = sum of P_r and P_m = complex random vector = probability of the resultant complex stochastic force in $\mathcal{C}$
$DOK = \lvert z\rvert^2$	the degree of our knowledge of the random system or experiment, it is the square of the norm of z
Chf	the chaotic factor of z
$MChf$	magnitude of the chaotic factor of z
N	number of random vectors = number of random atoms or particles or molecules
Z	the resultant complex random vector = $\sum_{j=1}^{N} z_j$
$DOK_Z = \frac{\lvert Z\rvert^2}{N^2}$	the degree of our knowledge of the whole stochastic system
$Chf_Z = \frac{Chf}{N^2}$	the chaotic factor of the whole stochastic system
$MChf_Z$	magnitude of the chaotic factor of the whole stochastic system
Z_U	the resultant complex random vector corresponding to a uniform random distribution
DOK_{Z_U}	the degree of our knowledge of the whole stochastic system corresponding to a uniform random distribution
Chf_{Z_U}	the chaotic factor of the whole stochastic system corresponding to a uniform random distribution
$MChf_{Z_U}$	the magnitude of the chaotic factor of the whole stochastic system corresponding to a uniform random distribution
$P_c\lvert_{Z_U}$	probability in the complex probability set $\mathcal{C}$ of the whole stochastic system corresponding to a uniform random distribution

Author details

Abdo Abou Jaoudé
Department of Mathematics and Statistics, Faculty of Natural and Applied Sciences, Notre Dame University-Louaize, Lebanon

*Address all correspondence to: abdoaj@idm.net.lb

References

[1] Ben-Chaim M. *Experimental Philosophy and the Birth of Empirical Science: Boyle, Locke and Newton*, Aldershot: Ashgate, ISBN 0-7546-4091-4. OCLC. 2004;53887772

[2] Agar, J. (2012). *Science in the Twentieth Century and beyond*, Cambridge: Polity Press, ISBN 978-0-7456-3469-2.

[3] Knudsen, J. M., Hjorth, P. (2012). *Elements of Newtonian Mechanics* (Illustrated Edition). Springer Science & Business Media. ISBN 978-3-642-97599-8.

[4] *MIT physics 8.01 lecture notes*. Archived 2013-07-09 at the Library of Congress Web Archives (PDF).

[5] Thornton, S. T., Marion, J. B. (2004). *Classical dynamics of particles and systems* (5th edition). Belmont, CA: Brooks/Cole. ISBN 978-0-534-40896-1.

[6] Jesseph, D. M. (1998). "*Leibniz on the Foundations of the Calculus: The Question of the Reality of Infinitesimal Magnitudes*". Perspectives on Science. Retrieved 31 December 2011.

[7] Alonso M, Finn J. *Fundamental University Physics*. Addison-Wesley; 1992

[8] Feynman, R. (1999). *The Feynman Lectures on Physics*. Perseus Publishing. ISBN 978-0-7382-0092-7.

[9] Feynman, R., Phillips, R. (1998). *Six Easy Pieces*. Perseus Publishing. ISBN 978-0-201-32841-7.

[10] Goldstein, H., Poole, C. P., Safko, J. L. (2002). *Classical Mechanics* (3rd Edition). Addison Wesley. ISBN 978-0-201-65702-9.

[11] Kibble, T. W. B., Berkshire, F. H. (2004). *Classical Mechanics* (5th Edition). Imperial College Press. ISBN 978-1-86094-424-6.

[12] Kleppner, D., Kolenkow, R. J. (1973). *An Introduction to Mechanics*. McGraw-Hill. ISBN 978-0-07-035048-9.

[13] Landau, L. D., Lifshitz, E. M. (1972). *Course of Theoretical Physics*, Vol. 1 – Mechanics. Franklin Book Company. ISBN 978-0-08-016739-8.

[14] Morin, D., (2008). *Introduction to Classical Mechanics: With Problems and Solutions* (1st edition). Cambridge: Cambridge University Press. ISBN 978-0-521-87622-3.

[15] Sussman, G. J., Wisdom, J. (2001). *Structure and Interpretation of Classical Mechanics*. MIT Press. ISBN 978-0-262-19455-6.

[16] O'Donnell, P. J. (2015). *Essential Dynamics and Relativity*. CRC Press. ISBN 978-1-4665-8839-4.

[17] Wikipedia, the free encyclopedia, *Classical Mechanics*. https://en.wikipedia.org/

[18] Newton, I., Machin, J. (1729). *Principia*. 1 (1729 translation edition).

[19] Browne, M. E. (July 1999). *Schaum's outline of theory and problems of physics for engineering and science* (Series: Schaum's Outline Series). McGraw-Hill Companies. ISBN 978-0-07-008498-8.

[20] Holzner, S. (December 2005). *Physics for Dummies.Wiley*, John & Sons, Incorporated. Bibcode:2005pfd..book.....H. ISBN 978-0-7645-5433-9.

[21] Greiner, W. (2003). *Classical mechanics: point particles and relativity*. New York: Springer. ISBN 978-0-387-21851-9.

[22] Zeidler, E. (1988). *Nonlinear Functional Analysis and its Applications IV: Applications to Mathematical Physics*.

New York: Springer. ISBN 978-1-4612-4566-7.

[23] Wachter, A., Hoeber, H. (2006). *Compendium of Theoretical Physics*. New York: Springer. ISBN 978-0-387-25799-0.

[24] Truesdell, C. A., Becchi, A., Benvenuto, E. (2003). *Essays on the history of mechanics: in memory of Clifford Ambrose Truesdell and Edoardo Benvenuto*. New York: Birkhäuser. ISBN 978-3-7643-1476-7.

[25] Lubliner, J. (2008). *Plasticity Theory* (PDF) (revised edition). Dover. Publications. ISBN 978-0-486-46290-5. Archived from the original (PDF) on 31 march 2010.

[26] Galili, I., Tseitlin, M. (2003). *"Newton's First Law: Text, Translations, Interpretations and Physics Education"*. Science & Education. 12 (1). Bibcode: 2003Sc&Ed..12...45G. doi:10.1023/A: 1022632600805. S2CID 118508770.

[27] Benjamin, C. (2001). "*4. Force and Motion*". *Newtonian Physics*. ISBN 978-0-9704670-1-0.

[28] *Isaac Newton, The Principia*, A new translation by I.B. Cohen and A. Whitman, University of California press, Berkeley 1999.

[29] Woodhouse, N. M. J. (2003). *Special relativity*. London/Berlin: Springer. ISBN 978-1-85233-426-0.

[30] Beatty, M. F. (2006). *Principles of engineering mechanics* Volume 2 of Principles of Engineering Mechanics: Dynamics-The Analysis of Motion. Springer. ISBN 978-0-387-23704-6.

[31] Thornton, M. (2004). *Classical dynamics of particles and systems* (5th Edition). Brooks/Cole. ISBN 978-0-534-40896-1.

[32] Plastino, A. R., Muzzio, J. C. (1992). *"On the Use and Abuse of Newton's Second Law for Variable Mass Problems"*. Celestial Mechanics and Dynamical Astronomy. 53 (3). Bibcode: 1992CeMDA..53..227P. doi:10.1007/BF00052611. ISSN 09232958. S2CID 122212239.

[33] Halliday, R. *Physics*. 1. ISBN 978-0-471-03710-1.

[34] Hannah, J., Hillier, M. J. (1971). *Applied Mechanics*. Pitman Paperbacks.

[35] Serway, R. A., Faughn, J. S. (2006). *College Physics*. Pacific Grove CA: Thompson-Brooks/Cole. ISBN 978-0-534-99724-3.

[36] Cohen, I. B. (2002). Harman, P. M., Shapiro, A. E. (editions). *The Investigation of Difficult Things: Essays on Newton and The History of The Exact Sciences in Honor of D.T. Whiteside*. Cambridge, UK: Cambridge university press. ISBN 978-0-521-89266-7.

[37] Stronge, W.J. (2004). *Impact mechanics*. Cambridge, UK: Cambridge University Press. ISBN 978-0-521-60289-1.

[38] Halliday, R. (1992). *Physics*, Volume 1 (4th Edition).

[39] Hellingman, C. (1992). "*Newton's Third Law Revisited*". Phys. Educ. 27 (2). Bibcode:1992PhyEd..27..112H. doi: 10.1088/0031-9120/27/2/011.

[40] Halliday R. *Physics* (Third Edition). John Wiley & Sons; 1977

[41] Fairlie G, Cayley E. *The life of a genius*. Hodder and Stoughton. 1965

[42] Cohen, I.B. (1995). *Science and the Founding Fathers: Science in the Political Thought of Jefferson, Franklin, Adams and Madison*. New York: W.W. Norton. ISBN 978-0-393-24715-2.

[43] Cohen, I.B. (1980). *The Newtonian Revolution: With Illustrations of the Transformation of Scientific Ideas*.

Cambridge, England: Cambridge University Press. ISBN 978-0-521-27380-0.

[44] Wikipedia, the free encyclopedia, *Newton's Laws of Motion*. https://en.wikipedia.org/

[45] Abou Jaoude A, El-Tawil K, Kadry S. *Prediction in complex dimension using Kolmogorov's set of axioms*. Journal of Mathematics and Statistics, Science Publications, vol. 2010;**6**(2):116-124

[46] Abou Jaoude, A. (2013)."*The Complex Statistics Paradigm and the Law of Large Numbers*", journal of mathematics and statistics, Science Publications, vol. 9(4), pp. 289-304.

[47] Abou Jaoude A. *The theory of complex probability and the first order reliability method*. Journal of Mathematics and Statistics, Science Publications, vol. 2013;**9**(4):310-324

[48] Abou Jaoude A. *Complex probability theory and prognostic*. Journal of Mathematics and Statistics, Science Publications, vol. 2014;**10**(1):1-24

[49] Abou Jaoude A. *The complex probability paradigm and analytic linear prognostic for vehicle suspension systems*. American Journal of Engineering and Applied Sciences, Science Publications, vol. 2015;**8**(1):147-175

[50] Abou Jaoude A. *The paradigm of complex probability and the Brownian motion*. Systems Science and Control Engineering, Taylor and Francis Publishers, vol. 2015;**3**(1):478-503

[51] Abou Jaoude A. *The paradigm of complex probability and Chebyshev's inequality*. Systems Science and Control Engineering, Taylor and Francis Publishers, vol. 2016;**4**(1):99-137

[52] Abou Jaoude A. *The paradigm of complex probability and analytic nonlinear prognostic for vehicle suspension systems*. Systems Science and Control Engineering, Taylor and Francis Publishers, vol. 2016;**4**(1):99-137

[53] Abou Jaoude A. *The paradigm of complex probability and analytic linear prognostic for unburied petrochemical pipelines*. Systems Science and Control Engineering, Taylor and Francis Publishers, vol. 2017;**5**(1):178-214

[54] Abou Jaoude A. *The paradigm of complex probability and Claude Shannon's information theory*. Systems Science and Control Engineering, Taylor and Francis Publishers, vol. 2017;**5**(1): 380-425

[55] Abou Jaoude A. *The paradigm of complex probability and analytic nonlinear prognostic for unburied petrochemical pipelines*. Systems Science and Control Engineering, Taylor and Francis Publishers, vol. 2017;**5**(1): 495-534

[56] Abou Jaoude A. *The paradigm of complex probability and Ludwig Boltzmann's entropy*. Systems Science and Control Engineering, Taylor and Francis Publishers, vol. 2018;**6**(1): 108-149

[57] Abou Jaoude A. *The paradigm of complex probability and Monte Carlo methods*. Systems Science and Control Engineering, Taylor and Francis Publishers, vol. 2019;7(1):407-451

[58] Abou Jaoude A. *Analytic prognostic in the linear damage case applied to buried petrochemical pipelines and the complex probability paradigm*. Fault Detection, Diagnosis and Prognosis, IntechOpen. 2020. DOI: 10.5772/intechopen.90157

[59] Abou Jaoude A. *The Monte Carlo techniques and the complex probability paradigm*. Forecasting in Mathematics - Recent Advances, New Perspectives and Applications, IntechOpen. 2020. DOI: 10.5772/intechopen.93048

[60] Abou Jaoude, A. (2020). "*The Paradigm of Complex Probability and Prognostic Using FORM*", London journal of research in science: Natural and formal (LJRS), London journals press, vol. 20(4), pp. 1-65. Print ISSN: 2631-8490, online ISSN: 2631-8504, DOI: 10.17472/LJRS, 2020.

[61] Abou Jaoude, A. (2020). "*The Paradigm of Complex Probability and The Central Limit Theorem*", London journal of research in science: Natural and formal (LJRS), London journals press, vol. 20(5), pp. 1-57. Print ISSN: 2631-8490, online ISSN: 2631-8504, DOI: 10.17472/LJRS, 2020.

[62] Benton W. *Probability*, *Encyclopedia Britannica*. Vol. 18, pp. 570-574. Chicago: Encyclopedia Britannica Inc; 1966

[63] Benton W. *Mathematical Probability*, *Encyclopedia Britannica*. Vol. 18, pp. 574-579. Chicago: Encyclopedia Britannica Inc; 1966

[64] Feller W. *An Introduction to Probability Theory and its Applications*. 3rd ed. New York: Wiley; 1968

[65] Walpole R, Myers R, Myers S, Ye K. *Probability and Statistics for Engineers and Scientists*. 7th ed. New Jersey: Prentice Hall; 2002

[66] Freund JE. *Introduction to Probability*. New York: Dover Publications; 1973

[67] Srinivasan SK, Mehata KM. *Stochastic Processes*. 2nd ed. New Delhi: McGraw-Hill; 1988

[68] Stewart I. *Does God Play Dice?* 2nd ed. Oxford: Blackwell Publishing; 2002

[69] Stewart I. *From Here to Infinity*. 2nd ed. Oxford: Oxford University Press; 1996

[70] Stewart I. *In Pursuit of the Unknown*. New York: Basic Books; 2012

[71] Barrow J. *Pi in the Sky*. Oxford: Oxford University Press; 1992

[72] Bogdanov I, Bogdanov G. *Au Commencement du Temps*. Paris: Flammarion; 2009

[73] Bogdanov I, Bogdanov G. *Le Visage de Dieu*. Paris : Editions Grasset et Fasquelle. 2010

[74] Bogdanov I, Bogdanov G. *La Pensée de Dieu*. Paris : Editions Grasset et Fasquelle. 2012

[75] Bogdanov I, Bogdanov G. *La Fin du Hasard*. Paris : Editions Grasset et Fasquelle. 2013

[76] Van Kampen NG. *Stochastic Processes in Physics and Chemistry*. Sydney, Elsevier: Revised and Enlarged Edition; 2006

[77] Bell ET. *The Development of Mathematics*. United States of America: New York, Dover Publications, Inc.; 1992

[78] Boursin J-L. *Les Structures du Hasard*. Paris: Editions du Seuil; 1986

[79] Dacunha-Castelle D. *Chemins de l'Aléatoire*. Paris: Flammarion; 1996

[80] Dalmedico-Dahan A, Chabert J-L, Chemla K. *Chaos Et Déterminisme*. Paris: Edition du Seuil; 1992

[81] Ekeland I. *Au Hasard*. In: *La Chance, la Science et le Monde*. Paris: Editions du Seuil; 1991

[82] Gleick J. *Chaos, Making a New Science*. New York: Penguin Books; 1997

[83] Dalmedico-Dahan A, Peiffer J. *Une Histoire des Mathématiques*. Paris: Edition du Seuil; 1986

[84] Gullberg J. *Mathematics from the Birth of Numbers*. New York: W.W. Norton & Company; 1997

[85] Vie SE. *Le Mystère des Mathématiques*. Numéro. 1999;**984**

[86] Davies P. *The Mind of God*. London: Penguin Books; 1993

[87] Gillies, D. (2000). *Philosophical Theories of Probability*. London: Routledge. ISBN 978-0415182768.

[88] Guillen M. *Initiation Aux Mathématiques*. Paris: Albin Michel; 1995

[89] Hawking S. *On the Shoulders of Giants*. London: Running Press; 2002

[90] Hawking S. *God Created the Integers*. London: Penguin Books; 2005

[91] Hawking S. *The Dreams That Stuff Is Made of*. London: Running Press; 2011

[92] Pickover C. *Archimedes to Hawking*. Oxford: Oxford University Press; 2008

[93] Abou Jaoude A. *The Computer Simulation of Monté Carlo Methods and Random Phenomena*. United Kingdom: Cambridge Scholars Publishing; 2019

[94] Abou Jaoude A. *The Analysis of Selected Algorithms for the Stochastic Paradigm*. United Kingdom: Cambridge Scholars Publishing; 2019

[95] Abou Jaoude, A. (August 1st 2004). Ph.D. Thesis in Applied Mathematics: *Numerical Methods and Algorithms for Applied Mathematicians*. Bircham International University. http://www.bircham.edu.

[96] Abou Jaoude, A. (October 2005). Ph.D. Thesis in Computer Science: *Computer Simulation of Monté Carlo Methods and Random Phenomena*. Bircham International University. http://www.bircham.edu.

[97] Abou Jaoude, A. (27 April 2007). Ph.D. Thesis in Applied Statistics and Probability: *Analysis and Algorithms for the Statistical and Stochastic Paradigm*. Bircham International University. http://www.bircham.edu.

[98] Stepić AI, Ognjanović Z. *Complex valued probability logics*. Publications De L'institut Mathématique, Nouvelle Série, tome. 2014;**95**(109):73-86. DOI: 10.2298/PIM1409073I

[99] Cox, D. R. (1955). "*A Use of Complex Probabilities in the Theory of Stochastic Processes*", mathematical proceedings of the Cambridge philosophical society, 51, pp. 313–319.

[100] Weingarten D. *Complex probabilities on R^N as real probabilities on C^N and an application to path integrals*. Physical Review Letters. 2002;**89**:335 http://dx.doi.org/10.1103/PhysRevLett.89.240201

[101] Youssef S. *Quantum mechanics as complex probability theory*. Modern Physics Letters A. 1994;**9**:2571-2586

[102] Fagin R, Halpern J, Megiddo N. *A logic for reasoning about probabilities*. Information and Computation. 1990;**87**: 78-128

[103] Bidabad B. *Complex Probability and Markov Stochastic Processes*. Proc: First Iranian Statistics Conference, Tehran, Isfahan University of Technology; 1992

[104] Ognjanović Z, Marković Z, Rašković M, Doder D, Perović A. *A probabilistic temporal logic that can model reasoning about evidence*. Annals of Mathematics and Artificial Intelligence. 2012;**65**:1-24

Chapter 3

Flooding Fragility Model Development Using Bayesian Regression

Alison Wells and Chad L. Pope

Abstract

Traditional component pass/fail design analysis and testing protocol drives excessively conservative operating limits and setpoints as well as unnecessarily large margins of safety. Component performance testing coupled with failure probability model development can support selection of more flexible operating limits and setpoints as well as softening defense-in-depth elements. This chapter discuses the process of Bayesian regression fragility model development using Markov Chain Monte Carlo methods and model checking protocol using three types of Bayesian p-values. The chapter also discusses application of the model development and testing techniques through component flooding performance experiments associated with industrial steel doors being subjected to a rising water scenario. These component tests yield the necessary data for fragility model development while providing insight into development of testing protocol that will yield meaningful data for fragility model development. Finally, the chapter discusses development and selection of a fragility model for industrial steel door performance when subjected to a water-rising scenario.

Keywords: fragility model development, Bayesian regression, Markov Chain Monte Carlo, fragility model checking, Bayesian p-value

1. Introduction

Traditional component pass/fail design analysis and testing protocol drives excessively conservative operating limits and setpoints as well as unnecessarily large margins of safety. Additionally, pass/fail testing tends to result in data shortcomings which must then be addressed using defense-in-depth elements. Contrarily, component performance testing and failure probability model development can support selection of more flexible operating limits and setpoints as well as softening defense-in-depth elements. The two major obstacles involved in developing a failure probability model, also known as a fragility model, center on devising an optimum component performance testing protocol so that meaningful data can be collected, and navigating the process of developing and testing an appropriate fragility model.

This chapter will first discuss the process of Bayesian regression fragility model development which includes model checking protocol. The foundation of fragility model development is Bayesian in nature where both data and parameters have probability distributions, and we seek a model that establishes a relationship between parameters and observables ultimately yielding a posterior probability distribution. That is, the Bayesian method requires an aleatory model, a prior distribution for the parameters of the aleatory model, and data associated with the aleatory model. Then, using Bayes Theorem, the posterior distribution for the model output can be obtained using Markov Chain Monte Carlo (MCMC) methods to address complicated integration. Multiple models are then developed, and a rigorous process is used to check model validity to help identify the most appropriate model. The model checking and comparison process uses multiple techniques including three types of Bayesian p-values.

With a firm foundation for fragility model development, checking, and selection established, the chapter then discusses component flooding performance experiments associated with industrial steel doors subjected to a rising water scenario. These component tests yield the necessary data for fragility model development while providing insight into development of testing protocol that will yield meaningful data for fragility model development. Finally, the chapter discusses the development and selection of a fragility model for industrial steel door performance when subjected to a rising water flood scenario.

2. Bayesian data analysis

Significant experience exists with fragility modeling focused on seismic fragility model determination. In a seismic fragility model, the single vertical ground acceleration variable is used to completely characterize the failure probability of structures or components of interest. However, other observable parameters may be important indicators for the potential of failure. Expanding upon the seismic example, these observables could include the detailed characteristics of the earthquake such as X, Y, and Z components of the ground motion; frequency of the waves; the age of the component; the anchorage of the component; the specifics of the component type; or any combination of the above.

Limitations found in these traditional fragility models include simplistic (single "driving" parameter) and excessive conservatism. For complex flooding fragility modeling requiring more observables, these issues will be avoided by moving to a more flexible, data-informed approach—Bayesian fragility modeling through phenomena-driven regression modeling. As stated by Box and Tiao, "Bayesian inference alone seems to offer the possibility of sufficient flexibility to allow reaction to scientific complexity free from impediment from purely technical limitation." [1].

From the Bayesian perspective, both data and parameters can have probability distributions, and the task of Bayesian analysis is to build a model for the relationship between parameters (θ) and observables (y), and then calculate the posterior probability. The Bayesian method, therefore, relies on three items: an aleatory model, a prior distribution for the parameter(s) of the aleatory model, and data associated with the aleatory model. An aleatory model pertains to stochastic or non-deterministic events, the outcome of which is described using probability. The posterior distribution for the model output function is developed in accordance with Bayes' Theorem [2], which is generally written as:

$$p(\theta|y) = \frac{p(\theta)p(y|\theta)}{p(y)} \tag{1}$$

where,

- $p(\theta|y)$:Posterior distribution, which is conditional upon data (y) that is known related to the hypothesis (θ);

- $p(\theta)$:Prior distribution, for knowledge of the hypothesis (θ) that is independent of data (y);

- $p(y|\theta)$:Likelihood, or aleatory model, representing the process or mechanism that provides data (y);

- $p(y)$:Marginal distribution, which serves as a normalization constant.

In summary, the above equation takes our prior knowledge about the parameters and updates this knowledge with the likelihood to observe the data for particular parameter values and gives the posterior probability. It essentially states: posterior $\propto$ prior $\times$ likelihood

This process combines everything that is known about a particular data set and model response to produce a posterior estimate of the output function's probability distribution.

Integration of functions plays an important role in Bayesian statistical analysis; however, explicit evaluation of these integrals is only possible for a limited number of special cases. Usually, problems will involve complex distributions and explicit evaluation is not possible. Traditionally, statisticians would be forced to use numerical integration or analytical approximation techniques. However, there are now several powerful software programs that exist for Bayesian inference. One of the most widely used by statistical practitioners is the BUGS (Bayesian inference Using Gibbs Sampling) family of programs. The most popular packages from the BUGS family are WinBUGS and OpenBUGS. There are several methods devised for construction and sampling complex Bayesian posterior distributions. BUGS software utilizes MCMC methods to determine the posterior [3].

MCMC is a general method based on randomly sampling values from a prior distribution to approximate the posterior distribution $p(\theta|y)$. The sampling is done sequentially, with the distribution of the sampled parameter depending on the value from the previous step only, forming a Markov chain [4]. Eventually the Markov chain will converge to a unique stationary distribution, the posterior distribution. Therefore, the key to MCMC method is the approximate distributions are improved at each step in the simulation, and after running the simulation long enough, converging to the posterior distribution.

3. Model checking and comparison

After constructing a probability model and computing posterior distributions for all estimated parameters, the next step of a Bayesian analysis includes checking that the model adequately represents the data and is plausible for the purpose for which the model will be used. There are multiple ways of assessing a model's performance. The approach selected is posterior predictive checking, a useful direct way of assessing the fit of a model to various aspects of the data. Additionally, residual tests are used for informal model criticism and outlier identification.

Posterior predictive checks are a primary form of Bayesian model checking used to assess the fit of the model to various aspects of the data. The procedure is based upon the following assumption: if a given model fits, then data simulated or replicated under the model should be comparable to the real-world observed data the model was fitted to [4]. In other words, the observed data should be plausible under the posterior predictive distribution. If any systematic differences occur between simulations and the data, it potentially indicates that model assumptions are not being met.

The model is checked for deviations from an assumed parameter form by means of test quantities or discrepancy functions, $T(y|\theta)$, that depend on both data (y) and parameters (θ). A check is made whether $T(y|\theta)$ is compatible with the simulated distribution of $T(y^{simulated}|\theta)$ by calculating a Bayesian p-value [4]. Regarding the choice of discrepancy functions, focus is given to diagnosing global lack of fit rather than discovering outliers; a task given to residual calculations. A summary of candidate discrepancy functions considered is provided in **Table 1**. Note, to avoid numerical errors for binomial models if $p = 0$ or 1, a small $\varepsilon = 0.00001$ is added in the expressions.

Note that ideally model checking should be based on new data, although in practice the same data is generally used for both developing and checking the model. This means Bayesian p-values based on these checks tend to be conservative [3]. However, this does not imply that posterior predictive checks lack value. Given that tests are conservative, small (less than 0.05) and large (greater than 0.95) p-values strongly suggest lack of fit. P-values closest to 0.5 indicate a high degree of predictive capability [2]. The concept of Bayesian p-value is graphically represented in **Figure 1**.

Residuals measure the discrepancy between the observed data and an assumed model. Informal tests based on Pearson and deviance residuals can be used to identify obvious assumption violations. Note that these analyses are generally carried out informally in Bayesian application, since all residuals depend on θ and have posterior distributions [6]. Therefore, they are not truly independent as required in unbiased application of goodness-of-fit tests.

Name	Definition	Binomial Expression
χ^2	$T(y,\theta) = \sum_i \frac{(y_i - E(y_i\|\theta))^2}{Var(y_i\|\theta)}$	$\sum_i \frac{(y_i - np_i)^2}{np_i(1-p_i)+\varepsilon}$
Likelihood ratio	$T(y,\theta) = 2\sum_i y_i \log\left(\frac{y_i}{E(y_i\|\theta)}\right)$	$2\sum_i y_i \log\left(\frac{y_i+\varepsilon}{np_i+\varepsilon}\right)$
Freeman-Tukey	$T(y,\theta) = \sum_i \left(\sqrt{y_i} - \sqrt{E(y_i\|\theta)}\right)^2$	$\sum_i \left(\sqrt{y_i} - \sqrt{np_i}\right)^2$

Table 1.
Discrepancy functions used for model checking [5].

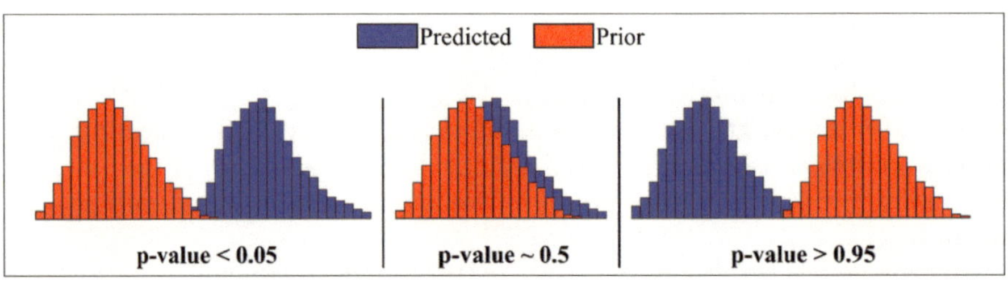

Figure 1.
Depiction of the Bayesian p-value predictability.

A standardized Pearson residual is defined as:

$$r_i = \frac{y_i - E(y_i|\theta)}{\sqrt{Var(y_i|\theta)}} \tag{2}$$

Where is the $E(y_i|\theta)$ expected value and $Var(y_i|\theta)$ is the variance. Since it is considered a function of random y_i for a fixed θ, Pearson residuals should generally take on values between −2.0 and 2.0 [6]. Values falling outside this range would represent outliers.

Residuals can also be based on a saturated version of the deviance., defined as:

$$D_s(\theta) = -2\log p(y|\theta) + 2\log p\left(y|\hat{\theta}_s(y)\right) \tag{3}$$

where $\hat{\theta}_s(y)$ are the saturated estimates. Models for which saturated deviance is appropriate, such as Poisson and binomial, the rule of thumb for a rough assessment of the fit is the mean saturated deviance should approximately equal sample size n [3].

Following model checking, comparisons can be made on the performance of alternative hypothesized models. It is not an uncommon occurrence for more than one probability model to provide an adequate fit to the data. These models may differ in prior specification, link function selection, or which explanatory variables are included in the regression, to name a few. Therefore, an analysis should not only examine models to see how they fail to fit reality but compare how sensitive the resulting posterior distributions are to arbitrary specifications using any number of model comparison or performance metrics.

There are a variety of Bayesian model comparison methods, including methods based on information criteria, which are measures of the relative fit. Deviance Information Criteria (DIC) is a measure of model fit that can be applied to Bayesian models and is applicable when the parameter estimation is done using techniques such as Gibbs sampling. It is particularly useful in Bayesian model selection problems where the posterior distributions of the model have been obtained by MCMC simulation. DIC is a generally straightforward computation, and no additional scripting is needed to calculate it in OpenBUGS, making it the comparison approach selected for this work.

As a rule of thumb, the model with the smallest DIC usually indicates the better fitting model. Note, however, only differences between models in DIC are important, not strictly absolute values. While it is not easy to define what constitutes an important difference, the following rough guide can be used for DIC comparison [3]:

- Differences greater than 10 can be used to rule out the model with the higher DIC.
- Differences between 5 and 10 are substantial.
- Differences less than 5, there is uncertainty about choice of model. Other methods may need to be considered, especially if models make different inferences.

Note that these considerations include negative values for the DIC, which occur in cases where the deviance is negative. It must also be noted that since DIC is a measure of relative fit, a model with the smallest DIC can still be a poor fit for the data [2].

4. Experiments

The objectives of component flooding experiments are to test individual component performance in flooding scenarios and acquire the necessary data to develop component fragility mathematical models. To conduct rising water experiments, the Portal Evaluation Tank (PET) was designed and built to facilitate testing.

Figure 2.
PET tank and piping.

Test	Depth (in.)	Flow rate (gal/min)	Temperature (F)	Notes
1S	46.1	1148	67.4	
2S	39.0	1130	63.3	
3S	37.1	1120	63.1	
4S	37.8	979	63.0	
5S	37.5	1133	63.0	
6S	37.6	604	63.0	
7S	37.7	593	63.0	
8S	37.1	598	63.1	
12S	44.5	975	64.0	
—	25.7	248	61.6	Non-Failure
—	17.0	117	59.0	Non-Failure
—	27.4	285	59.3	Non-Failure
—	30.9	397	59.4	Non-Failure
—	32.3	484	59.6	Non-Failure
—	24.3	247	60.2	Non-Failure
—	34.8	593	60.7	Non-Failure
—	37.5	696	61.0	Non-Failure
—	38.0	734	61.2	Non-Failure
13S	41.4	1025	61.3	

Table 2.
Steel door performance results [5].

The PET is a steel semi-cylindrical tank with a height and diameter of 8 ft. Its design includes a 62.4 ft^2 opening for installation of components to be tested, a front water tray with a 90-degree v-notch weir and the ability to hold up to 2,000-gal of water. The PET is connected through 12 in. PVC pipes to a 60 HP pump, which is located inside an 8,000 gal water reservoir, to support variable inlet flow rates up to ~4,500 gpm. Additionally, the design of PET, once filled, can rely on the pump and pressure and air relief values to provide hydrostatic head to simulate depths up to 20 ft. The PET, along with piping, is shown in **Figure 2**.

Accompanying instrumentation and measurements included electromagnetic flowmeters for upstream and downstream flow rates and two pressure transducers for averaged water depths and temperature. The PET can also measure small leakage rates that do not exceed the v-notch weir barrier using an ultrasonic depth sensor. The top of the PET is also equipped with pressure and air relief valves and a digital pressure gauge to measure pressures for simulated hydrostatic head once the PET is filled.

The components tested were industrial steel doors oriented to swing outwards, away from the tank interior. A strengthened wall was built to support the doorframe, ensuring stability. The aim of these experiments was to test the door to failure only and not the supporting wall structure. The experimental approach subjected each steel door to a water rising scenario until catastrophic failure of the door occurred or the leakage rate equalized with the filling rate. A compiled summary of the steel door results, including non-failure tests, are given in **Table 2**.

5. Model development

Having conducted the flooding experiments and collected observational data on door failures, models where developed that analyzed the fragility of components using explanatory variables. An explanatory variable is a type of independent variable that is possibly predictive of a component's fragility in a regression analysis. For the probability of door failure during a flooding event, water depth, flow rate, and temperature may be leading indicators of failure and information about these explanatory variables is incorporated into the Bayesian inference.

The mathematical modeling uses the discrete binomial distribution to represent failure of a door installed in the PET during a rising water flood event. This is a commonly used model for failure on demand with key parameters p, the probability of failure on demand, and trials $n = 1$ (only a single door is potentially challenged during testing). The fragility model in this case looked at seven possibilities: each of the variables alone driving the model to failure, a combination of two variables driving the model to failure, and all three variables driving the model to failure. The above cases are modeled as:

$$Logit(p) = intercept + aD \tag{4a}$$

$$Logit(p) = intercept + bF \tag{4b}$$

$$Logit(p) = intercept + cT \tag{4c}$$

$$Logit(p) = intercept + aD + bF \tag{4d}$$

$$Logit(p) = intercept + aD + cT \tag{4e}$$

$$Logit(p) = intercept + bF + cT \tag{4f}$$

$$Logit(p) = intercept + aD + bF + cT \tag{4g}$$

where a, b, and c are the coefficients of the covariate parameters represented as D, F, and T for depth, flow rate, and temperature respectively. Since parameter p represents a probability, it must be constrained between 0 and 1 with a link function. The logit function was selected, which is defined as:

$$Logit(p) = \ln\left(\frac{p}{1-p}\right) \tag{5}$$

While the logit function should transform the parameter p onto an appropriate scale, in practice this was not always true from the special case of $n = 1$. Periodically, the sampler from the prior distribution selects illogical or extreme values. This can cause errors such as numerical overflow or, within the logistic regression, results in negative parameter values that cannot be log transformed. The improper value prompted a binomial calculation that OpenBUGS is unable to perform, causing the run to crash. It should also be noted that subtle differences in programs could resolve some of these problems. Not all available programs, for instance, use the same sampling approach. A similar model setup in R or JAGS could run without additional considerations for the case of $n = 1$.

A robust solution focuses on the parameter that fails to meet specifications. The binomial probability of failure, p, needs to take on values between 0 and 1 for OpenBUGS to perform the calculation, as referenced earlier. This requirement can be achieved by restricting p using built-in scalar functions, max and min. They are defined and operate as follows:

- max(e1, e2) e1 if e1 > e2; e2 otherwise,
- min(e1, e2) e1 if e1 < e2; e2 otherwise.

For the probability of failure to be properly scaled, the following criteria need to hold true:

- return 1 if probability of failure is greater than 1,
- return 0 if probability of failure is less than 0,
- otherwise p.

The quantity *p.bound[i] - > max(0, min(1, p[i]))* performs all three listed criteria. Inserting p.bound into the model script restricts the probability to lie between 0 and 1 and prevents OpenBUGS from crashing [7]. A logistic link function can now be used when $n = 1$ for all regression models.

The water temperature data was included as an explanatory variable with the expectation that it would be eliminated as part of the Bayesian analysis. To address the possibility of temperature as a failure influence, centering was used on the covariates. Interpreting coefficients in models with interactions can be simplified by subtracting the mean, $\overline{x} = N^{-1}\sum x_i$, of each input variable x_i. For example, the temperature T in Eq. (4c) would be subtracted by $\overline{T}$ and the following logistic regression would be fit:

$$Logit(p) = intercept + c\left(T - \overline{T}\right) \tag{6}$$

where the data is now centered at zero. The main effects of using explanatory variables are now interpretable based on comparison to the mean of the data.

Coefficients that stay relatively the same compared to the un-centered results indicate low predictability, while large predictive differences are leading indicators of component failure.

Looking at the steel door data, however, leads to a different discovery. **Table 3** gives the results for the standard models, and **Table 4** gives the results when centering is applied. Depth's predictive difference is greater than flow rate, but the highest is temperature. Additionally, temperature has the smallest DIC between the three models.

To understand why temperature appears to be the leading indicator of failure, the steel door data, along with its collection process, must be examined. Of the nineteen test results recorded in **Table 2**, the first nine tests all resulted in door failures. These nine tests were conducted exclusively during the spring. The remainder of the tests, nine non-failure and one failure, where conducted in a single day during the winter when the reservoir water was cooler. The results could mean that warmer water temperatures cause steel doors to fail in flooding events, implying a correlation of variables observed together. It is noted, however, that correlation does not necessarily mean causation. The relationship could have alternative explanations, such as a third-cause fallacy, where a spurious correlation is mistaken for causation. A spurious correlation is a relationship in which events or variables are associated, but not causally related, due to the presence of a third factor [8]. Seasonal weather changing the interior temperature of the laboratory is a hidden third factor. Therefore, steel door flooding failure and water temperature may be correlated with each other only because they are correlated with the weather when testing was conducted. By conducting all non-failure tests in the cooler winter conditions and majority of failures in the warmer spring, an unintentional bias was introduced into the temperature data. This bias, that temperature impacts failure, becomes apparent when looking at the centering comparison.

There is another means of verifying the introduced bias in temperature by looking at the residuals. Pearson residuals should take on values between −2.0 and 2.0. Any data point with values outside this range represent an outlier. If there is a bias introduced from when the tests were conducted, the last data point, a failure during winter testing, should be considered an outlier. **Figure 3** shows the residual box plot for the temperature regression model. Note that the last data point has an outlier residual value of 3.53 ± 6.037, confirming the bias.

Model	Mean	Standard Dev.	97.5% Interval	DIC
Depth	1.66	0.91	(0.42, 3.89)	13.86
Flow Rate	0.013	0.006	(0.004, 0.028)	16.0
Temperature	2.56	0.88	(1.10, 4.51)	8.294

Table 3.
Coefficient results for standard logit regression model for steel doors [5].

Model	Mean	Standard Dev.	97.5% Interval	DIC
Depth	2.05	1.26	(0.46, 5.21)	14.39
Flow Rate	0.013	0.006	(0.005, 0.028)	15.98
Temperature	7.85	4.69	(2.04, 19.74)	8.98

Table 4.
Coefficient results for centered logit regression models for steel doors [5].

Since the steel door temperature data is biased, it is dropped from consideration as an explanatory variable for now. In experiments, controlling and extensively testing the relationship between dependent and independent variables can identify spurious correlation. For component flooding experiments, steps could be taken to control the temperature of the reservoir water. If future testing corrects for this bias, temperature data could again be considered as part of the Bayesian analysis for steel doors. Of the remaining depth and flow rate data, centering simplified interpreting coefficients and indicated depth as a significant indicator of failure.

Development of the logistic regression models so far has been directly interpreting the failure response given some predictor(s) data. It is also possible to interpret indirectly by incorporating an additional random variability. These models assume that besides the observed variables, there could be an unobserved variable or random effects. Therefore, the probability of the binomial distribution is allowed to adjust by some small amount, λ_i, for each observation.

A script was written where logistic regression equations contain a random or latent effect. In the case of the depth model, previously given by Eq. (4a), it would now be defined as follows:

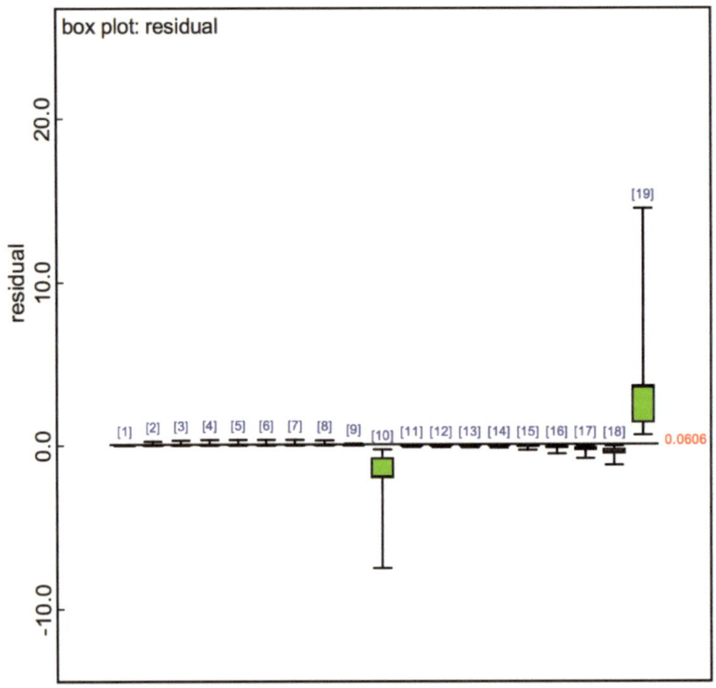

Figure 3.
Box plot of the temperature regression model residuals using steel door data [5].

Model	χ^2	Likelihood Ratio	Freeman-Tukey	DIC
Depth	0.97	0.97	0.97	0.41
Flow Rate	0.99	0.99	0.99	0.13
Depth. Flow Rate	0.99	0.99	0.99	0.08

Table 5.
Depth, flow and combined p-values and DIC [5].

```
#Bound Binomial Model using Logit Regression: Final
#Steel Door Data

model{

for(i in 1:tests){
failure[i] ~ dbin(p.bound[i], numtested)
p.bound[i] < - max(0, min(1, p[i]))

#Regression Model
logit(p[i]) < - int. + depth*WDepth[i]

failure.rep[i] ~ dbin(p.bound[i], numtested)

#Fit Assessment: Pearson Residuals Posterier Predective check (Bayesian P-Value)
residual[i] < - (failure[i] - (numtested*p.bound[i]))/sqrt(numtested*p.bound[i]*(1-p.bound
[i]) + 0.00001)
residual.rep[i] < - (failure.rep[i] - (numtested*p.bound[i]))/sqrt(numtested*p.bound[i]*(1-p.bound
[i]) + 0.00001)
sq.[i] < - pow(residual[i], 2)
sq.rep[i] < - pow(residual.rep[i], 2)

#Fit Assessment: Likelihood Statistic Posterier Predective check (Bayesian P-Value)
like.obs[i] < - failure[i]*log((failure[i] + 0.00001)/(numtested*p.bound[i] + 0.00001))
like.rep[i] < - failure.rep[i]*log((failure.rep[i] + 0.00001)/(numtested*p.bound[i] + 0.00001))

#Fit Assessment: Freeman-Tukey Statistic Posterier Predective check (Bayesian P-Value)
diff.obs[i] < - pow(sqrt(failure[i]) - sqrt(numtested*p.bound[i]), 2)
diff.rep[i] < - pow(sqrt(failure.rep[i]) - sqrt(numtested*p.bound[i]), 2)

prop[i] < - failure[i]/numtested
Ds[i] < - 2*numtested*(prop[i]*log((prop[i] + 0.00001)/(p.bound[i] + 0.00001))
+ (1-prop[i])*log((1-prop[i] + 0.00001)/((1-p.bound[i]) + 0.00001)))

phat[i] < - failure[i]/numtested

}

chisq.obs < - sum(sq[])
chisq.rep <- sum(sq.rep[])
p.chisq <- step(chisq.rep - chisq.obs)

likelihood.obs < - sum(like.obs[])
likelihood.rep <- sum(like.rep[])
p.likelihood <- step(likelihood.rep - likelihood.obs)

freeman.obs < - sum(diff.obs[])
freeman.rep <- sum(diff.rep[])
p.freeman <- step(freeman.rep - freeman.obs)

dev.sat < - sum(Ds[])

#Prior Distributions
int. ~ dnorm(0, .000001)
depth ~ dnorm(0, .000001)
}

data
list(
tests = 19,
numtested = 1,
failure = c(1, 1, 1, 1, 1, 1, 1, 1, 1, 0, 0, 0, 0, 0, 0, 0, 0, 0, 1),
WDepth = c(46.1, 39.0, 37.1, 37.8, 37.5, 37.6, 37.7, 37.1, 44.5, 25.7, 17.0, 27.4, 30.9, 32.3, 24.3, 34.8, 37.5,
```

```
38.0, 41.4),
WFlow = c(1148, 1130, 1120, 979, 1133, 604, 593, 598, 975, 248, 117, 285, 397, 484, 247, 593, 696, 734,
1025)
)

inits

#Depth
list(int = −28, depth = 4, flow = 0, temp = 0, failure.rep = c(0,0,0,0,0,0,0,0,0,0,0,0,0,0,0,0,0,0,0))
list(int = −122, depth = 0, flow = 0, temp = 0, failure.rep = c(0,0,0,0,0,0,0,0,0,0,0,0,0,0,0,0,0,0,0))
```

Table 6.
OpenBUGS script [5].

Parameter	Depth	Flow Rate	Depth, Flow Rate
intercept	75.68	−8.51	−72.5
a (depth coeff.)	2.05	—	1.83
b (flow rate coeff.)	—	0.01	0.007
Sat. deviance	12.88	14.29	13.31
Chi-squared	0.19	0.26	0.14
Likelihood ratio	0.38	0.36	0.29
Freeman-Tukey	0.33	0.23	0.21

Table 7.
Summary posterior estimates of logistic regression parameters and Bayesian p-values using steel door data [5].

$$Logit(p) = intercept + aD + \lambda_i \tag{7}$$

with $\lambda_i \sim N(0, \sigma^2)$ and unknown variance. A prior distribution is specified for σ. More variability is accounted for by allowing the probability to vary on an observation-by-observation bases.

The resulting p-values and DIC for the depth, flow rate, and combined regression models are given in **Table 5**. The larger p-values (all greater than 0.95) strongly suggest lack of fit. The regression models without variability are favorable over the inclusion of unobserved effects for their better fit.

The final OpenBUGS script for the depth regression model, prior distributions, and dispersed initial values is shown in **Table 6**. Included are the script for the three Bayesian p-value calculations and the saturated deviance.

The mean values calculated for the applicable parameters in the outward swinging steel door fragility models and corresponding Bayesian p-values are shown in **Table 7**. The saturated deviance for all three models compared with the data sample size suggests that all three models fit adequately. The DIC is nearly the same for all three models, the smallest belonging to the depth model by a non-significant amount. The model with only depth as an explanatory variable has the closest Bayesian p-value using the likelihood ratio (0.38). It also has the slightly closer average p-value compared to 0.5 than the regression model with only flow rate and the combined model with both variables. Given the results, the model with only depth is recommended for predictive analyses.

With depth selected as the explanatory variable regression model, the parameters in **Table 7** are used with the fragility model to calculate the failure probability for a steel door as a function of water depth. The probability p is given by:

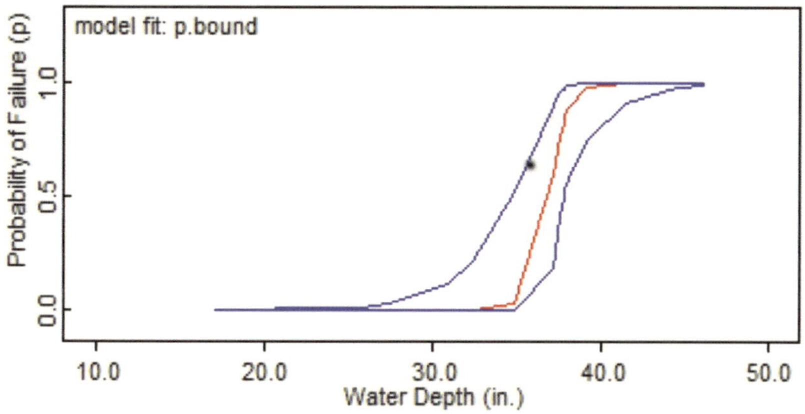

Figure 4.
Fragility curve showing probability of failure versus water depth. Blue curves represent the 95% credible intervals [5].

$$p = \frac{1}{e^{-(-75.68+2.05x)} + 1} \tag{8}$$

where x is the given water depth. **Figure 4** shows the plot of failure probability versus water depth with 95% credible intervals. It should be noted that the mean, shown in red, is close to the bound at low probabilities. This is due to a couple of non-failure tests reaching water depths greater than some observed failure depths, bringing the mean near the credible interval at low fragility probabilities.

6. Conclusion

Component failure probability models provide a pathway for selection of more flexible operating limits and setpoints. Model development requires component performance data and an effective process for probability model selection and checking. Using Bayesian methodology, prior knowledge about model parameters can be updated with the knowledge of the likelihood to observe data for parameter values giving a posterior probability. In short, the process combines everything that is known about a particular data set and model response to produce a posterior estimate of the output function's probability distribution. Integration of these functions is necessary and can be accomplished through MCMC methods.

Bayesian model checking is used to assess the fit of the model to various aspects of the data using the assumption that if a given model fits, then data simulated or replicated under the model should be comparable to the real-world observed data. If any systematic differences occur between simulations and the data, it potentially indicates that model assumptions are not being met. The model is also checked for deviations by means of test quantities or discrepancy functions that depend on both data and parameters by calculating a Bayesian p-value. The DIC can also be used as a measure of model fit that can be applied to Bayesian models and is applicable when the parameter estimation is done using techniques such as Gibbs sampling. It is particularly useful in Bayesian model selection problems where the posterior distributions of the model have been obtained by MCMC simulation.

Application of the data collection, model development, and model checking process was carried out for the performance of steel doors subjected to water rise flooding conditions. The resulting fragility model provides a carefully developed representation of the failure probability as the flood depth changes. The model can then be used in more comprehensive probabilistic flooding analyses rather than simply using an empirically derived pass-fail water depth for steel doors subjected to water rise flooding scenarios. The overall result of using the rigorously developed fragility model is a more robust representation of how components will perform when subjected to challenges such as flooding. With an improved representation of overall performance available, necessary limits and controls can then be selected without undue conservatism.

Acknowledgements

Funding support for the PET construction and experiments and fragility model development was provided to Idaho State University by the US Department of Energy Light Water Reactor Sustainability Program through Contract Number 154652.

Author details

Alison Wells[1] and Chad L. Pope[2*]

1 Idaho National Laboratory, Idaho Falls, Idaho, USA

2 Idaho State University, Pocatello, Idaho, USA

*Address all correspondence to: popechad@isu.edu

References

[1] Box G, Tiao G. Bayesian Inference in Statistical Analysis. John Wiley & Sons Wiley, 1992.

[2] Kelly D, Smith C. Bayesian Inference for Probabilistic Risk Assessment: A Practitioner's Guidebook. Springer, 2011.

[3] Lunn D, Jackson C, Best N, Thomas A, Spiegelhalter D. The BUGS Book: A Practical Introduction to Bayesian Analysis: CRC Press, 2013.

[4] Gelman A, Carlin J, Stern H, Dunson D, Vehtari A, Rubin D. Bayesian Data Analysis. Texts in Statistical Science. CRC Press, 3rd edition, 2014.

[5] Wells A. 2020. Assessing Nuclear Power Plant Component Fragility in Flooding Events Using Bayesian Regression Modeling with Explanatory Variables [Doctoral Dissertation]. Pocatello: Idaho State University

[6] Conn P, Johnson D, Williams P, Melin S, Hooten M. A guide to Bayesian model checking for ecologists. Ecological Monographs, 88(4):526–542, 2018.

[7] Gelman A, Hill J. Data Analysis Using Regression and Muiltilevel/Hierarchical Models. Cambridge University Press, 2007.

[8] Burns W. Spurious Correlations, 1997.

Chapter 4

Markov Chain Monte Carlo in a Dynamical System of Information Theoretic Particles

Tokunbo Ogunfunmi and Manas Deb

Abstract

In Bayesian learning, the posterior probability density of a model parameter is estimated from the likelihood function and the prior probability of the parameter. The posterior probability density estimate is refined as more evidence becomes available. However, any non-trivial Bayesian model requires the computation of an intractable integral to obtain the probability density function (PDF) of the evidence. Markov Chain Monte Carlo (MCMC) is a well-known algorithm that solves this problem by directly generating the samples of the posterior distribution without computing this intractable integral. We present a novel perspective of the MCMC algorithm which views the samples of a probability distribution as a dynamical system of Information Theoretic particles in an Information Theoretic field. As our algorithm probes this field with a test particle, it is subjected to Information Forces from other Information Theoretic particles in this field. We use Information Theoretic Learning (ITL) techniques based on Rényi's α-Entropy function to derive an equation for the gradient of the Information Potential energy of the dynamical system of Information Theoretic particles. Using this equation, we compute the Hamiltonian of the dynamical system from the Information Potential energy and the kinetic energy. The Hamiltonian is used to generate the Markovian state trajectories of the system.

Keywords: Hamiltonian Monte Carlo (HMC), information theoretic learning, Kernel density estimator (KDE), Markov chain Monte Carlo, Parzen window, Rényi's entropy, information potential

1. Introduction

Bayesian learning involves estimating the PDF of a model parameter from the likelihood function and the prior probability of the parameter. Bayesian inference incorporates the concept of belief where the parameter estimate is refined as more data or evidence becomes available. The posterior PDF of the model parameter θ with the PDF of the evidence X denoted as $P(X)$, is expressed by the following well-known Bayes' equation:

$$P(\theta|X) = \frac{P(X|\theta)P(\theta)}{P(X)} \tag{1}$$

$P(X)$ is the integral of the probability of all possible values of θ weighted by the likelihood function:

$$P(X) = \int_{\theta} P(X|\theta)P(\theta)d\theta \tag{2}$$

This is an intractable integration for most non-trivial Bayesian inference problems and makes it impossible to compute the posterior probability. The MCMC algorithm described in [1] provides a solution to this problem by directly generating samples of the posterior PDF without computing this intractable integral. The shape of the posterior PDF and other statistics can be inferred from these samples.

The MCMC algorithm requires knowledge of a function that is proportional to the unknown posterior PDF. It uses this function to generate sample proposals of the unknown PDF. Usually, this function is the product of the likelihood function and the prior probability. In practical applications, one often encounters a system whose outputs are observable, but the process within the system that generated these outputs are unknown. We present a novel perspective on the MCMC method to solve these types of practical problems, where instead of generating the samples of the unknown PDF, it uses the samples of the unknown distribution to estimate the PDF. In this chapter we use the Hamiltonian MCMC (HMC) method described in [2–4] and ITL concepts to show how the samples of the unknown distribution can be viewed as Information Theoretic particles of a dynamical system. The sample space of the given probability distribution is explored by computing trajectories corresponding to the state transition of this dynamical system. The evolution or state transition of the dynamical system is governed by equations which use the total energy or the Hamiltonian of the system of Information Theoretic particles. Each such particle has an inherent Information Potential by virtue of its position with respect to the other particles of the system. The system of Information Theoretic particles creates an Information Field which enables each particle to exert an Information Force on the other particles. We use ITL techniques [5] based on Rényi's α-Entropy function to derive an equation for the gradient of the Information Potential energy of this dynamical system. This equation is one of the main contributions of our work and it is used to compute the Hamiltonian of the system to explore the probability space of the Information Theoretic particles.

In this work, we implement an iterative PDF estimator of an unknown sample distribution, using the HMC method. At every iteration of the estimator, the HMC generates samples such that the mutual information between the generated samples and the given unknown distribution is large. To do this, it uses the Information Potential, the Information Force and the kinetic energy of an Information Theoretic "probe" particle. To compute the Information Potential and the Information Force, the algorithm uses a non-parametric Kernel Density Estimator (KDE). The bandwidth of the KDE determines how close the generated samples are from the unknown sample distribution. At the end of each iteration, the Kullback–Leibler (K–L) divergence of the samples generated by the estimator from the given distribution is computed. The iteration continues until the K-L divergence falls below a specified threshold. We have derived an equation to adapt the kernel bandwidth for each iteration, based on the invariant point theorem. Before starting the next iteration, this equation is used to adapt the kernel bandwidth before generating the next set of samples.

An important application of our algorithm is in machine learning where sometimes the dataset is either too large to fit in the memory of a computer or too small to obtain an accurate inference model. The dataset can be resampled to the desired size using the PDF estimator and the HMC equations derived in this chapter.

The sections in this chapter are organized in the following manner: In Section 2 we review the MCMC algorithm. Section 3 provides an overview of the Hamiltonian MCMC algorithm. Rényi's Entropy and the concept of Information Theoretic particles are introduced in Section 4. In Section 5 we show how the Hamiltonian MCMC algorithm can be used with Information Theoretic particles and derive a key equation for the system potential gradient. Section 6 describes a method to iteratively estimate the PDF of the target distribution using HMC. In this section we derive an equation to adapt the Information Potential energy estimator bandwidth for each iteration. The simulation results of the HMC algorithm on a system of Information Theoretic particles are listed in Section 7 and we summarize our conclusions in Section 8 of this chapter.

2. Review of the MCMC algorithm

The core principle underlying MCMC techniques is that an ergodic, reversible Markov chain reaches a stationary state. MCMC models the sampling from a distribution as an ergodic and reversible Markov process. When this process reaches a stationary state, the probability distribution of the states of the Markov chain becomes invariant and matches the given probability distribution. The sampling operation in the MCMC is a Markov process that satisfies the following detailed balance equation:

$$\pi_i \mathrm{P}(X_{t-1} = i, X_t = j) = \pi_j \mathrm{P}(X_{t-1} = j, X_t = i) \quad \forall i, j \tag{3}$$

In the detailed balance equation, π_i and π_j are the stationary probability distribution of being in states i and j respectively and $X_0, X_1, X_2, \ldots X_t \ldots$ are a sequence of random variables at discrete time indices $0, 1, 2, \ldots t-1, t, \ldots$. The Monte Carlo part of the MCMC algorithm is used to generate random "proposal" samples from a known probability distribution $Q(X)$. The proposal sample for the next time step of the MCMC algorithm is dependent on the current proposal sample and the transition probability for the new sample is enforced by an acceptance function. The proposal distribution is usually symmetric to ensure the reversibility of the Markov chain:

$$Q(x_t | x_{t-1}) = Q(x_{t-1} | x_t) \tag{4}$$

Symmetric distributions like the Gaussian distribution or the Uniform distribution centered around the current sample value can be used to generate the proposal sample. There are cases where asymmetric distributions are used but we will focus on symmetric distributions to illustrate our algorithm, without any loss of generality.

To lay the groundwork for the HMC, we review the simple Metropolis-Hastings (MH) MCMC [6] in this section. The simplest MH algorithm is the Random-Walk MH which uses a symmetrical proposal distribution. It comprises of the following 3 parts:

1. Generate a proposal sample for the posterior probability from a known symmetric distribution. The new proposal sample is based on the current proposal sample: $x_{proposal} \sim Q(x_i | x_{i-1})$. For example, if $Q(X)$ is a Gaussian distribution, it is centered at sample x_{i-1} to generate sample x_i

2. Calculate the acceptance probability by passing this sample through the posterior density function using:

$$P(\theta|X) = \frac{1}{Z}P(X|\theta)P(\theta)$$
$$\text{where } Z = \int_{\theta} P(Z|\theta)P(\theta)d\theta \tag{5}$$

3. Accept the candidate sample with probability α or reject it with probability $1 - \alpha$ where α is defined in (Eq. (8))

If the proposal density function is symmetric, we have:

$$Q\left(x_{i-1}|x_{proposal}\right) = Q\left(x_{proposal}|x_{i-1}\right) \tag{6}$$

The acceptance function is derived as follows:

$$\begin{aligned}
\frac{P\left(\theta|X = x_{proposal}\right)}{P\left(\theta|X = x_{i-1}\right)} \frac{Q\left(x_{i-1}|x_{proposal}\right)}{Q\left(x_{proposal}|x_{i-1}\right)} &= \frac{\cancel{1/Z}\, P\left(X = x_{proposal}|\theta\right)P(\theta)}{\cancel{1/Z}\, P\left(X = x_{i-1}|\theta\right)P(\theta)} \frac{\cancel{Q\left(x_{i-1}|x_{proposal}\right)}}{\cancel{Q\left(x_{proposal}|x_{i-1}\right)}} \\
&= \frac{P\left(X = x_{proposal}|\theta\right)P(\theta)}{P\left(X = x_{i-1}|\theta\right)P(\theta)} \\
&= \frac{P\left(X = x_{proposal}, \theta\right)}{P\left(X = x_{i-1}, \theta\right)}
\end{aligned} \tag{7}$$

It is evident from (Eq. (7)) that since the acceptance function is a ratio of the posterior probability, the intractable integral to compute the value of Z is completely bypassed. The acceptance probability of a sample proposal of the MH-MCMC is:

$$\alpha = \min\left\{1, \frac{P(X = x_{proposal}, \theta)}{P(X = x_{i-1}, \theta)}\right\} \tag{8}$$

The transition probability of each state of the Markov chain is defined by the acceptance probability. In the stationary state, the product of the Markov chain state probability and the transition probability matrix remains stationary and matches the posterior PDF of the model parameter. The sample points x_i generated by this MCMC in the stationary state of the Markov chain are therefore the sample points of the posterior PDF.

3. The Hamiltonian MCMC algorithm

Instead of the random-walk method of the Metropolis-Hastings algorithm, this MCMC technique uses Hamiltonian dynamics to sample from the posterior PDF. The random-walk method of the Metropolis-Hastings algorithm is inefficient and converges slowly to the target posterior distribution. Instead of randomly generating "proposal" samples from a known probability distribution, the Hamiltonian method uses the dynamics of a physical system to generate these samples. This enables the system to explore the target posterior probability space more efficiently, which in turn results in faster convergence compared to random-walk methods.

Hamiltonian dynamics is a concept borrowed from statistical mechanics where the energy of a dynamic system changes from potential energy to kinetic energy and back. The Hamiltonian represents the total energy of the system, which for a closed system, is the sum of its potential and kinetic energy.

As described in [2, 3], Hamiltonian dynamics operates on an N dimensional position vector $\mathbf{q}$ and an N dimensional momentum vector $\mathbf{p}$ and the dynamic system is described by the Hamiltonian $H(\mathbf{q}, \mathbf{p})$. The partial derivatives of the Hamiltonian define how the system evolves with time:

$$\begin{aligned} \frac{dq_i}{dt} &= \frac{\partial H}{\partial p_i} \qquad\qquad i = 1, 2, \ldots, N \\ \frac{dp_i}{dt} &= -\frac{\partial H}{\partial q_i} \end{aligned} \tag{9}$$

Given the state of the system at time t, these equations can be used to determine the state of the system at time $t + T$ where $T = 1, 2, 3, \ldots$. For the time evolution of the dynamical system, we use the following Hamiltonian:

$$H(\mathbf{q}, \mathbf{p}) = U(\mathbf{q}) + K(\mathbf{p}) \tag{10}$$

In (10), $U(\mathbf{q})$ is the potential energy and $K(\mathbf{p})$ is the kinetic energy of the system. The position vector $\mathbf{q}$ corresponds to the model parameter and the PDF of $\mathbf{q}$ is the target posterior PDF that we want to estimate. The potential energy of the Hamiltonian system is expressed as the negative log of the probability of $\mathbf{q}$:

$$U(\mathbf{q}) = -\log\left(P(\mathbf{q})\right) \tag{11}$$

To relate the Hamiltonian $H(\mathbf{q}, \mathbf{p})$ to the target posterior probability, we use a basic concept from statistical mechanics known as the canonical ensemble. If there are several microstates of a physical system contained in the vector $\boldsymbol{\theta}$ and there is an energy function $E(\boldsymbol{\theta})$ defined for these microstates, then the canonical probability distribution of the microstates is expressed as:

$$p(\theta) = \frac{1}{Z} e^{-\frac{E(\theta)}{T}} \tag{12}$$

where T is the temperature of the system and the variable Z is a normalizing constant called the partition function. Z scales the canonical probability distribution such that it sums to one. For a system described by Hamiltonian dynamics, the energy function is:

$$E(\boldsymbol{\theta}) = H(\mathbf{q}, \mathbf{p}) = U(\mathbf{q}) + K(\mathbf{p}) \tag{13}$$

In MCMC, the Hamiltonian is an energy function of the states of both the position $\mathbf{q}$ and the momentum $\mathbf{p}$. Therefore, the canonical probability distribution of a Hamiltonian system can be expressed as:

$$\begin{aligned} P(\mathbf{q}, \mathbf{p}) &= \frac{1}{Z} e^{-\frac{H(\mathbf{q}, \mathbf{p})}{T}} \\ &= \frac{1}{Z} e^{-\frac{U(\mathbf{q}) + K(\mathbf{p})}{T}} \\ &= \frac{1}{Z} \exp\left(-\frac{U(\mathbf{q})}{T}\right) \exp\left(-\frac{K(\mathbf{p})}{T}\right) \end{aligned} \tag{14}$$

This equation shows that $\mathbf{q}$ and $\mathbf{p}$ are independent and each have canonical distributions with energy functions $U(\mathbf{q})$ and $K(\mathbf{p})$. The probability density of $\mathbf{q}$ is the posterior probability density of the model parameter θ and is the product of the likelihood function of $\boldsymbol{\theta}$ given the data $\mathbf{D}$ and the prior probability of $\boldsymbol{\theta}$. An important point to note here is that the momentum variable $\mathbf{p}$ has been introduced in the probability distribution in (Eq. (14)) so that we can use Hamiltonian dynamics. Since $\mathbf{p}$ is independent of $\mathbf{q}$, we can choose any distribution for this variable. In our HMC algorithms use a zero-mean multivariate Gaussian distribution for the momentum vector $\mathbf{p}$. The temperature $T = 1$ in this discussion on the HMC.

The kinetic energy of the dynamical system for a unit mass is expressed as:

$$K(\mathbf{p}) = \frac{1}{2}\mathbf{p}^T\mathbf{p} \tag{15}$$

On applying the Hamiltonian partial derivatives in (9) to the definition of the HMC in (10) we get the following differential equations which describe the time evolution of the dynamical system:

$$\begin{aligned} \frac{d\mathbf{q}}{dt} &= \frac{\partial H}{\partial \mathbf{p}} = \frac{\partial[U(\mathbf{q}) + K(\mathbf{p})]}{\partial \mathbf{p}} = \frac{\partial}{\partial \mathbf{p}}\left(\frac{1}{2}\mathbf{p}^T\mathbf{p}\right) = \mathbf{p} \\ \frac{d\mathbf{p}}{dt} &= -\frac{\partial H}{\partial \mathbf{q}} = -\frac{\partial[U(\mathbf{q}) + K(\mathbf{p})]}{\partial \mathbf{q}} = -\frac{\partial U(\mathbf{q})}{\partial \mathbf{q}} \end{aligned} \tag{16}$$

Since the Hamiltonian equations for the time evolution of the system are differential equations, computer simulation of the HMC must discretize time. A popular scheme to implement this discretization is the "Leapfrog" algorithm [4]. The HMC algorithm uses the leapfrog algorithm to update the momentum and the position while computing the trajectory towards the next sample proposal in the distribution. The Leapfrog integrator has 2 main advantages:

1. It is time reversible. A Leapfrog integration by N steps in the forward direction and then in the backward direction results in the same starting position

2. It is symplectic in nature. In other words, it conserves the energy of dynamical systems

The steps of the Hamiltonian MCMC algorithm are:

1. At every time step t, determine a trajectory of the system potential and kinetic energy. To do that, generate a random value from a standard normal distribution for the momentum variable.

2. Execute the Leapfrog algorithm to update the position and momentum variables according to the differential equations in (Eq. (16)). This determines the trajectory of the system towards the next sample proposal

3. Compute the potential and kinetic energy $(U(\mathbf{q}_{t-1}), K(\mathbf{p}_{t-1}))$ of the system at the beginning of the trajectory and at the end $\left(U\left(\mathbf{q}_{proposed}\right), K\left(\mathbf{p}_{proposed}\right)\right)$ of the proposed trajectory

4. Calculate the acceptance probability of the new trajectory using the following ratio of probabilities:

$$\beta = \min\left\{1, \frac{P\left(\mathbf{q}_{proposal}, \mathbf{p}_{proposal}\right)}{P\left(\mathbf{q}_{t-1}, \mathbf{p}_{t-1}\right)}\right\}$$

$$= \min\left\{1, \frac{\frac{1}{Z}\exp\left(-U\left(\mathbf{q}_{proposal}\right)\right)\exp\left(-K\left(\mathbf{p}_{proposal}\right)\right)}{\frac{1}{Z}\exp\left(-U\left(\mathbf{q}_{t-1}\right)\right)\exp\left(-K\left(\mathbf{p}_{t-1}\right)\right)}\right\} \tag{17}$$

$$= \min\left\{1, \exp\left(\begin{array}{c}\left(U\left(\mathbf{q}_{t-1}\right) + K\left(\mathbf{p}_{t-1}\right)\right) - \\ \left(U\left(\mathbf{q}_{proposal}\right) + K\left(\mathbf{p}_{proposal}\right)\right)\end{array}\right)\right\}$$

5. Generate a random number $u \sim Uniform(0, 1)$ to accept or reject the proposal

$if\,(\beta > u)then$

$\mathbf{q}_t \leftarrow \mathbf{q}_{proposal}$ //accept the proposed trajectory

$else$

$\mathbf{q}_t \leftarrow \mathbf{q}_{t-1}$ //reject the proposed trajectory

$endif$

4. Rényi's entropy and Information Theoretic particles

The concept of Information Theoretic particles comes from Alfréd Rényi's pioneering work on generalized measures of entropy and information [7]. At the core of Rényi's work is the concept of generalized mean or the Kolmogorov-Nagumo (K-N) mean [8–10]. For numbers $x_1, x_2, \ldots x_N$, the K-N mean is expressed as:

$$\psi^{-1}\left(\frac{1}{N}\sum_{i=1}^{N}\psi(x_i)\right) \tag{18}$$

where, $\psi(.)$ is the K-N function. This function is continuous and strictly monotonic implying that it has an inverse. In the general theory of means, the quasi-linear mean of a random variable X which takes the values $x_1, x_2, \ldots x_N$ with probabilities $p_1, p_2, \ldots p_N$ is defined as:

$$E_\psi[X] = \langle X\rangle_\psi = \psi^{-1}\left(\sum_{k=1}^{N} p_k\psi(x_k)\right) \tag{19}$$

From the theorem on additivity of quasi-linear means [11], if $\psi(.)$ is a K-N function and $c$is a real constant, then:

$$\psi^{-1}\left(\sum_{k=1}^{N} p_k\psi(x_k + c)\right) = \psi^{-1}\left(\sum_{k=1}^{N} p_k\psi(x_k)\right) + c \tag{20}$$

if and only if $\psi(.)$ is either linear or exponential.

4.1 Rényi's entropy

Consider a random variable X which takes the values $x_1, x_2, \dots x_N$ with probabilities $p_1, p_2, \dots p_N$. The amount of information generated when X takes the value x_k is given by the Hartley [12] information measurement function $I(x_k)$:

$$I(x_k) = \log_2\left(\frac{1}{p_k}\right) \text{ bits} \tag{21}$$

The expected value of $I(x_k)$ yields the expression for Shannon's entropy [13]:

$$H(X) = \sum_{k=1}^{N} p_k I(x_k) = \sum_{k=1}^{N} p_k \log_2\left(\frac{1}{p_k}\right) \tag{22}$$

Rényi replaced the linear mean in (Eq. (22)) with the quasi-linear mean in (Eq. (19)) to obtain a generalized measure of information:

$$H_\psi(X) = \psi^{-1}\left(\sum_{k=1}^{N} p_k \psi\left(\log_2\left(\frac{1}{p_k}\right)\right)\right) \tag{23}$$

For $H_\psi(X)$ to satisfy the additivity property of independent events, it must satisfy $\langle X + c\rangle_\psi = \langle X\rangle_\psi + c$ where c is a constant. From (Eq. (20)), this implies that $\psi(x) = cx$ (linear) or $\psi(x) = c2^{(1-\alpha)x}$ (exponential). Setting $\psi(x) = cx$ reduces (Eq. (23)) to the linear mean and yields Shannon entropy equation. Substituting $\psi(x) = c2^{(1-\alpha)x}$ and the corresponding inverse function $\psi^{-1} = \frac{1}{(1-\alpha)} \log_2$ in (Eq. (23)) yields the expression for Rényi's α–entropy:

$$H_\alpha(X) = \frac{1}{(1-\alpha)} \log_2\left(\sum_{k=1}^{N} p_k^\alpha\right) \quad \alpha > 0 \text{ and } \alpha \neq 1 \tag{24}$$

Rényi's α–entropy equation is therefore a general expression for entropy and comprises of a family of entropies for different values of the parameter α. Shannon's entropy is a special case of Rényi's entropy in the limit as $\alpha \to 1$. The argument of the logarithm function in (Eq. (24)) is called the Information Potential. The α-Information Potential is expressed as:

$$V_\alpha(X) = \sum_{k=1}^{N} p_k^\alpha \tag{25}$$

Substituting (Eq. (25)) in (Eq. (24)), we get the following expression for Rényi's entropy in terms of the Information Potential:

$$H_\alpha(X) = \frac{1}{(1-\alpha)} \log_2(V_\alpha(X)) \tag{26}$$

The Information Potential in (Eq. (25)) can be written as the expected value of the PDF of the sample distribution raised to $\alpha - 1$:

$$V_\alpha(X) = \sum_{k=1}^{N} p_k^\alpha = \sum_{k=1}^{N} p_k p_k^{\alpha-1} = E\left[p_k^{\alpha-1}\right] \tag{27}$$

For $\alpha = 2$ in (Eq. (24)), we get Rényi's quadratic entropy, which has the useful property that it allows us to compute the entropy directly from the samples. The equations for Rényi's Quadratic Entropy (QE) and Quadratic Information Potential (QIP) are obtained by substituting $\alpha = 2$ in (Eqs. (26) and (27)):

$$\begin{aligned} H_2(X) &= \frac{1}{(1-2)} \log_2(V_2(X)) = -\log_2(V_2(X)) \\ \text{where } V_2(X) &= E\left[p_k^{2-1}\right] = E\left[p_k\right] \end{aligned} \tag{28}$$

The QIP is therefore the expected value of the PDF of the given data samples.

4.2 Rényi's quadratic information potential (QIP) estimator

From (Eq. (28)), it is evident that to compute the QIP we need to know the PDF of the given data samples. In practical applications an analytical expression of the PDF is rarely available. Therefore, the QIP computation involves a non-parametric estimator of the PDF directly from the samples [14]. The Parzen-Rosenblatt window estimator [15, 16] is a non-parametric way to estimate the PDF of a random variable from its sample values. This estimator places a kernel function with its center at each of the samples. The resulting output values are averaged over all the samples to estimate the PDF. The laws governing the interaction of the Information Theoretic particles is defined by the shape of the kernel. We use a Gaussian kernel, since this kernel when placed over the samples, behaves like an Information Theoretic field whose strength decays with increasing distance between the samples. Just like a charge in space creates an electric field, the samples of a probability distribution behave like Information Particles with unit charge. Information particles exert Information Forces on other particles through this Information Theoretic field.

For scalar samples $x_1, x_2, \ldots x_N$, the Parzen window PDF estimator with a Gaussian kernel is expressed as:

$$\hat{p}(x) = \frac{1}{N} \sum_{i=1}^{N} G_\sigma(x - x_i) \tag{29}$$

where $G_\sigma(u)$ is the following standard univariate Gaussian kernel:

$$G_\sigma(u) = \frac{1}{\sqrt{2\pi\sigma^2}} \exp\left[-\frac{1}{2}\left(\frac{u}{\sigma}\right)^2\right] \tag{30}$$

σ is the kernel bandwidth of the estimator and it must be carefully chosen to obtain an accurate and unbiased estimate of the PDF. The Parzen window estimator of a multivariate PDF for vector samples $\mathbf{x}_1, \mathbf{x}_2, \ldots \mathbf{x}_N$ of dimension d is expressed as:

$$\hat{p}(\mathbf{x}) = \frac{1}{N} \sum_{i=1}^{N} G_\mathrm{C}(\mathbf{x} - \mathbf{x}_i) \tag{31}$$

where $G_\mathrm{C}(u)$ is the following standard multivariate Gaussian kernel:

$$G_\mathrm{C}(\mathbf{u}) = \frac{1}{\sqrt{(2\pi)^d |\mathbf{C}|}} \exp\left[-\frac{1}{2}\mathbf{u}^T \mathbf{C}^{-1} \mathbf{u}\right] \tag{32}$$

d is the dimension of the input vector $\mathbf{u}$, $\mathbf{C}$ is the $d \times d$ covariance matrix and $|\mathbf{C}|$ is the determinant of the covariance matrix. For the multivariate PDF case, the kernel bandwidth $\mathbf{C}$ must be carefully chosen to obtain an accurate and unbiased estimate of the PDF.

Rényi's quadratic entropy for a continuous random variable is expressed as:

$$H_2(X) = -\log \int_{-\infty}^{\infty} p^2(x)dx \tag{33}$$

Substituting $\hat{p}(x)$ from (Eq. (29)) for $p(x)$ in the above equation as described in [5], we get the following equation for the QE estimator:

$$\hat{H}_2(X) = -\log\left[\frac{1}{N^2}\sum_{i=1}^{N}\sum_{j=1}^{N} G_{\sigma\sqrt{2}}(x_j - x_i)\right] \tag{34}$$

where:

$$G_{\sigma\sqrt{2}}(u) = \frac{1}{\sqrt{2\pi(\sigma\sqrt{2})^2}} \exp\left[-\frac{1}{2}\left(\frac{u}{\sigma\sqrt{2}}\right)^2\right] \tag{35}$$

The equation for the QE estimator shows that we can compute the QE estimate directly from the samples of a distribution without knowing its PDF, by applying the Parzen-Rosenblatt kernel on these samples. From (Eqs. (28) and (34)), the QIP estimator can be expressed as:

$$\hat{V}_2(X) = \frac{1}{N^2}\sum_{i=1}^{N}\sum_{j=1}^{N} G_{\sigma\sqrt{2}}(x_j - x_i) \tag{36}$$

4.3 Information potential energy and the information force of information theoretic particles

The total QIP energy estimate of the system is given by (Eq. (36)). The QIP energy estimate of sample x_j due to the Information Potential field of a single sample x_i is:

$$\hat{V}_2(x_j; x_i) = G_{\sigma\sqrt{2}}(x_j - x_i) \tag{37}$$

The Quadratic Information Potential energy estimate of scalar sample x_j in the Information Field created by all the samples $x_i \in \mathbb{R}$, for $i = 1, 2, \dots N$ is defined as the average of $\hat{V}_2(x_j; x_i)$ taken over all the samples x_i:

$$\begin{aligned}\hat{V}_2(x_j) &= \frac{1}{N}\sum_{i=1}^{N} G_{\sigma\sqrt{2}}(x_j - x_i) \\ &= \frac{1}{N}\frac{1}{\sqrt{2\pi}(\sigma\sqrt{2})}\sum_{i=1}^{N} \exp\left[-\frac{1}{2}\left(\frac{x_j - x_i}{\sigma\sqrt{2}}\right)^2\right]\end{aligned} \tag{38}$$

If the samples are d dimensional vectors, then the Quadratic Information Potential energy estimate of vector sample $\mathbf{x}_j$ in the Information Potential Field created by all vector samples $\mathbf{x}_i \in \mathbb{R}^d$, for $i = 1, 2, \ldots N$ is defined as:

$$\hat{V}_2(\mathbf{x}_j) = \frac{1}{N}\sum_{i=1}^{N} G_{2\mathbf{C}}(\mathbf{x}_j - \mathbf{x}_i) \tag{39}$$

where:

$$G_{2\mathbf{C}} = \frac{1}{\sqrt{(2\pi)^d |\mathbf{C}| (2^d)}} \sum_{i=1}^{N} \exp\left[-\frac{1}{2}(\mathbf{x}_j - \mathbf{x}_i)^T (2\mathbf{C})^{-1} (\mathbf{x}_j - \mathbf{x}_i)\right] \tag{40}$$

From (Eqs. (39) and (40)) we can re-write the QIP energy estimate for vector samples of d dimensions as:

$$\hat{V}_2(\mathbf{x}_j) = \frac{1}{N}\frac{1}{\sqrt{(2\pi)^d |\mathbf{C}| (2^d)}} \sum_{i=1}^{N} \exp\left[-\frac{1}{2}(\mathbf{x}_j - \mathbf{x}_i)^T (2\mathbf{C})^{-1} (\mathbf{x}_j - \mathbf{x}_i)\right] \tag{41}$$

To obtain the Quadratic Information Force estimate on scalar sample x_j due to the Information Potential field of sample x_i, we take the derivative of the Quadratic Information Potential energy estimate:

$$\begin{aligned}
\hat{F}_2(x_j; x_i) &= \frac{\partial}{\partial x_j}\hat{V}_2(x_j; x_i) = \frac{\partial}{\partial x_j} G_{\sigma\sqrt{2}}(x_j - x_i) \\
&= \frac{\partial}{\partial x_j}\left[\frac{1}{\sqrt{2\pi}(\sigma\sqrt{2})} \exp\left[-\frac{1}{2}\left(\frac{x_j - x_i}{\sigma\sqrt{2}}\right)^2\right]\right] \\
&= \frac{1}{\sqrt{2\pi}(\sigma\sqrt{2})} \exp\left[-\frac{1}{2}\left(\frac{x_j - x_i}{\sigma\sqrt{2}}\right)^2\right]\left(-\frac{1}{2(2\sigma^2)}\right)\frac{\partial}{\partial x_j}(x_j - x_i)^2 \\
&= \frac{1}{\sqrt{2\pi}(\sigma\sqrt{2})} \exp\left[-\frac{1}{2}\left(\frac{x_j - x_i}{\sigma\sqrt{2}}\right)^2\right]\left(-\frac{1}{2(2\sigma^2)}\right)[2(x_j - x_i)] \\
\hat{F}_2(x_j; x_i) &= \left(\frac{1}{2\sigma^2}\right)\frac{1}{\sqrt{2\pi}(\sigma\sqrt{2})} \exp\left[-\frac{1}{2}\left(\frac{x_j - x_i}{\sigma\sqrt{2}}\right)^2\right][(-1)(x_j - x_i)] \\
&= \left(\frac{1}{2\sigma^2}\right)\frac{1}{\sqrt{2\pi}(\sigma\sqrt{2})} \exp\left[-\frac{1}{2}\left(\frac{x_j - x_i}{\sigma\sqrt{2}}\right)^2\right](x_i - x_j) \qquad (42) \\
&= \left(\frac{1}{2\sigma^2}\right) G_{\sigma\sqrt{2}}(x_j - x_i)(x_i - x_j)
\end{aligned}$$

The Quadratic Information Force on scalar sample x_j in the Information Potential Field created by all the samples $x_i \in \mathbb{R}$, for $i = 1, 2, \ldots N$ is defined as the average of $\hat{F}_2(x_j; x_i)$ taken over all the samples x_i:

$$\begin{aligned}\hat{F}_2(x_j) &= \frac{1}{N(2\sigma^2)}\sum_{i=1}^{N} G_{\sigma\sqrt{2}}(x_j - x_i)(x_i - x_j) \\ &= \frac{1}{(2N\sigma^2)}\frac{1}{\sqrt{2\pi}(\sigma\sqrt{2})}\sum_{i=1}^{N}\exp\left[-\frac{1}{2}\left(\frac{x_j - x_i}{\sigma\sqrt{2}}\right)^2\right](x_i - x_j)\end{aligned} \tag{43}$$

If the samples are d dimensional vectors, then the Quadratic Information Force on vector sample $\mathbf{x}_j$ in the Information Potential Field created by all samples $\mathbf{x}_i \in \mathbb{R}^d$, for $i = 1, 2, \ldots N$ is defined as:

$$\begin{aligned}\hat{F}_2(\mathbf{x}_j) &= \frac{1}{(2^d N|\mathbf{C}|)}\sum_{i=1}^{N} G_{2\mathbf{C}}(\mathbf{x}_j - \mathbf{x}_i)(\mathbf{x}_i - \mathbf{x}_j) \\ &= \frac{1}{(2^d N|\mathbf{C}|)}\frac{1}{\sqrt{(2\pi)^d|\mathbf{C}|(2^d)}} \\ &\times \sum_{i=1}^{N}\exp\left[-\frac{1}{2}(\mathbf{x}_j - \mathbf{x}_i)^T(2\mathbf{C})^{-1}(\mathbf{x}_j - \mathbf{x}_i)\right](\mathbf{x}_i - \mathbf{x}_j)\end{aligned} \tag{44}$$

5. Hamiltonian MCMC with information theoretic particles

The expression for the potential energy in the Hamiltonian function (Eq. (11)) is similar to the expression for Rényi's quadratic entropy (Eq. (28)). This is consistent with the principles of statistical mechanics where the entropy is related to the dissipation of the potential energy of the system. Based on this intuition from statistical mechanics, we replace the PDF of the position vector $\mathbf{q}$ in (Eq. (11)) with the QIP energy estimator as follows:

$$U(\mathbf{q}_j) = -\log\left[P(\mathbf{q}_j)\right] = -\log\left[\hat{V}_2(\mathbf{q}_j)\right] \tag{45}$$

The change in momentum of the jth Information Theoretic particle in the dynamical system is equal to the negative potential energy gradient defined in (Eq. (16)). This can be expressed in terms of the QIP energy estimator as:

$$\frac{d\mathbf{p}}{dt} = -\frac{dU(\mathbf{q}_j)}{d\mathbf{q}_j} = -\frac{d\log\left[P(\mathbf{q}_j)\right]}{d\mathbf{q}_j} = -\frac{d\log\left[\hat{V}_2(\mathbf{q}_j)\right]}{d\mathbf{q}_j} \tag{46}$$

From the above expression, we derive the expression for the Hamiltonian system's negative potential gradient in terms of the Information Potential and the Information Force as follows:

$$-\frac{d\log\left[\hat{V}_2(\mathbf{q}_j)\right]}{d\mathbf{q}_j} = -\frac{d}{d\mathbf{q}_j}\log\left[\begin{array}{l}\frac{1}{N}\frac{1}{\sqrt{(2\pi)^d|\Sigma|(2^d)}}\times \\ \sum_{i=1}^{N}\exp\left[-\frac{1}{2}(\mathbf{q}_j - \mathbf{q}_i)^T(2\Sigma)^{-1}(\mathbf{q}_j - \mathbf{q}_i)\right]\end{array}\right]$$

$$
\begin{aligned}
-\frac{d\log\left[\hat{V}_2\left(\mathbf{q}_j\right)\right]}{d\mathbf{q}_j} &= -\frac{d}{d\mathbf{q}_j}\log\left[\frac{1}{N}\frac{1}{\sqrt{(2\pi)^d|\Sigma|\left(2^d\right)}}\right] \\
&\quad -\frac{d}{d\mathbf{q}_j}\log\left[\sum_{i=1}^{N}\exp\left[-\frac{1}{2}\left(\mathbf{q}_j-\mathbf{q}_i\right)^T(2\Sigma)^{-1}\left(\mathbf{q}_j-\mathbf{q}_i\right)\right]\right] \\
&= -\frac{1}{\sum_{i=1}^{N}\exp\left[-\frac{1}{2}\left(\mathbf{q}_j-\mathbf{q}_i\right)^T(2\Sigma)^{-1}\left(\mathbf{q}_j-\mathbf{q}_i\right)\right]} \\
&\quad \times\sum_{i=1}^{N}\frac{d}{d\mathbf{q}_j}\exp\left[-\frac{1}{2}\left(\mathbf{q}_j-\mathbf{q}_i\right)^T(2\Sigma)^{-1}\left(\mathbf{q}_j-\mathbf{q}_i\right)\right] \\
&= -\frac{1}{\sum_{i=1}^{N}\exp\left[-\frac{1}{2}\left(\mathbf{q}_j-\mathbf{q}_i\right)^T(2\Sigma)^{-1}\left(\mathbf{q}_j-\mathbf{q}_i\right)\right]} \\
&\quad \times\frac{1}{\left(2^d|\Sigma|\right)}\sum_{i=1}^{N}\exp\left[-\frac{1}{2}\left(\mathbf{q}_j-\mathbf{q}_i\right)^T(2\Sigma)^{-1}\left(\mathbf{q}_j-\mathbf{q}_i\right)\right]\left(\mathbf{q}_i-\mathbf{q}_j\right) \\
&= -\frac{\hat{F}_2\left(\mathbf{q}_j\right)}{\hat{V}_2\left(\mathbf{q}_j\right)}
\end{aligned}
\tag{47}
$$

This result shows that the gradient of the potential energy of the Hamiltonian system of Information Particles is just the Information Force estimate normalized by the Information Potential energy estimate. This also shows that the Information Force vector influences the trajectory of sample proposals in the HMC algorithm. This equation is one of the important contributions of our work. Our simulation of the HMC of a dynamical system of Information Theoretic particles uses this potential energy gradient equation to evolve the system over time.

5.1 Quality of the information potential energy estimator

As described in [5], the Information Potential energy estimator is a kernel estimator of the 2-norm of the underlying PDF of the Information Particles. Just like a PDF estimator, we can define metrics to describe the quality of the Information Potential energy estimator. The Mean Integrated Square Error (MISE) is an important metric used to assess the quality of an estimator. This is expressed as:

$$
\begin{aligned}
MISE\left[\hat{V}_2\left(q_j\right)\right] &= E\left[\int\left(\hat{V}_2\left(q_j\right)-V_2\left(q_j\right)\right)^2 dq\right] \\
&= \int E\left\{\hat{V}_2\left(q_j\right)-E\left[\hat{V}_2\left(q_j\right)\right]\right\}^2 dq+\int\left\{E\left[\hat{V}_2\left(q_j\right)\right]-V_2\left(q_j\right)\right\}^2 dq \\
&= \int Variance\left(\hat{V}_2\left(q_j\right)\right)dq+\int Bias^2\left(\hat{V}_2\left(q_j\right)\right)dq
\end{aligned}
\tag{48}
$$

The bias and variance of the Information Potential estimator can be derived as follows:

Bias:

$$E\left[\hat{V}_2\left(q_j\right)\right] - V_2\left(q_j\right) = E\left[\frac{1}{N}\sum_{i=1}^{N} G_{\sigma\sqrt{2}}\left(q_j - q_i\right)\right] - V_2\left(q_j\right) \\ = E\left[G_{\sigma\sqrt{2}}\left(q_j - q_i\right)\right] - V_2\left(q_j\right) \tag{49}$$

Since the Gaussian kernel is symmetric under the expectation operation:

$$G_{\sigma\sqrt{2}}\left(q_j - q_i\right) = G_{\sigma\sqrt{2}}\left(q_i - q_j\right) \tag{50}$$

Substituting this in (Eq. (49)) and using the definition of $G_{\sigma\sqrt{2}}$ from (Eq. (35)):

$$E\left[\hat{V}_2\left(q_j\right)\right] - V_2\left(q_j\right) == E\left[G_{\sigma\sqrt{2}}\left(q_i - q_j\right)\right] - V_2\left(q_j\right) \\ = \frac{1}{\sigma\sqrt{2}} E\left[G\left(\frac{q_i - q_j}{\sigma\sqrt{2}}\right)\right] - V_2\left(q_j\right) \\ = \frac{1}{\sigma\sqrt{2}}\int G\left(\frac{s - q_j}{\sigma\sqrt{2}}\right) V_2(s) ds - V_2\left(q_j\right) \tag{51}$$

In the above equation $s$is the dummy variable of integration. Let $y = \frac{s-q_j}{\sigma\sqrt{2}}$.This implies that $dy = \frac{ds}{\sigma\sqrt{2}}$. Substituting this in (Eq. (51)), we get:

$$E\left[\hat{V}_2\left(q_j\right)\right] - V_2\left(q_j\right) = \int G(y) V_2\left(q_j + \sigma\sqrt{2}y\right) dy - V_2\left(q_j\right) \tag{52}$$

When $\sigma\sqrt{2}$ is small, we can write the Taylor series expansion of $V_2\left(q_j + \sigma\sqrt{2}y\right)$ as:

$$V_2\left(q_j + \sigma\sqrt{2}y\right) = V_2\left(q_j\right) + \sigma\sqrt{2}yV_2'\left(q_j\right) + \frac{1}{2}2\sigma^2 y^2 V_2''\left(q_j\right) + o\left(\sigma^2\right) \tag{53}$$

Substituting this in (Eq. (52)), we get:

$$E\left[\hat{V}_2\left(q_j\right)\right] - V_2\left(q_j\right) \\ = \int G(y)\left[V_2\left(q_j\right) + \sigma\sqrt{2}yV_2'\left(q_j\right) + \sigma^2 y^2 V_2''\left(q_j\right) + o\left(\sigma^2\right)\right] dy - V_2\left(q_j\right) \\ = V_2\left(q_j\right)\int G(y)dy + \sigma\sqrt{2}V_2'\left(q_j\right)\int yG(y)dy + \sigma^2 V_2''\left(q_j\right)\int y^2 G(y)dy + o\left(\sigma^2\right) - V_2\left(q_j\right) \\ = V_2\left(q_j\right)(1) + \sigma\sqrt{2}V_2'\left(q_j\right)(0) + \sigma^2 V_2''\left(q_j\right)\int y^2 G(y)dy + o\left(\sigma^2\right) - V_2\left(q_j\right) \\ = \sigma^2 V_2''\left(q_j\right)\int y^2 G(y)dy + o\left(\sigma^2\right) \tag{54}$$

This result implies that as the kernel bandwidth $\sigma \to 0$ the bias of the Information Potential energy estimator for sample q_j reduces at the rate of $O(\sigma^2)$. From the above equation it is also evident that the main reason for the bias is the second derivative of the true Information Potential energy (i.e., the rate of curvature of the true PDF of the samples). In other words, if the true PDF of the samples has a sharp spike, the bias of the Information Potential energy estimator will increase. The Information Potential energy estimator tends to smooth out sharp curvatures or spikes in the PDF which increases bias. The amount of smoothness is governed by the bandwidth parameter σ.

Variance:

$$\begin{aligned} E\left\{\left[\hat{V}_2\left(q_j\right)\right]^2\right\} - \left\{E\left[\hat{V}_2\left(q_j\right)\right]\right\}^2 &= E\left\{\left[\hat{V}_2\left(q_j\right)\right]^2\right\} - \frac{1}{N}\left(V_2\left(q_j\right) + Bias\right)^2 \\ &= E\left\{\left[\hat{V}_2\left(q_j\right)\right]^2\right\} + O\left(N^{-1}\right) \\ &= E\left\{\left[\frac{1}{N}\sum_{i=1}^{N} G_{\sigma\sqrt{2}}\left(q_j - q_i\right)\right]^2\right\} + O\left(N^{-1}\right) \\ &= E\left\{\left[\frac{1}{N}\sum_{i=1}^{N} G_{\sigma\sqrt{2}}\left(q_i - q_j\right)\right]^2\right\} + O\left(N^{-1}\right) \\ &= \frac{1}{N}E\left\{\left[G_{\sigma\sqrt{2}}\left(q_i - q_j\right)\right]^2\right\} + O\left(N^{-1}\right) \\ &= \frac{1}{2N\sigma^2}\int G^2\left(\frac{s - q_j}{\sigma\sqrt{2}}\right) V_2(s)ds + O\left(N^{-1}\right) \end{aligned} \tag{55}$$

Let $y = \frac{s-q_j}{\sigma\sqrt{2}}$.This implies that $dy = \frac{ds}{\sigma\sqrt{2}}$. Substituting this in (Eq. (55)), we get:

$$\begin{aligned} &E\left\{\left[\hat{V}_2\left(q_j\right)\right]^2\right\} - \left\{E\left[\hat{V}_2\left(q_j\right)\right]\right\}^2 = \\ &\frac{1}{N\sigma\sqrt{2}}\int G^2(y)V_2\left(q_j + \sigma\sqrt{2}y\right)dy + O\left(N^{-1}\right) \end{aligned} \tag{56}$$

When $\sigma\sqrt{2}$ is small, we can write the Taylor series expansion of $V_2\left(q_j + \sigma\sqrt{2}y\right)$ as:

$$V_2\left(q_j + \sigma\sqrt{2}y\right) = V_2\left(q_j\right) + \sigma\sqrt{2}yV_2'\left(q_j\right) + o(\sigma) \tag{57}$$

Substituting this in (Eq. (56)), we get:

$$\begin{aligned} &E\left\{\left[\hat{V}_2\left(q_j\right)\right]^2\right\} - \left\{E\left[\hat{V}_2\left(q_j\right)\right]\right\}^2 \\ &= \frac{1}{N\sigma\sqrt{2}}\int G^2(y)\left[V_2\left(q_j\right) + \sigma\sqrt{2}yV_2'\left(q_j\right) + o(\sigma)\right]ds + O\left(N^{-1}\right) \\ &= \frac{1}{N\sigma\sqrt{2}}V_2\left(q_j\right)\int G^2(y) + o\left(\frac{1}{N\sigma\sqrt{2}}\right) \end{aligned} \tag{58}$$

This result shows that as the number of samples $N \to \infty$ and kernel bandwidth $\sigma \to \infty$, the variance of the Information Potential energy estimator for the sample

q_j reduces at the rate of $O\left(\frac{1}{N\sigma\sqrt{2}}\right)$. However, as $\sigma \to 0$, the variance of the estimator increases. The result also shows that the variance of the estimator is large where the value of the Information Potential energy $V_2\left(q_j\right)$ (i.e., true probability of the sample) is also large. This happens when there are many Information Particles closer together.

5.2 The Kernel bandwidth parameter and the information potential energy estimator bias-variance trade-off

We have shown that the Gaussian kernel bandwidth σ directly influences the bias and variance of the Information Potential energy estimator. This in turn affects the sample distribution of the PDF estimate generated by the Hamiltonian MCMC. From (Eq. (54)) it is evident that the bias of the estimator reduces when we decrease the kernel bandwidth σ. However, (Eq. (58)) clearly shows that the decreasing σ increases the variance of the estimator. Therefore, we must choose an optimum bandwidth which minimizes both the systematic error (bias) and the random error (variance) of the Information Potential energy estimator. An iterative algorithm to converge to the optimum kernel bandwidth is described in the following section.

5.3 Computational complexity of the information potential energy estimator

From (Eqs. (38) and (41)) it may appear that the complexity of computing the Information Potential is $O(N^2)$. However, as described in [5], the Information Potential can be written as a symmetric positive Gramm Matrix which can be approximated using the incomplete Cholesky decomposition (ICD) as an $N \times D$ matrix where $D \ll N$. Using this technique, the time complexity for computing the Information Potential reduces to $O(ND^2)$ and the space complexity reduces to $O(ND)$.

6. Maximum-likelihood iterative algorithm to adapt the kernel bandwidth of the information potential energy estimator

There are many iterative kernel bandwidth adaptation techniques available in the literature. We present a simple iterative technique to illustrate how MCMC with Hamiltonian of Information Theoretic Particles can be used to adjust the bandwidth parameter of the iterative PDF estimator. Here, we have chosen to minimize the Kullback–Leibler (K-L) divergence between the samples of the estimated PDF and the target sample distribution as the criteria for adapting the kernel bandwidth of the Information Potential energy and Information Force estimator. As described in [17], this is equivalent to maximizing the likelihood that the estimated PDF samples output by the MCMC, has the same distribution as the target samples.

The ML estimate of the optimum kernel bandwidth $\mathbf{C}_{ML}$ for vector Information Particle samples $\mathbf{q}_j$ is the solution to the following log-likelihood maximization problem:

$$\mathbf{C}_{ML} = \arg\max_{\mathbf{C}} \sum_{j=1}^{N} \log\left[\hat{V}\left(\mathbf{q}_j|\mathbf{C}\right)\right] \tag{59}$$

Using (Eq. (41)) in the summation of the above equation, we get:

$$\sum_{j=1}^{N} \log\left[V\left(\mathbf{q}_j|\mathbf{C}\right)\right] = \sum_{j=1}^{N} \log\left[\frac{1}{N-1}\sum_{\substack{i=1\\i\neq j}}^{N} G_{2\mathbf{C}}\left(\mathbf{q}_j - \mathbf{q}_i\right)\right]$$

$$= \sum_{j=1}^{N} \log\left[\begin{array}{l}\frac{1}{(N-1)}\frac{1}{\sqrt{(2\pi)^d\left(2^d\right)|\mathbf{C}|}}\times \\ \sum_{\substack{i=1\\i\neq j}}^{N} \exp\left[-\frac{1}{2}\left(\mathbf{q}_j - \mathbf{q}_i\right)^T (2\mathbf{C})^{-1}\left(\mathbf{q}_j - \mathbf{q}_i\right)\right]\end{array}\right] \tag{60}$$

To maximize the above equation, we take the derivative and equate it to 0. This gives us the following update equation for scalar Information Theoretic particles:

$$\sigma_{t+1}^2 = \left[\frac{1}{2N(N-1)}\sum_{j=1}^{N}\frac{1}{\hat{V}\left(q_j\right)}\sum_{\substack{i=1\\i\neq j}}^{N} G_{\sigma\sqrt{2}}\left(q_j - q_i\right)\left(q_j - q_i\right)^2\right]_t \tag{61}$$

In the above equation, σ_{t+1} is the kernel bandwidth at iteration $t+1$. It is updated with the result of the right-hand side of the equation obtained at time t. This kernel bandwidth update equation (Eq. (61)) is in the form of a fixed-point (or invariant point) equation. This equation is like the equation in [18] except for the factor of $1/2$. For vector Information Theoretic particles, the kernel bandwidth update equation is:

$$\mathbf{C}_{t+1} = \left\{\frac{1}{2N(N-1)}\sum_{j=1}^{N}\frac{1}{\hat{V}\left(\mathbf{q}_j\right)}\sum_{\substack{i=1\\i\neq j}}^{N} G_{2\mathbf{C}}\left(\mathbf{q}_j - \mathbf{q}_i\right)\left[\left(\mathbf{q}_j - \mathbf{q}_i\right)\left(\mathbf{q}_j - \mathbf{q}_i\right)^T\right]\right\}_t \tag{62}$$

In this equation, $\mathbf{C}$ is the kernel bandwidth matrix and can have unequal elements along its diagonal or non-zero off-diagonal elements. If the kernel bandwidth matrix is constrained to an identity matrix multiplied by a scaling factor, the kernel bandwidth matrix update equation can be expressed as:

$$\mathbf{C}_{t+1} = \left\{\frac{1}{2N(N-1)}\sum_{j=1}^{N}\frac{1}{V\left(\mathbf{q}_j\right)}\sum_{\substack{i=1\\i\neq j}}^{N} G_{2\mathbf{C}}\left(\mathbf{q}_j - \mathbf{q}_i\right)\left\|\left(\mathbf{q}_j - \mathbf{q}_i\right)\right\|^2\right\}_t \tag{63}$$

From the fixed- or invariant-point theorem, the range over which the fixed-point bandwidth update equations will converge to a unique solution is:

$$\left[\frac{\min\left(q_j - q_i\right)^2}{2}, Trace\left\{E\left[\mathbf{q}\mathbf{q}^T\right]\right\}\right] \tag{64}$$

In the above equation, q_i, q_j are information particles from the target sample distribution and $\mathbf{q}$ is the column vector of all the target information particles. From the fixed-point theorem, this fixed-point equation will converge to a unique solution if $|f'(\sigma^2)| < 1$.

7. Simulation results

The potential energy surface, which is the plot of (Eq. (41)), of a Hamiltonian system of Information Theoretic particles for a bivariate Gaussian distribution is shown in **Figure 1**. From this figure it is evident that the potential energy surface of the Hamiltonian system has larger values when the Information Theoretic particles are sparse and is lowest at the bottom of the bowl-shaped surface where the particles have the highest density.

The momentum variable of the HMC algorithm occasionally moves the "probe" particle to a higher energy level but the Hamiltonian system has the tendency to fall back to its lowest energy level along the bowl-shaped surface. As a result, the HMC tends to sample the given target distribution more often where the density of the Information Theoretic particles is the largest.

Figure 2 shows the potential energy gradient of the same bivariate Gaussian distribution. This is the plot of (Eq. (47)) for this distribution. Each surface in this figure is one component of the potential energy gradient. Each surface tilts towards the corresponding mean value $\mu = [-5, 6]$ of the bivariate Gaussian distribution. The figure shows that the potential energy gradient of the Hamiltonian system is lowest near the mean of the distribution and is highest further away from the mean. The time evolution trajectory of the Hamiltonian system lies on this surface.

The iterative PDF estimate of a bivariate Gaussian distribution with $\mu = [-5, 6], \Sigma = [3, 0; 0, 4]$ using MCMC with 3 different kernel bandwidths is shown in **Figure 3**.

From **Figure 3**, it is evident that the MCMC algorithm based on the Hamiltonian of Information Theoretic particles accurately estimates the PDF of the target distribution. The sample points generated by the HMC algorithm covers most of the target samples in this figure. This figure shows that our intuition of comparing the Entropy to the system's potential and also using the Information Potential in the derivation of the potential gradient (Eq. (47)) of the Hamiltonian system of Information Theoretic particles, was correct.

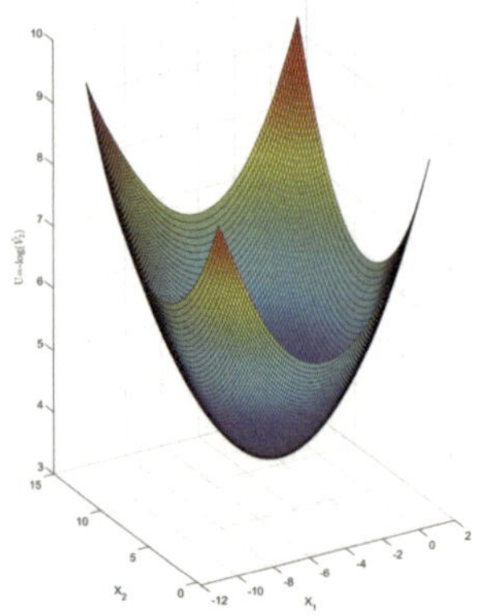

Figure 1.
Potential energy surface of the Hamiltonian system of a bivariate Gaussian ($\mu = [-5, 6], \Sigma = [3, 0; 0, 4]$) *distribution of Information Theoretic particles.*

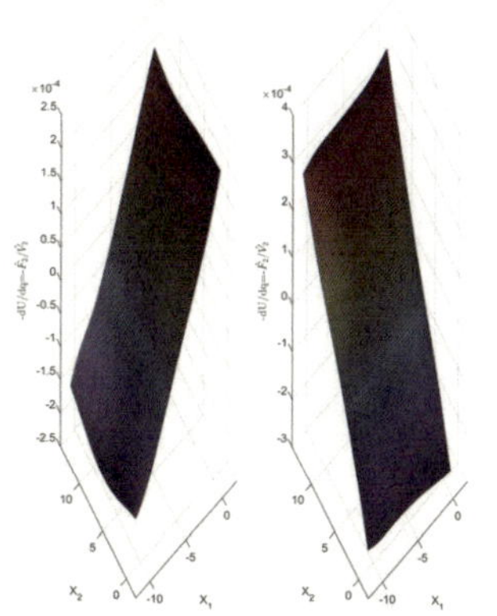

Figure 2.
Vector components of the potential energy gradient of the Hamiltonian system of a bivariate Gaussian ($\mu = [-5, 6], \Sigma = [3, 0; 0, 4]$) *distribution of Information Theoretic particles.*

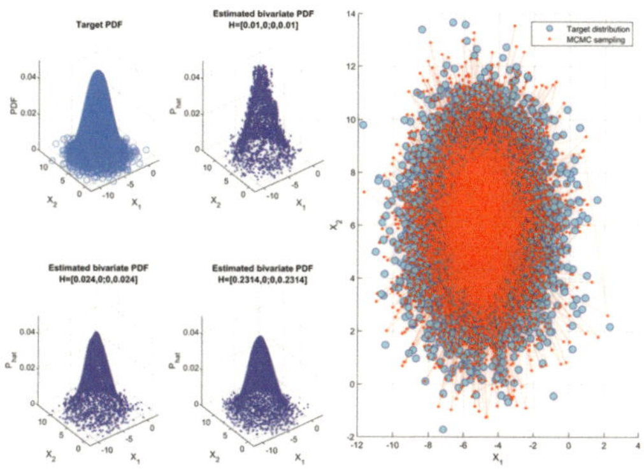

Figure 3.
The left-hand side figure shows the iterative PDF estimate of a bivariate Gaussian distribution ($\mu = [-5, 6], \Sigma = [3, 0; 0, 4]$) *with the MCMC method using the Hamiltonian of Information Theoretic particles. The right-hand side figure shows that the samples generated by the HMC method mostly overlaps the samples of the target distribution.*

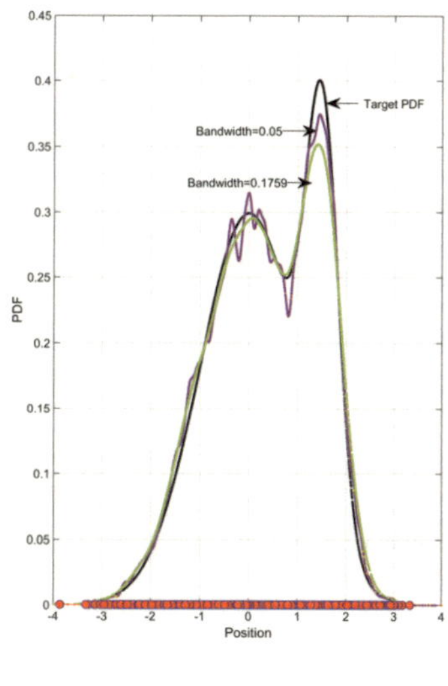

Figure 4.
Iterative estimation of the PDF of a bivariate Gaussian mixture distribution with the MCMC method using the Hamiltonian of Information Theoretic particles.

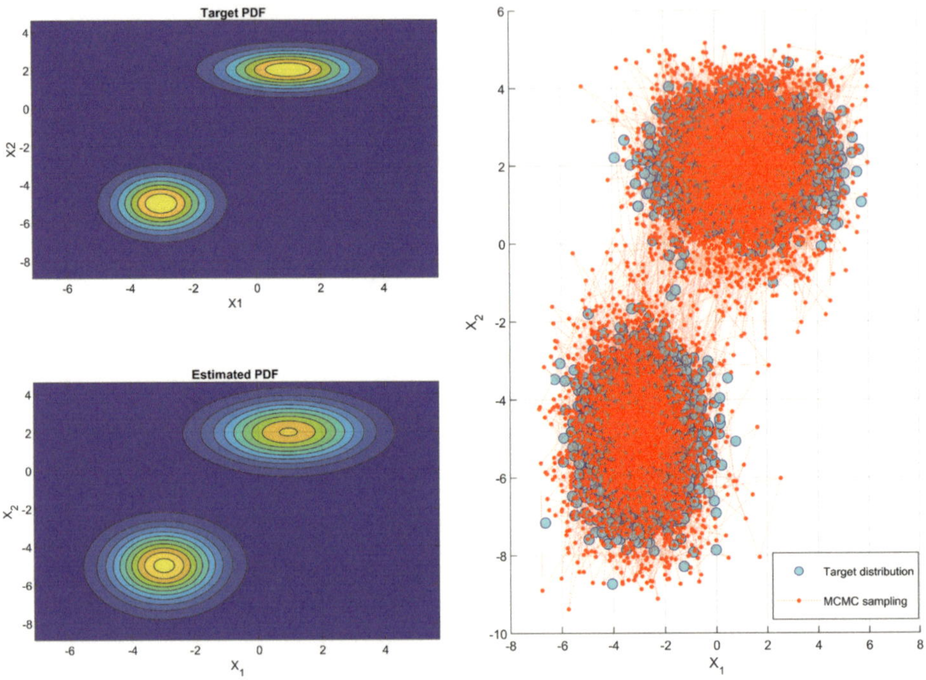

Figure 5.
Contour plots of the target PDF and the estimated PDF of the bivariate Gaussian mixture distribution. Samples generated by the MCMC algorithm using the Hamiltonian of Information Theoretic particles.

Our HMC algorithm using Information Theoretic particles also works well for Gaussian mixture distributions. **Figure 4** shows that our MCMC algorithm using the Hamiltonian of Information Theoretic particles can be used to iteratively estimate the PDF of different multivariate distributions.

Figure 5 shows that the contour plot of the estimated PDF matches closely to the target PDF. The corresponding samples generated by the HMC algorithm traverses the two clusters of the bivariate Gaussian mixture distribution and covers most of the samples of the target distribution.

8. Conclusion

We have proposed a novel perspective on the MCMC method where we used it to iteratively estimate the PDF of a given target sample distribution. We have shown that the samples of a probability distribution can be viewed as Information Particles in an Information Field. These particles have Information Potential energy and are subject to Information Forces by virtue of their position in the field. The concept of Information Potential energy fits perfectly within the framework of the Hamiltonian of a dynamical system. We have derived an important result that the gradient of the potential energy of the Hamiltonian system of Information Particles is just the Information Force estimate normalized by the Information Potential energy estimate.

Our simulation results show that our intuition of comparing Rényi's Quadratic Entropy equation with the Hamiltonian potential energy equation to derive the equation for the potential gradient of a dynamical system of Information Theoretic particles was correct. Using this equation, we were able to accurately estimate univariate and multivariate PDFs. Based on the fixed- or invariant-point theorem, we also derived an equation to iteratively update the bandwidth parameter of the Information Potential and Information Force estimators.

In machine learning applications the dataset is sometimes resampled to the appropriate size before starting the learning operation. Our algorithm can be used to view the data samples as Information Theoretic particles and resample it using the HMC described in this chapter.

Author details

Tokunbo Ogunfunmi* and Manas Deb
Signal Processing Research Lab (SPRL), Department of Electrical and Computer Engineering, Santa Clara University, CA, USA

*Address all correspondence to: togunfunmi@scu.edu

References

[1] R. Neal, "Probabilistic Inference using Markov Chain Monte Carlo methods, Technical Report CRG-TR-93-1," Department of Computer Science, University of Toronto, Toronto, 1993.

[2] R. Neal, "An improved acceptance procedure for the hybrid Monte Carlo algorithm, "*Journal of Computational Physics,* vol. 111, no. 1, pp. 194–203, 1994.

[3] M. Betancourt, "A Conceptual Introduction to Hamiltonian Monte Carlo, "arXiv: 1701.02434 [stat.ME], 2018.

[4] R. Neal, "MCMC using Hamiltonian Dynamics, "in *Handbook of Markov Chain Monte Carlo*, CRC Press, 2011, pp. 113–162.

[5] J. Principe, Information Theoretic Learning, New York: Springer, 2010.

[6] W. K. Hastings, "Monte Carlo Sampling Methods Using Markov Chains and Their Applications, "*Biometrika,* vol. 57, no. 1, pp. 97–109, 1970.

[7] A. Rényi, "On measures of entropy and information, "*Proceedings of the 4th Berkeley symposium on math, statistics and probability,* vol. 1, pp. 547–561, 1961.

[8] Z. Makó and Z. P. M. D. 7. 4. & Páles, "On the equality of generalized quasiarithmetic means.," *Publicationes Mathematicae, Debrecen,* vol. 72, pp. 407–440, 2008.

[9] A. N. Kolmogorov, "Sur la notion de la moyenne," *Atti Accad. Naz. Lincei. Rend. 12:9,* pp. 388–391, 1930.

[10] M. Nagumo, "Über eine klasse von mittelwerte," *Japanese Journal of Mathematics,* vol. 7, pp. 71–79, 1930.

[11] G. H. Hardy, L. J. E. and P. G., Inequalities, Cambridge, 1934.

[12] R. V. L. Hartley, "Transmission of Information," *Bell System Technical Journal,* vol. 7, p. 535, 1928.

[13] C. E. Shannon, "A mathematical theory of communication," *Bell System Technical Journal,* p. 535, 1928.

[14] M. Deb and T. Ogunfunmi, "Using Information Theoretic Learning techniques to train neural networks," in *51st Asilomar conference on signals, systems and computers*, 2017.

[15] E. Parzen, "On Estimation of a Probability Density Function and Mode," *Annals of Mathematical Statistics,* vol. 33, no. 3, pp. 1065–1076, 1962.

[16] M. Rosenblatt, "Remarks on some Nonparametric Estimates of a Density Function," *Annals of Mathematical Statistics,* vol. 27, no. 3, pp. 832–837, 1956.

[17] T. Cover and J. Thomas, Elements of Information Theory, Wiley & Sons, 2012.

[18] J. M. Leiva-Murillo and A. Artés-Rodriguez, "Fixed point algorithm for finding the optimal covariance matrix in kernel density modeling," *IEEE International Conference on Acoustics, Speech and Signal Procesing,* vol. 5, 2006.

Chapter 5

Monte Carlo and Medical Physics

Omaima Essaad Belhaj, Hamid Boukhal and El Mahjoub Chakir

Abstract

The different codes based on the Monte Carlo method, allows to make simulations in the field of medical physics, so the determination of all the magnitudes of radiation protection namely the absorbed dose, the kerma, the equivalent dose, and effective, what guarantees the good planning of the experiment in order to minimize the degrees of exposure to ionizing radiation, and to strengthen the radiation protection of patients and workers in clinical environment as well as to respect the 3 principles of radiation protection ALARA (As Low As Reasonably Achievable) and which are based on: -Justification of the practice -Optimization of radiation protection -Limitation of exposure.

Keywords: Radioprotection, Monte Carlo, medical physics, simulation

1. Introduction

The monte Carlo method is a method of the family of algorithmic methods, it makes it possible to solve statistical problems and contribute to the analysis of data based on stochastic processes, thus it allows to evaluate the maturity risk and these probabilities thus the appointment of this method refers to the random side used at the casino of Monte Carlo located in MONACO.

The scope of the Monte Carlo method is very broad and covers all fields of nuclear medicine and particle transport, namely applications in radiotherapy-brachytherapy, scintigraphic imaging, shielding, dosimetry, PET, gamma camera, etc.

This choice of use of this method is not due to chance, but on the one hand because it allows to simulate the behavior of different particles and to deduce the average behavior of all particles according to the law of large numbers and the central limit theorem, so it can handle coupled and 3D problems with complex geometry and it can answer specific questions (average flow in a volume, Absorbed dose, Kerma, Hp(10), etc.), and its major advantages and determining the sources of errors, on the other hand, this method makes it possible to comply with the laws of ALARA (As Low As Reasonably Achievable) radiation protection which are:

- Justification of the practice
- Limitation of exposure
- Optimization of radiation protection

All this through simulation and modeling of the experience by the Monte Carlo method before practicing it in order to be able to determine the limits of the doses absorbed by the staff and the patient, as well as to develop and determine the correction methods to improve the quantification images and results, as well as the design of radiation detection equipment.

Monte Carlo uses software and calculation codes requiring a better knowledge of the input data of the problem (source, energy, angle and distribution, spatial and temporal dependence), and of the geometry which must represent the real situation, as well. The materials constituting the system to be studied, type of calculation envisaged, criteria for stopping the simulation, these parameters are defined and entered by the operator, and the basic nuclear data will be taken directly from data libraries (effective diffusion section, adsorption, fission, etc.).

Among these codes we cite GEANT4, GAMOS, GATE, FLUKA, PENELOPE... generally written in the C ++ language. So, Monte Carlo is a reliable method and offers complete and very close to reality solutions, which cover all the needs in nuclear medicine.

2. Analogous simulation of photon transport by the Monte Carlo method

2.1 Generality of photons

Most visualization techniques exploit radiation, photonic or otherwise, the intensity of which can be measured in total flux, while gamma radiation used in nuclear medicine is exploited at the level of its smallest indivisible component, the "photon", and to detect this gamma radiation, a scintillation detector is generally used, the sensitive cell of which is a crystal which has the property of producing a small burst of light when it is touched by a photon, a photomultiplier tube associated with this crystal transforms this spark into an electrical pulse whose amplitude is proportional to the energy of the radiation. The total number of photons detected during a given time interval, or count rate, is the measurement of the radioactivity present in the field of the detector. It's this ability to count the number of individual photons that make medicine nuclear energy provides quantitative results [1].

So, photons ($m_{photon} = 0$, $q_{photon} = 0$), are electromagnetic radiations characterized by their energy and their origin, they can be produced by the following phenomena:

- de-excitation of the nucleus following a modification of its structure;
- de-excitation of the electronic procession;
- bremsstrahlung production;
- dematerialization or positron annihilation

There are two types of photons:
X-rays:

- Products during the rearrangements of the electronic procession, usually caused by the collision of electrons on atoms, [10 keV; 100 keV].
- Produced by linear accelerators used in radiotherapy external, [6 MeV; 25 MeV].

Ray γ:

Gamma radiation generally accompany α and β decays, their energy ranges from 60 keV at 3 MeV.

Whatever the origin of the photon, its behavior in matter will be identical.

The photons interact directly with the electrons of matter, and then regenerate various effects depending on the disappearance or not of this one, we find the photo-electric effect, creation of pairs which are due to a disappearance of the photon, the coherent and incoherent diffusion due to the non-disappearance of the photon.

2.2 Simple flowchart

The method to generate histories is to query the probability distributions that describe the problem. This concept is called sampling. The method chosen for Sampling these probability distributions depends on the nature of the distribution [2].

The analogous simulation of particle transport using the Monte Carlo method allows the particle to be followed in its actual path, and the particle is sampled by following the following steps (**Figure 1**):

1. Sampling of the "random distance" collision distance (average free path).

2. transport of the particle to the point of interaction taking into account the constraints of geometry.

3. Selection of the collided nucleus (heterogeneous environment).

4. selection of the type of interaction.

If we have

- **Absorption**, the particle's story ends, and the system will repeat steps 1 to 4 with a new particle.
- **Diffusion**, modification of the trajectory, of the energy, and calculation of the angle of diffusion, the system will repeat steps from 1 to 4 until reaching E_{seuil}.

2.2.1 Collision distance

This is the distance the particle traveled before it interacted. Or a particle that will undergo an interaction at a point x, at a distance l in a given volume.

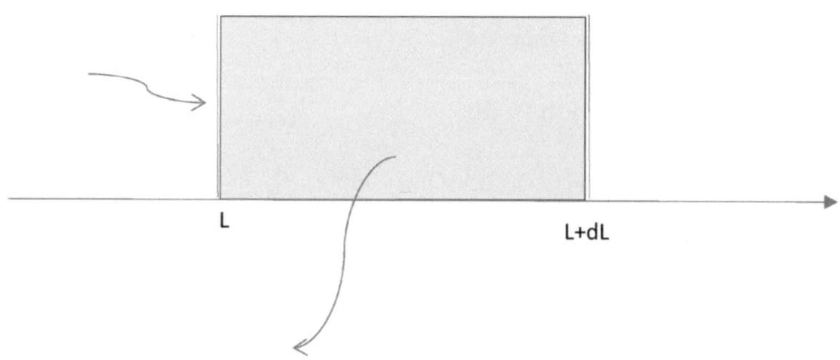

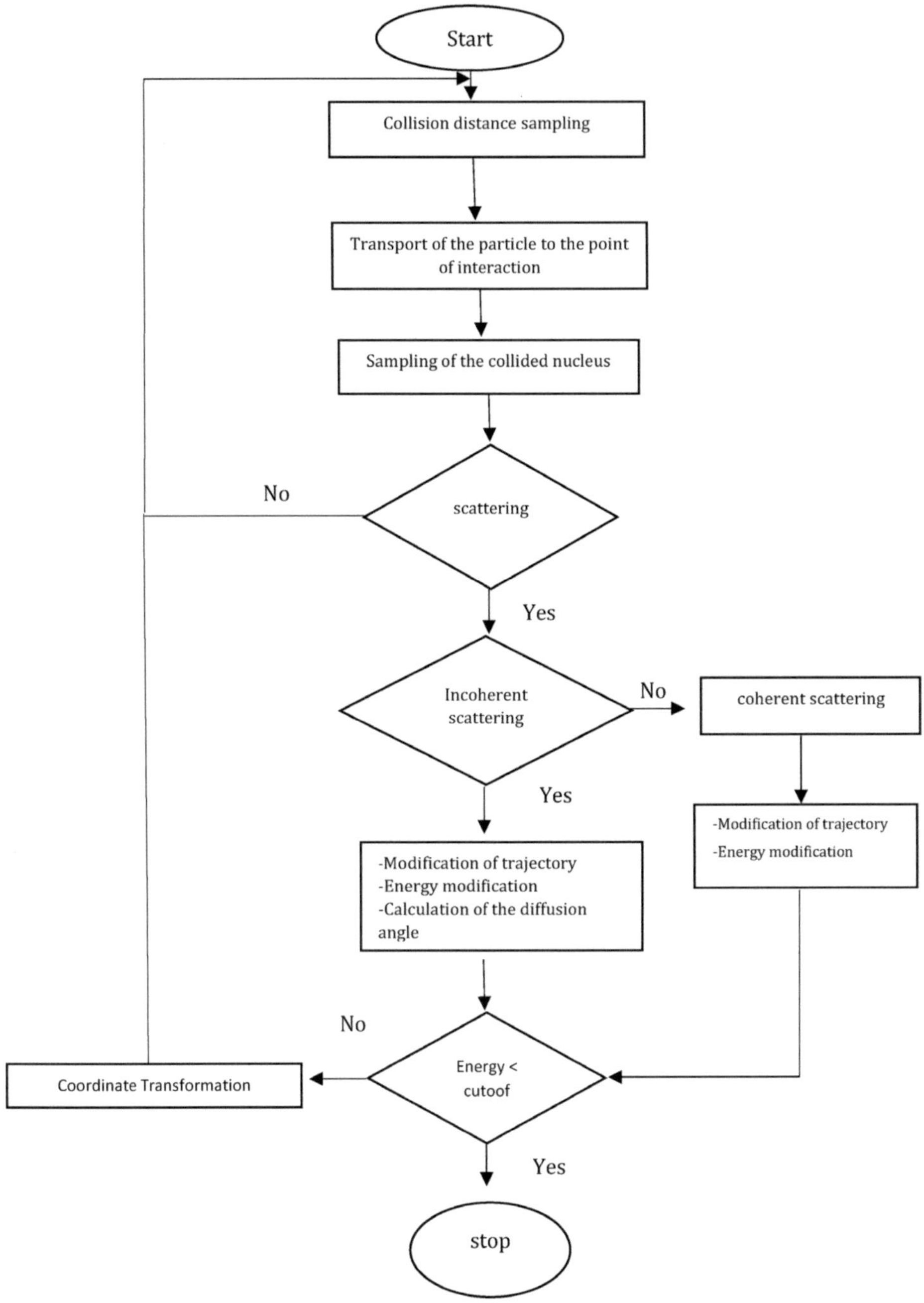

Figure 1.
Flowchart of photon transport in Monte Carlo.

The probability of interaction in dl:

$$P(L)dL = \mu(h\nu)e^{-\mu(h\nu)L}\,dL \tag{1}$$

The cumulative function is calculated by:

$$F(L) = \int_0^L \mu(h\nu)\, e^{-\mu(h\nu)s} ds \tag{2}$$

We draw a random number ε, and we look for L such that:

$$ɥ = F(L) \rightarrow L = F^{-1}(ɥ) \tag{3}$$

So,

$$\boxed{L = -\frac{\ln(\varepsilon)}{\mu(h\nu)}} \tag{4}$$

where $\mu(h\nu)$ is the linear attenuation coefficient for the material at photon energy $h\nu$.

2.2.2 Selection of the collided nucleus

Which nuclide is subject to collision in the case of a mixture environment?

We must just generate a random number ε between 0 and 1 and compare it to the cumulative probabilities (**Figures 2** and **3**).

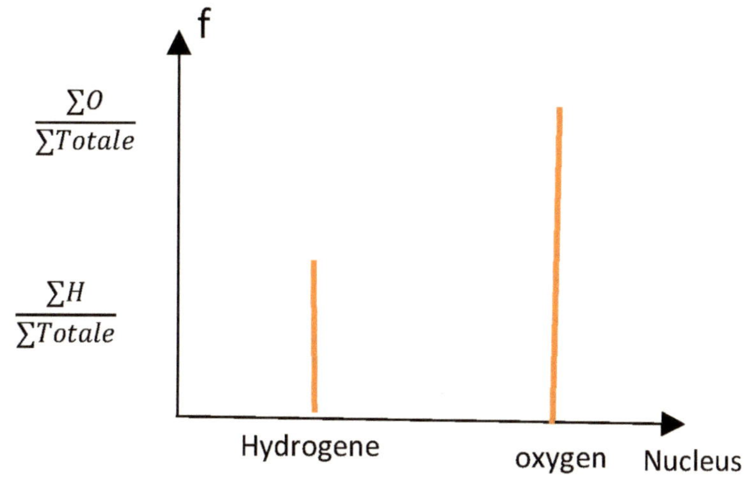

Figure 2.
Density of probability.

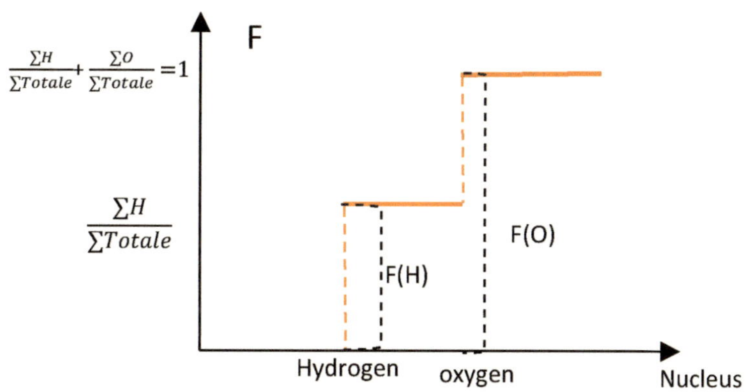

Figure 3.
Inversion of the CDF for selection of the collided nucleus.

Example (H_2O):

if
- $\varepsilon < \sum H/\sum Totale$ (5) ⟷ interaction on hydrogen
- $\varepsilon > \sum O/\sum Totale$ (6) ⟷ interaction on oxygen

with $\sum_H = N_H \sigma_H$ (7) and $\sum_o = N_o \sigma_o$ (8)

Macroscopic cross section — Microscopic cross section

2.2.3 Choice of the type of interaction

The choice of the interaction type is done in the same way of choosing of the collided nucleus.

if
- $\varepsilon \leqslant \sigma_{photoelecric}(h\nu)/\mu(h\nu)$ (9) photoelectric absorption
- $\varepsilon < (\sigma_{photoelecric}(h\nu) + \sigma_{coh}(h\nu))/\mu(h\nu)$ (10) incoherent scattering
- $\sigma_{photoelecric}(h\nu)/\mu(h\nu) < \varepsilon \leqslant \sigma_{photoelecric}(h\nu) + \sigma_{coh}(h\nu))/\mu(h\nu)$ (11) coherent scattering

with $\sigma_{photoelecric}$: cross section of photoelectric effect
σ_{coh}: cross section of coherent scattering effect
σ_{incoh}: cross section of incoherent scattering effect

2.2.4 Choice of angle and scattering energy

In the case of photoelectric effect, the history of the photon ends, and the program begins with another sample, except if there is emission of a fluorescence X-ray in this case The direction of the emitted photon in the laboratory system is given by.

$$\vec{\Omega} = (\sin\theta\ \cos\Phi, \sin\theta\ \sin\Phi, \cos\theta) \tag{12}$$

(θ and Φ, are polar and azimuthal angles, respectively), with the azimuthal angle sampled from,

$$\Phi = 2\pi\varepsilon_1 \tag{13}$$

and the polar angle sampled from,

$$\theta = \arccos\left(1 - \varepsilon_2(1 - \cos\theta max)\right) \tag{14}$$

with $\theta max = \pi$

When the scattering is incoherent the energy hν' of the scattered photon is taken as the Compton energy:

$$h\nu' = h\nu/\left[1 + \left(h\nu/m_e c^2\right)(1 - \cos\theta)\right] \tag{15}$$

where

$m_e c^2$: the energy equivalence of the electron rest mass (511 keV);

θ: scattering angle.

It is common to neglect sampling coherent scattering angles [3], this may be justified in many situations. However, in diagnostic radiology the neglect of coherent scattering is a poor approximation [4].

Sampling of scattering angles can be done with the rejection method, from the total (incoherent more coherent) scattering cross-section [5], or separately, through method of Klein-Nishina cross-section and the classical Thompson scattering cross-section and corrected for the use of incorrect scattering cross-sections by applying a weight factor to the photon [6], or with and the distribution function techniques.

2.2.5 Weight of particle

Ideally, every real particle in a physical problem should be simulated by a fictitious particle in monte Carlo, in fact, to limit the duration of the simulations and improve computational efficiency [7], a monte Carlo particle does not exactly simulate a physical particle but rather represents a number w of particles physical. The number w is the weight, this weight represents the importance that is assigned to a particle. By default, the weight of each particle is 1 and the energy deposited by a particle equals the product of its energy by its weight.

The benefit of giving weight to a particle is to favor certain physical processes over others. It is necessary to simulate more particle with a low weight to decrease the uncertainty, if we have a sample N of σ (uncertainty) and we want to decrease σ to $\frac{\sigma}{\sqrt{n}}$, so we have to multiply N by $\sqrt{n}^2$.

3. Monte Carlo code "GAMOS"

There are several Monte Carlo calculation codes in the field of nuclear medicine, among these codes there is the famous GEANT4 code which is a very powerful and flexible toolkit, for medical applications, but the use of this code does not is not easy, it requires strong knowledge of C ++ language, and details of GEANT4, Most GEANT4 users are researchers who always want to know what is going on in simulation.

The thing that drove Pedro Acre to develop an easy-to-use framework based on the GEANT4 code, allowing to deal with medical physics problems with minimal knowledge of GEANT4 and no need for C ++, thus it provides all the necessary functionalities. to deal with a subject of medical physics while avoiding complicated coding, it is the GAMOS code (Geant4-Based Architecture for Medicine-Oriented Simulation).

The minimum set necessary to compile a project is to select a geometry, a physics list and a generator, to run N events to do this, the geometry will be written in a (.geom) extension file, the name of the geometry, the choice of the physics list, the generators, and the other functionality will be written in an input file of .in extension, therefore output items will be displayed in a file with an .out extension and errors in a file with the gamos_error.log extension.

3.1 Installation Gamos 6.2.0 under Ubuntu

Tap on terminal:

- mkdir gamos-6.2.0
- cd gamos-6.2.0
- run: wget http://fismed.ciemat.es/GAMOS/download/GAMOS.6.2.0/download_scripts.sh
- run: sh download_scripts.sh
- cd scripts
- bash./installMissingPackages.Ubuntu.20.04.sh
- run: sh installGamos.sh home/your home directory/gamos-6.2.0
- to make sure that Gamos has been installed correctly, you need to compile the tutorials
- to compile the tutorials tape:
    - cd ~/gamos-6.0.0/GAMOS.6.0.0
    - source ~/gamos-6.2.0/GAMOS.6.2.0/config/confgamos.sh
    - cd Tutorials
    - . /runAll

3.2 Creation of geometry and input file

3.2.1 Geometry file

There are 3 ways to describe your geometry:

- Using a text file.
- Using one of the geometry examples provided by Gamos.
- Using C ++.

In this case we will be concerned with the use of a geometry from a text file, the extension of this file must be (.geom), the 1st step is to create a mother volume that will generate all the other volumes, so that the particles do not escape from the mother volume, and to finish the history of the particles that will come out of this volume, then build the other volumes from the following tags:

a. Materials

:ISOT: For isotopes
:ELEM: For elements

:ELEM_FROM_ISOT: For element composed of several isotopes Material mixtures by weight, volume or number of atoms

:MIXT: For material made of a mixture of elements or materials it can be:

:MIXT_BY_WEIGHT
:MIXT_BY_NATOMS
:MIXT_BY_VOLUME

Examples

```
:ISOT     Cs137(Name)       55(Z) 137(A) 136,907(atomic mass)
:ELEM   Hydrogen(Name)  H(Symbol) 1.( Z) 1.0078 (A).

:ELEM  Water(Name)   1. (density)  2(Number of components).
          Hydrogen       2*1.0078/ (2*1.0078 + 15.999)
          Oxygene        1.0078/(2*1.0078 + 15.999)
```

Geant4 provides a list of predefined materials, whose compositions correspond to the definition of NIST. Among them you can find all the simple elements, you can use these materials when building a volume in GAMOS without needing to redefine them on your geometry file.

The elements can be found in ~ / *gamos-6.2.0 / GAMOS.6.2.0 / data / NIST_elements.txt*

Materials consisting of a mixture of elements or materials can be found in ~ / *gamos-6.2.0 / GAMOS.6.2.0 / data / NIST_materials.txt*

Other materials common in medical physics are also predefined in the files ~ / *gamos-6.2.0 / GAMOS.6.2.0 / data / NIST_materials.txt and /PET_materials.txt.*

b. Volume

: **VOLU**, can be BOX, TUBE, TUBS, CONE, CONS, PARA, TRAP, SPHERE, ORB, TORUS, POLYCONE, ELLIPTICALTUBE … .

For more details on List of solid parameters see the manual [8].

: **PLACE** mean the placement of the volume in relation to the parent volume, according to which rotation Matrix, and which coordinates x, y, z.

: **PLACE_PARAM** is the placement of several copies of a volume along a line.

c. Rotation matrix

:ROTM, a rotation matrix is interpreted as the rotation that should be applied to the volume in the reference system, it can be defined in three ways:

3 rotation angles around X,Y,Z
6 theta and phi angles of X,Y,Z axis
9 matrix values (XX, XY, XZ, YX, YY, YZ, ZX, ZY, ZZ)

Example

```
:ROTM     R000    0.      0.      0.
:VOLU     world   BOX   400.    400.   400.   G4_AIR
:VOLU     myvol   BOX   200.    200.   200.   G4_Pb
:PLACE    myvol   1      world   R00    0.      0.        0.
```

There are other features offered by Gamos such as Visibility, Color and transparency, Check overlaps … . For more details return to manual.

3.2.2 Input file

The first lines of this file are to appeal to the geometry file using the following commands:

/ gamos / setParam GmGeometryFromText: FileName test.geom

/ gamos / geometry GmGeometryFromText

Then, determine one of the lists found on physical Gamos and functions adabpted to your problem, such as:

/ gamos / physicsList GmEMPhysics: for gammas, electrons and positrons, as well as for optical photons.

/ gamos / physicsList HadrontherapyPhysics: this list relates to hadron-therapy.

/ gamos / physicsList GmEMExtendedPhysics: this list concerns subatomic particles, besons, leptons, mesons, barions, ions.

/ gamos / physicsList GmDNAPhysics: This physics list defines the physical processes and models to simulate the interactions of very low energy electrons (down to 7 eV) in water.

.... .

Then, you have to determine your generator by writing the following command.

/ gamos / generator GmGenerator

The generator allows you to choose the type of particle and combine any number of single particles or isotopes decaying into e +, e-, g, as well as choosing which distributions of time, energy, position and direction by the following commands:

3.2.2.1 Particle source

For a single source particle:

/gamos/generator/addSingleParticleSource SOURCE_NAME PARTICLE_NAME ENERGY.

For an isotope source:

/ gamos / generator / addIsotopeSource SOURCE_NAME ISOTOPE_NAME ACTIVITY.

...

3.2.2.2 Time distributions

There are 3 choices, Constant time, Time changing at constant interval, Decay time:

/ gamos / generator / timeDist SOURCE_NAME GmGenerDistTimeConstant TIME .

/gamos/generator/timeDist SOURCE_NAME GmGenerDistTimeConstantChange TIME_INTERVAL TIME_OFFSET.

/gamos/generator/ timeDist SOURCE_NAME GmGenerDistTimeDecay ACTIVITY LIFETIME.

...

3.2.2.3 Energy

It can be, Constant, BetaDecay, Gaussian, RandomFlat ...

/gamos/generator/energyDist SOURCE_NAME GmGenerDistEnergyConstant ENERGY.

/ gamos / generator / energyDist SOURCE_NAME GmGenerDistEnergy-BetaDecay.

/ gamos / generator / energyDist SOURCE_NAME GmGenerDistEnergy-Gaussian MEAN SIGMA.

/ gamos / generator / energyDist SOURCE_NAME GmGenerDistEnergyRandomFlat MIN_ENERGY MAX_ENERGY.

...

3.2.2.4 Position

It can be, at point, in a Geant4 volume, in a user defined volume, in steps along a line,in square, in a disc, in the voxels of a phantom(materials, structure ...) ... .

/gamos/generator/positionDist SOURCE_NAME GmGenerDistPositionPoint POS_X POS_Y POS_Z.

/gamos/generator/positionDist SOURCE_NAME GmGenerDistPositionInG4Volumes LV_NAME1 LV_NAME2.

/gamos/generator/positionDist SOURCE_NAME GmGenerDistPositionInUserVolumes POS_X POS_Y POS_Z ANG_X ANG_Y ANG_Z SOLID_TYPE SOLID_DIMENSIONS.

/gamos/generator/positionDist SOURCE_NAME GmGenerDistPositionLineSteps POS_X POS_Y POS_Z DIR_X DIR_Y DIR_Z STEP.

/gamos/generator/positionDist SOURCE_NAME GmGenerDistPositionSquare HALF_WIDTH POS_X POS_Y POS_Z DIR_X DIR_Y DIR_Z.

/gamos/generator/positionDist SOURCE_NAME GmGenerDistPositionDisc RADIUS POS_X POS_Y POS_Z DIR_X DIR_Y DIR_Z.

/gamos/generator/positionDist SOURCE_NAME GmGenerDistPositionPhantomVoxels.

It is also possible to create distributions where several of the four variables, are generated at the same time, so that they are related.

By using this minimum of commands described above we can run an example and also visualize the geometry by VRML, OpenGL and ASCII with this command:

/control/execute PATH_TO_MY_GAMOS_DIRECTORY/examples/visVRML2FILE.in

The main way to extract information of what is happening and modify the running conditions is **user action,**

/gamos/userAction MyUserAction

This user action feature allows you to add filters, and classifier in order to follow and focus on the particles as well as the processes you are interested in with this command:

For filters

/gamos/userAction USER_ACTION FILTER_NAME

/gamos/filter FILTER_NAME FILTER_CLASS PARAMETER_1 PARAMETER_2

For classifiers

/gamos/userAction USER_ACTION CLASSIFIER_NAME

/gamos/scoring/assignClassifier2Scorer CLASSIFIER_NAME SCORER_NAME

Next step consiste of attaching a sensitive detector to a volume, which used to creating hits (deposits of energy) each time a track traverses a sensitive volume and loses some energy.

Finally you can creat a score to calculate many quantities with or without error, in one or several volumes, for each scored quantity one of several filters can be used, only particles in a given volumen, and results can be displayed in a file or as a histogram.

3.3 Application

3.3.1 Attenuation study for photons of different energy (shielding)

To strengthen radiation protection, nuclear activities must be carried out in accordance with the fundamental principles to ensure the protection of man and the environment against the harmful effects of exposure to ionizing radiation, and since we are talking about medical nuclear medicine, The use of ionizing radiation for medical purposes contributes significantly to the exposure of the population. After natural exposure, this practice presents the first source of exposure of artificial origin, it is therefore recommended to control the doses, and to minimize the time, it is suggested to simulate the experiment before the practice.

We will be interested in the photon, in a midst, the fluence of photons decreases exponentially with the thickness of the material crossed. So, we have:

$$\Phi(x) = \Phi\ (0)e^{-\mu x} \tag{16}$$

As the Kerma is at any point proportional to the fluence of photons:

$$K(x) = K\ (0)e^{-\mu x} \tag{17}$$

For against, the absorbed dose is proportional to the photon fluence that when the electronic equilibruim is achieved in the material medium. In this case:

$$D(x) = D\ (0)e^{-\mu x} \tag{18}$$

Therefore the dose rate decreases exponentially with the thickness of the material traversed, so to protect ourselves from external exposure we must move far from the source and protect ourselves with shielding, where the role of simulation comes in determining the thickness necessary to attenuate these photons.

The attenuation study of a parallel beam of photons of diffetent energy which moves away from a distance of 1 m from the plate of material with Gamos is made by this user action: /gamos/userAction SHNthValueLayerUA, and we must stopped secondary particle, so that they will not be counted as particles coming out of the shield layers by this command /gamos/userAction GmKillAtStackingActionUA GmSecondaryFilter. This commands, allows us to studying penetration, and to establish the role of the shielding.

3.3.2 For photons of 150 Kev

Figures 4 and 5.

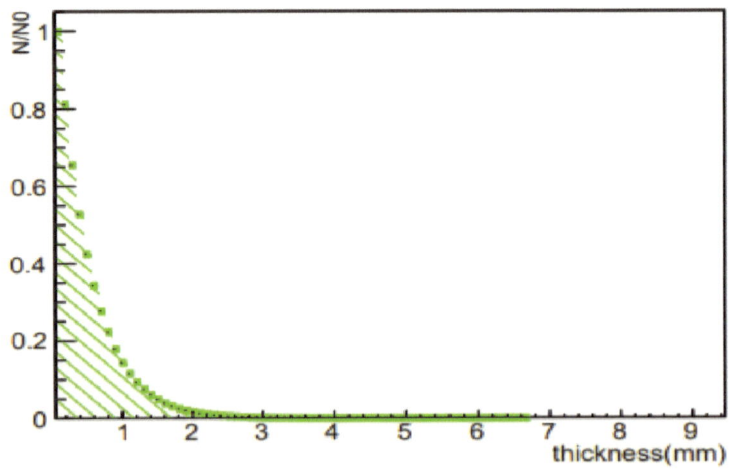

Figure 4.
Attenuation of E = 150 Kev photons by Pb.

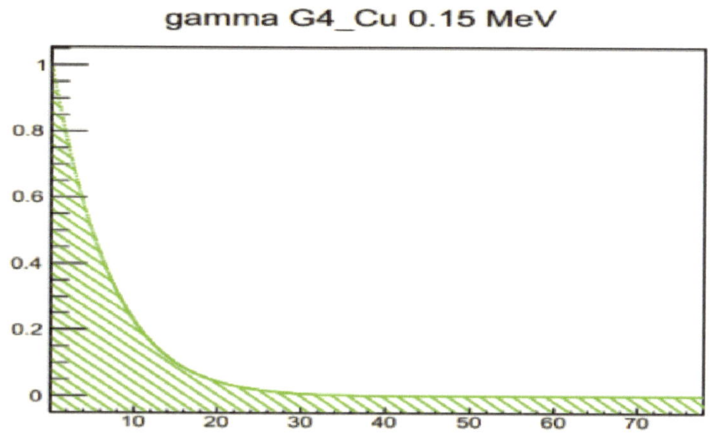

Figure 5.
Attenuation of E = 150 Kev photons by Cu.

3.3.3 For photon of 511 Kev

Figures 6 and 7.

3.3.4 For photon of 2 Mev

Figures 8 and **9**.

3.3.5 Interpretation

(**Figures 4–9**).

From the curves we can see that the element suitable for attenuating and absorbing photons is lead (you can try other materials than Cu), because it is less expensive, and it allows to have an optimal thickness compared to other materials, as well as for the facility of its control of the parameters of aquatic chemistry.

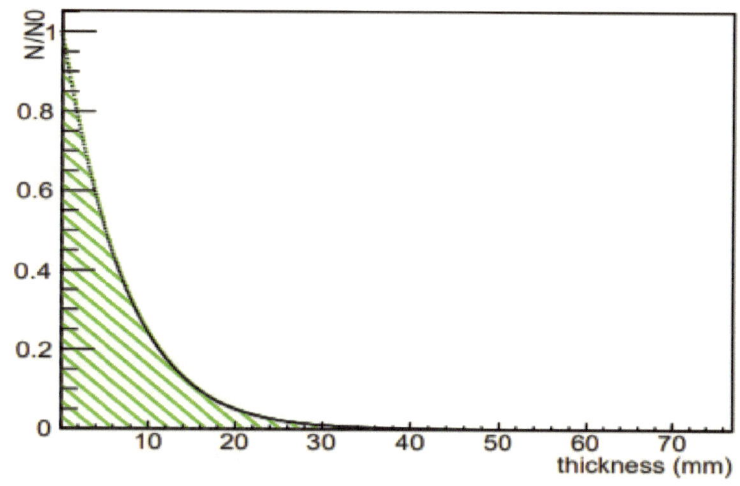

Figure 6.
Attenuation of E = 511 Kev photons by Pb.

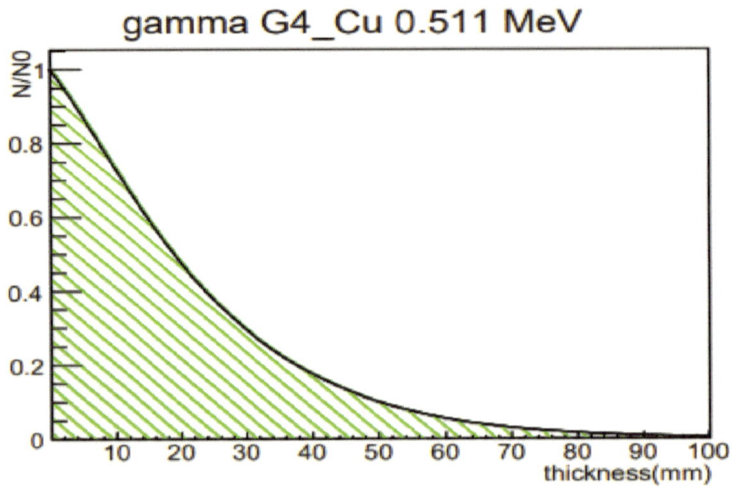

Figure 7.
Attenuation of E = 511 Kev photons by Cu.

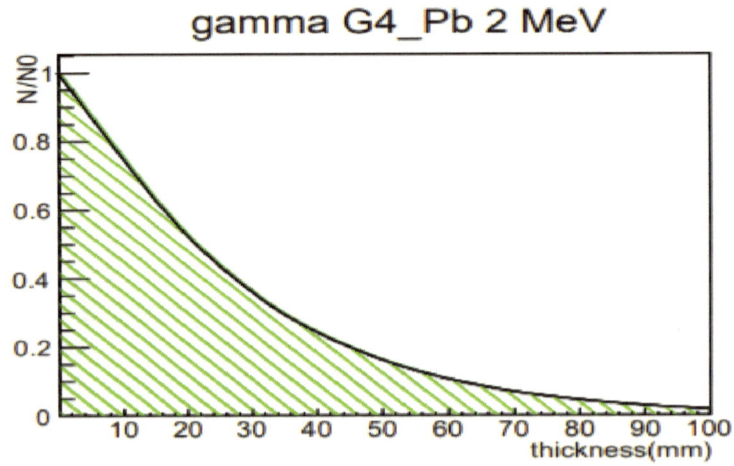

Figure 8.
Attenuation of E = 2Mev photons by Pb.

We can also Make an histogram of the energy spectrum of photons of energy 1 Mev and other types of particles that traverse the plate by using Gamos filters, and classifiers, with *GmClassifierByParticle:DataList FinalKineticEnergy* (**Figures 10–13**).

So we can remove the number of interactions that occur along the path of the particle in the material with this command */gamos/userAction GmCountProcessesUA*

Shielding calculations made by hand are often approximations, the most accurate are those performed by simulation, The thickness of the shielding needed depends on: Radiation Energy, the shield material, and Radiation intensity, So the lower the energy of the gamma rays, the easier it is to shield of them, and high energy gamma rays sometimes determine shielding requirements.

For gamma rays, the higher the atomic number of the shield material, the greater the attenuation of the radiation.

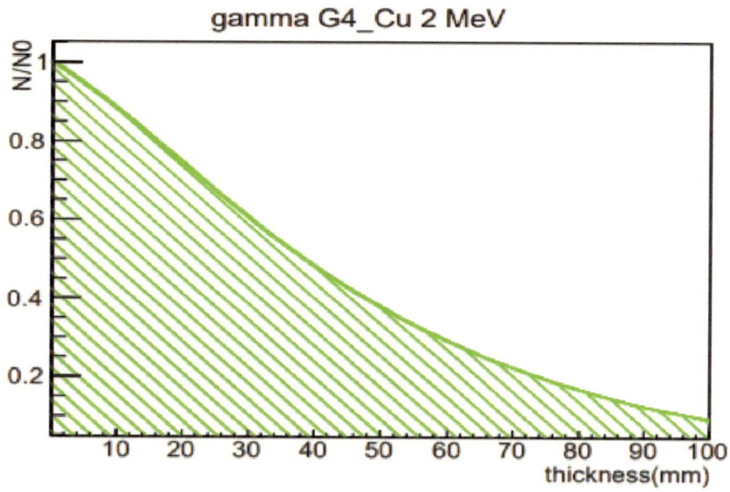

Figure 9.
Attenuation of E = 2Mev photons by Cu.

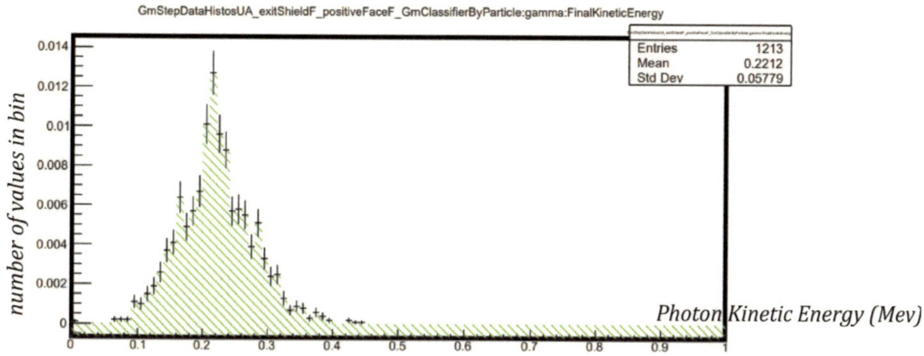

Figure 10.
Energy spectrum of photons that traverse the plate.

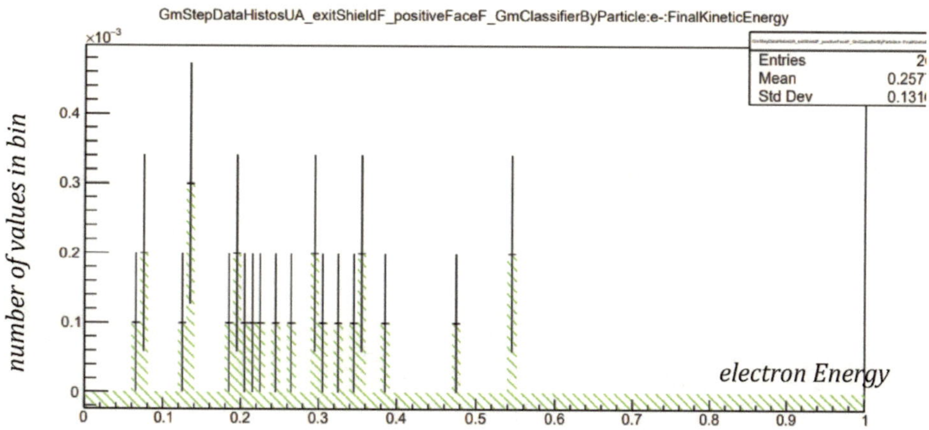

Figure 11.
Energy spectrum of electron.

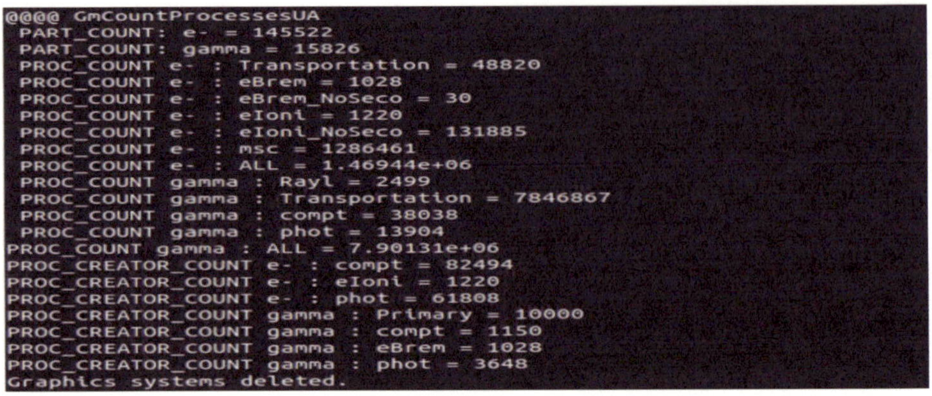

```
@@@@ GmCountProcessesUA
 PART_COUNT: e- = 145522
 PART_COUNT: gamma = 15826
 PROC_COUNT e- : Transportation = 48820
 PROC_COUNT e- : eBrem = 1028
 PROC_COUNT e- : eBrem_NoSeco = 30
 PROC_COUNT e- : eIoni = 1220
 PROC_COUNT e- : eIoni_NoSeco = 131885
 PROC_COUNT e- : msc = 1286461
 PROC_COUNT e- : ALL = 1.46944e+06
 PROC_COUNT gamma : Rayl = 2499
 PROC_COUNT gamma : Transportation = 7846867
 PROC_COUNT gamma : compt = 38038
 PROC_COUNT gamma : phot = 13904
PROC_COUNT gamma : ALL = 7.90131e+06
PROC_CREATOR_COUNT e- : compt = 82494
PROC_CREATOR_COUNT e- : eIoni = 1220
PROC_CREATOR_COUNT e- : phot = 61808
PROC_CREATOR_COUNT gamma : Primary = 10000
PROC_CREATOR_COUNT gamma : compt = 1150
PROC_CREATOR_COUNT gamma : eBrem = 1028
PROC_CREATOR_COUNT gamma : phot = 3648
Graphics systems deleted.
```

Figure 12.
Number and type of processes that occur when a number $N = 10^4$ of photons passes through the material.

3.4 PET scanner

PET imaging combined with CT scanning allows two examinations to be performed simultaneously: the PET examination studies the biological activity of organs, while the CT examination studies the anatomy and morphology of organs. The objective of this examination is to detect anomalous organic activities, by injecting the patient with low-level radioactive glucose, the radiation dose is very low, and does not represent a risk for the patient and his entourage. The injected product is a weakly radioactive marker (derived from glucose marked by Fluorine 18), which will be fixed on the organs, with a preference for the organs that work more, the radioactive marker to highlight the biological activity. This examination is performed on a hybrid machine consisting of two devices:

- PET scanner that records the radiation emissions of the injected product.
- X-ray CT which allows to obtain anatomical images

Quantitative reconstruction of PET (Positron emission tomography) data with GAMOS (Monte Carlo) needs to have knowledge of the scanner geometry (**Figure 13**). Both typical clinical and preclinical scanners use a block-type geometry. Many rectangular blocks of crystals are arrayed in regular polygons. Some of these polygons are arranged along the axis of the scanner, and Monte Carlo simulation remains an essential tool to help design new medical imaging devices, and to know what is happening in the simulation.

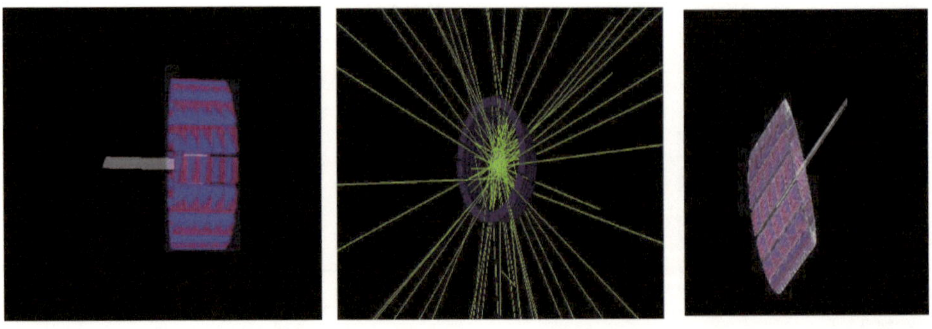

Figure 13.
Geometry realized by GAMOS. Three-dimensional (3D) acquisition with block description, and ^{18}F source.

With this user action we can obtain information about the physics process that occur in a PET scanner with all particle:

/gamos/userAction GmCountTracksUA

/gamos/userAction GmCountProcessesUA

/gamos/userAction GmHistosGammaAtSD

*/gamos/analysis/histo1Max *Energy* 1*MeV*

*/gamos/analysis/histo1Max *Pos* 200*mm*

/gamos/userAction GmTrackDataHistosUA GmPrimaryFilter

We will obtain results about track (**Figure 14**) and procces that occur for each particle in terminal and we can save them in file, for example

/gamos/userAction GmHistosGammaAtSD, this command give us an idea about the interaction of original gammas in the sensitive detector (**Figure 15**).

So we have 901 total number of events:

55.66% of 'original' gammas reaching one sensitive detector,

94.21% with photoelectric interaction in SD,

37.67%with photoelectric interaction and no Compton interactions,

38,51, with photoelectric interaction and one Compton interaction,

16.29% with photoelectric interaction and two Compton interaction,

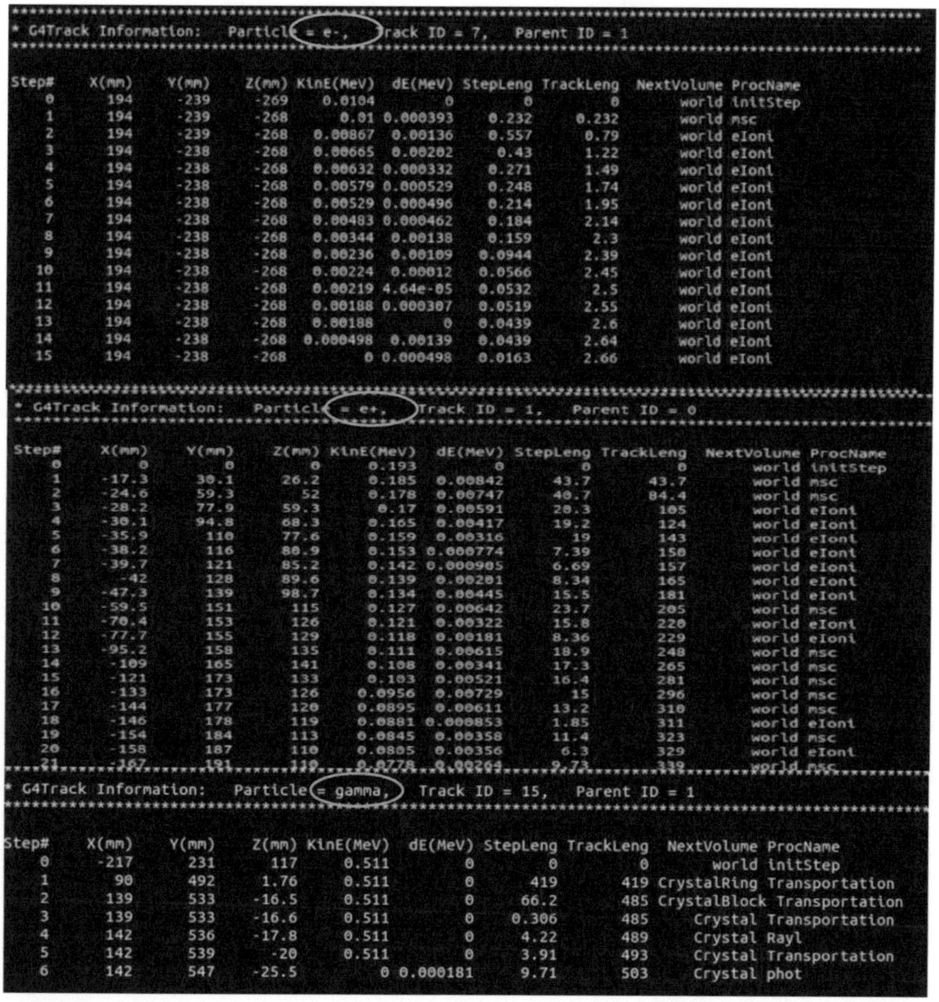

Figure 14.
Information about track of the job in interactive running.

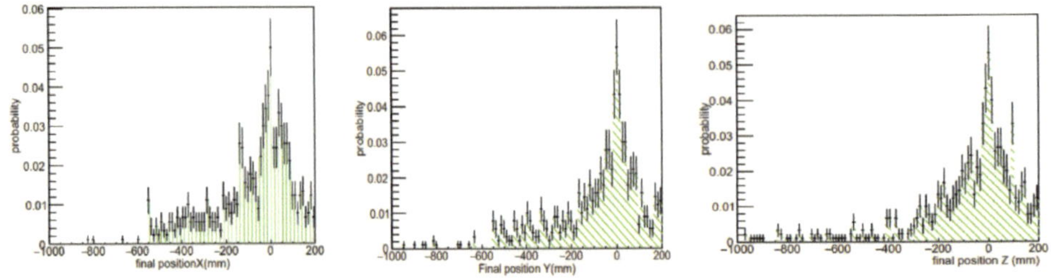

Figure 15.
Final position X ,Y,Z obtain with GAMOS.

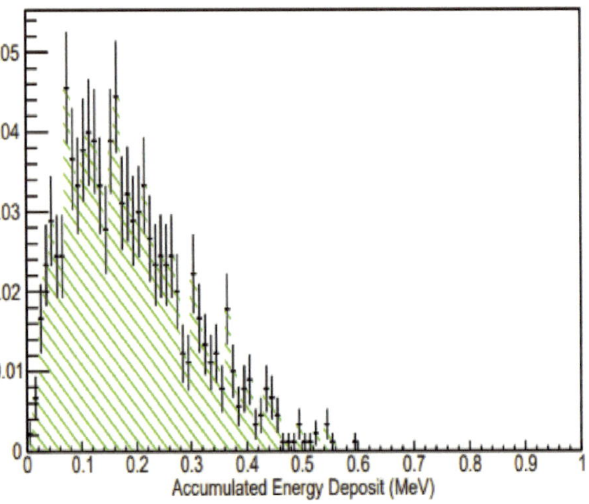

Figure 16.
Accumulated Energy deposit.

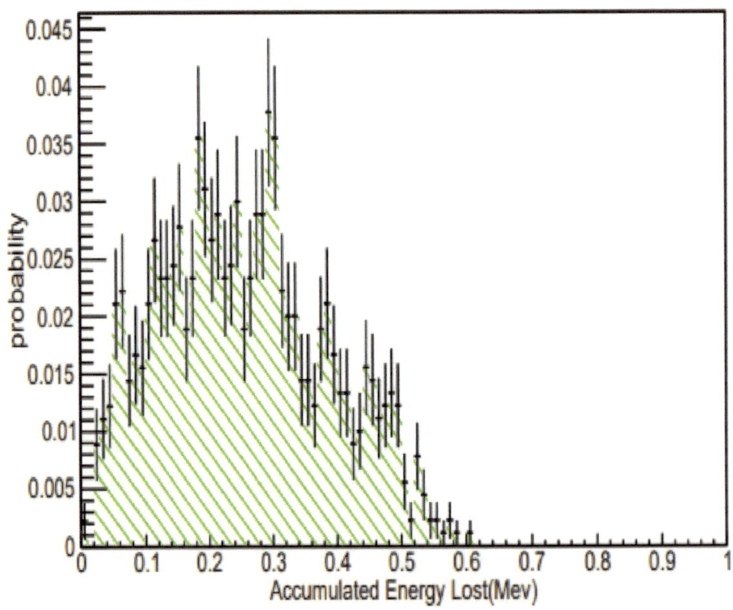

Figure 17.
Accumulated energy lost.

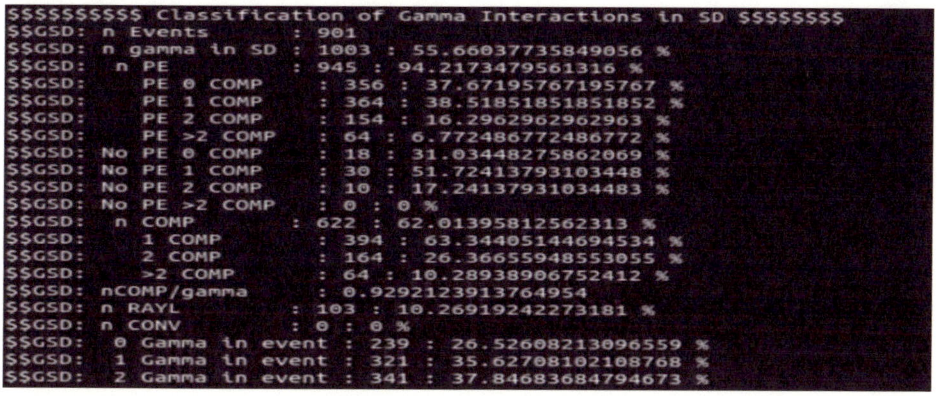

```
$$$$$$$$$$ Classification of Gamma Interactions in SD $$$$$$$$
$$GSD: n Events            : 901
$$GSD: n gamma in SD  : 1003 : 55.66037735849056 %
$$GSD:   n PE              : 945 : 94.2173479561316 %
$$GSD:      PE 0 COMP        : 356 : 37.67195767195767 %
$$GSD:      PE 1 COMP        : 364 : 38.51851851851852 %
$$GSD:      PE 2 COMP        : 154 : 16.2962962962963 %
$$GSD:      PE >2 COMP       : 64 : 6.772486772486772 %
$$GSD: No PE 0 COMP          : 18 : 31.03448275862069 %
$$GSD: No PE 1 COMP          : 30 : 51.72413793103448 %
$$GSD: No PE 2 COMP          : 10 : 17.24137931034483 %
$$GSD: No PE >2 COMP         : 0 : 0 %
$$GSD:   n COMP            : 622 : 62.01395812562313 %
$$GSD:      1 COMP           : 394 : 63.34405144694534 %
$$GSD:      2 COMP           : 164 : 26.36655948553055 %
$$GSD:      >2 COMP          : 64 : 10.28938906752412 %
$$GSD: nCOMP/gamma           : 0.9292123913764954
$$GSD: n RAYL              : 103 : 10.26919242273181 %
$$GSD: n CONV                : 0 : 0 %
$$GSD:   0 Gamma in event  : 239 : 26.52608213096559 %
$$GSD:   1 Gamma in event  : 321 : 35.62708102108768 %
$$GSD:   2 Gamma in event  : 341 : 37.84683684794673 %
```

Figure 18.
Accumulated energy lost.

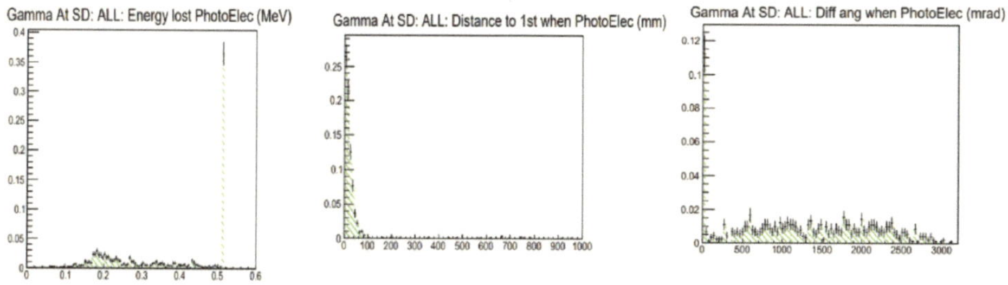

Figure 19.
Information about interaction of gammas in the sensitive detector.

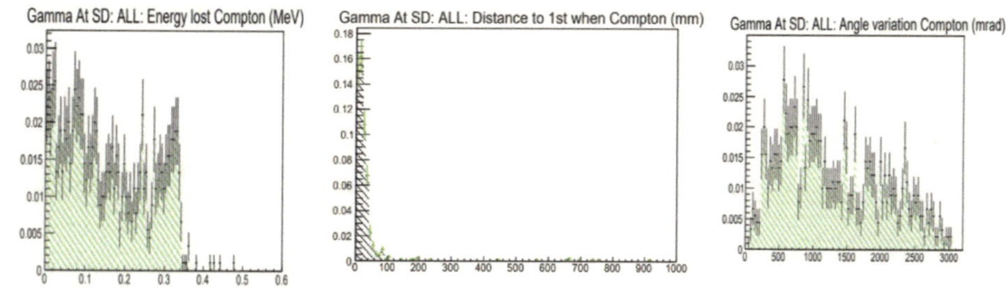

Figure 20.
Informations about Compton effect.

6.7% with photoelectric interaction and more than two Compton interaction,
31,03% with no photoelectric interaction and no Compton interaction.
Also we can get more details from the histrograms for examples:

Figures 16–20 show the different parameters and details after reaction generated by Gamos based on random number generator (Section 2.2.4).

4. Conclusion

To conclude, Monte Carlo simulation facilitates the experiment and minimizes the time to process the phenomena, so we can go as far as studying the treatment using proton therapy because if we use a cancer treatment with photons, there will be the maximum amount of X-rays, so there will be a maximum dose delivered to the area to be treated (tumor volume), but also there will be a certain level of dose around this volume which are the organs at risk, and which are the tissues that should not be irradiated and which are, so the great advantage of protons is to have a delicate dose, almost zero once the target is reached, because the protons deposit their energy locally, so they generate less complication, and allows to decrease the risk of having a radiation-induced carcinogenesis. Proton therapy has an extremely important indication, which makes it possible to prevent proton therapy, and the Monte Carlo method remains a very powerful tool that makes it possible to improve research in this field by going as far as microdosimetry.

Author details

Omaima Essaad Belhaj[1*], Hamid Boukhal[1] and El Mahjoub Chakir[2]

1 Laboratory of Radiation and Nuclear Systemes, University Abdelmalek Essaadi, Tetouan, Morocco

2 Laboratory of Materials Subatomic Physics, University Ibn Tofail Kenitra, Morocco

*Address all correspondence to: omaimaessaad.belhaj@gmail.com

References

[1] "La visualisation et demain," 1986.

[2] O. N. Vassiliev, *Monte Carlo Methods for Radiation Transport: Fundamentals and Advanced Topics*. 2016.

[3] S.M. *Seltzer, Calculated response of intrinsic Germanium detectors to narrow beams of photons with energies up to - 300 keV*, Nucl. Instr. Methods 188 (1981).

[4] G. Alm *Carlsson, C.A. Carlsson, K-F. Berggren and R. Ribberfors, Calculation of scattering cross-sections for increased accuracy in diagnostic radiology: I. Energy broadening of Compton scattered photons*, Med. Phys. 9 (1982).

[5] DR. Dance, The Monte Carlo calculation of integral radiation dose in xeromammography, Phys. Med. Biol. 25 (1980)

[6] C.S. Chen, K. Doi, C. Vyborny, H-*P. Chan* and G. Holje, Monte Carlo simulation studies of detectors used in the measurement of diagnostic X-ray spectra, Med. Phys. 7 (1980).

[7] "Jean-No¨ el Badel 9 septembre 2009," 2009.

[8] GAMOS Collaboration, "GAMOS User 's Guide, release 6.1.0," p. 336, 2019.

Chapter 6

Reliability and Comparison of Some GEANT4-DNA Processes and Models for Proton Transportation: An Ultra-Thin Layer Study

Gabriela Hoff, Raquel S. Thomaz, Leandro I. Gutierres, Sven Muller, Viviana Fanti, Elaine E. Streck and Ricardo M. Papaleo

Abstract

This chapter presents a specific reliability study of some GEANT4-DNA (version 10.02.p01) processes and models for proton transportation considering ultra-thin layers (UTL). The Monte Carlo radiation transport validation is fundamental to guarantee the simulation results accuracy. However, sometimes this is impossible due to the lack of experimental data and, it is then that the reliability evaluation takes an important role. Geant4-DNA runs in an energy range that makes impossible, nowadays, to perform a proper microscopic validation (cross-sections and dynamic diffusion parameters) and allows very limited macroscopic reliability. The chemical damage cross-sections reliability (experiment versus simulation) is a way to verify the consistency of the simulation results which is presented for 2 MeV incident protons beam on PMMA and PVC UTL. A comparison among different Geant4-DNA physics lists for incident protons beams from 2 to 20 MeV, interacting with homogeneous water UTL (2 to 200 nm) was performed. This comparison was evaluated for standard and five other optional physics lists considering radial and depth profiles of deposited energy as well as number of interactions and stopping power of the incident particle.

Keywords: Geant4-DNA, Monte Carlo methods, Proton transportation, Ultra-thin layer, Software reliability

1. Introduction

The Monte Carlo toolkit Geant4 [1–3] was developed as a general-purpose transportation toolkit. This toolkit has a framework that extends the transport process to model the early biological damage induced by ionizing radiation at cellular and sub-cellular scale [4–6], the so called Geant4-DNA [4–6], that makes possible to simulate the physical–chemical and chemical processes for water radiolysis, the molecular geometries, and the damage quantification. This framework can simulate energies from 10 eV to 100 MeV for protons, enabling the simulation of particles' interaction using discrete models at nanoscale. It is also known and well informed on Geant4 manual "Guide For Physics Lists" that simulation of transport for energies below 1 keV reduces significantly the accuracy of the transport models [7]. However, to get a simulation in the necessary scale for Geant4-DNA it is inevitable to consider energies below 1 keV. It allows to simulate, depending on the interacting particle and energy range, the following processes (applying different possible models): elastic scattering, ionization, excitation, electron capture, nuclear scattering, charge increase and decrease, attachment and vibrational excitation [8].

The validation (macroscopic and microscopic) of the results by comparing the Geant4-DNA cross sections or simulated quantities to experimental data is still extremely limited, considering the energy range used by Geant4-DNA, which makes important to be careful on generalizing the simulation results.

In this chapter simulations results of 2 MeV kinetic energy protons impinging on homogeneous water ultra-thin layers (UTLs), using different physics lists (including Geant4-DNA running on version 10.02.p01) are presented. The reliability evaluation was performed considering chemical damage cross section (CDCS) and stopping power (SP). The comparison among different Geant4 recommended physics lists was based on radial and depth deposited energy profiles, number of interactions and SP.

2. The experimental and simulation definitions

In this section the experimental setup, the developed application for the simulation and the results are presented. The strategy used to evaluate the reliability of the simulation for each physics list was performed comparing simulated-calculated to experimental CDCS, and simulated SP to NIST database. The physics lists evaluation was based on the comparison of the results of interaction files generated, that registered several information on each simulation step.

2.1 The experimental setup for chemical damage cross section estimation

The experimental setup used on reliability evaluation was defined to collect the CDCSs using polymer ultra-thin films. High-grade poly(methyl methacrylate) (PMMA) with density 1.190 g/cm^3 and poly(vinyl chloride)(PVC) with density 1.406 g/cm^3 powder were dissolved and spun onto polished silicon (Si) wafers. Homogeneous ultra-thin films, with thicknesses from 4 nm to 200 nm, and very low roughness ($\sim$0.3 nm RMS) were obtained. The films were bombarded by 2 MeV H^+ in vacuum at a HVEE 3 MV Tandetron (Porto Alegre, Brazil) with a set of fluences ranging from 10^{14} ions/cm^2 to 2.8x10^{15} ions/cm^2. X-ray photo-electron spectroscopy (XPS) was performed on the irradiated samples at Universit de Namur, Belgium, to evaluate bond-breaking cross sections of C=O and C-Cl bonds as a function of the thickness of the polymer. The radiolytic efficiency is usually estimated measuring CDCSs for different transformation processes induced by radiation such as bond-breaking [9–11]. These CDCSs for bond-breaking represent the energy loss by length (dE/dx) [12, 13] and are based on the number of specific bond-breaking at the end of an irradiation process. Additional information about the experimental data collection can be found at [14].

2.2 The Monte Carlo simulation

This Geant4-DNA application (version 10.02.p01) was developed considering a protons beam impinging normally on the entrance surface of an ultra-thin layer (UTL) of water in a semi-infinite configuration of 500 nm per 500 nm with thicknesses from 2 nm to 200 nm and a water 500 nm substract. The simulated protons beams were monodirectional and monochromatic with initial kinetic energies of 2 MeV, 5 MeV, 10 MeV and 20 MeV.The CDCS evaluation was performed only for the 2 MeV incident protons beam, while the SP one was performed for 2 MeV, 5 MeV, 10 MeV and 20 MeV protons beams. For each UTL and beam energy 10^5 histories were simulated, taking into account a cut-off of 1 nm for secondary particle generation. According to Geant4-DNA official webpage the Geant4-DNA processes are all discrete; as such, they simulate explicitly all interactions and do not use any production cut, so this 1 nm cut will have no effect on the Geant4-DNA Physics results [8]. The class G4EmDNAPhysics (henceforth named DNA) and the other five available physics lists (named DNAopt1 to DNAopt5) were evoked for each setup configuration for both reliability and comparison studies. So, to enlighten the physics lists evoked to transport protons and electrons, the processes and models used on Geant4-DNA classes are presented on **Figures 1** and 2.

To simulate the processes and models above cited, additional electromagnetic physics builders are needed and, to support the simulation, the Livermore physics list was implemented by default [15].

Since Geant4-DNA only simulates standard liquid water as interaction material, the only way to explore situations close to the experimental setup was by altering the water density. So, different CDCSs were simulated-calculated using water with different densities. In addition to standard liquid water, composed by 2 hydrogen and 1 oxygen with density of 1 g/cm^3, a "dense water" of the same composition but with a 1.190 g/cm^3 density was considered.

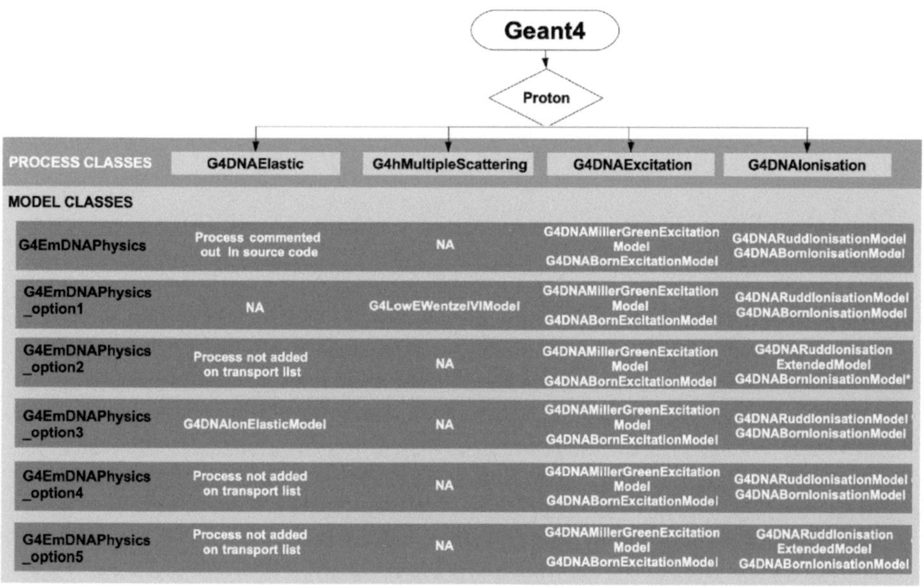

Figure 1.
*Scheme of the processes and models for protons transport and different physics lists. The symbol * indicates that the flag "SelectFasterComputation" was activated. The G4DNAChargeDecrease class always evoke the G4DNADingfelderChargeDecreaseModel class, so it was nod added to the scheme.*

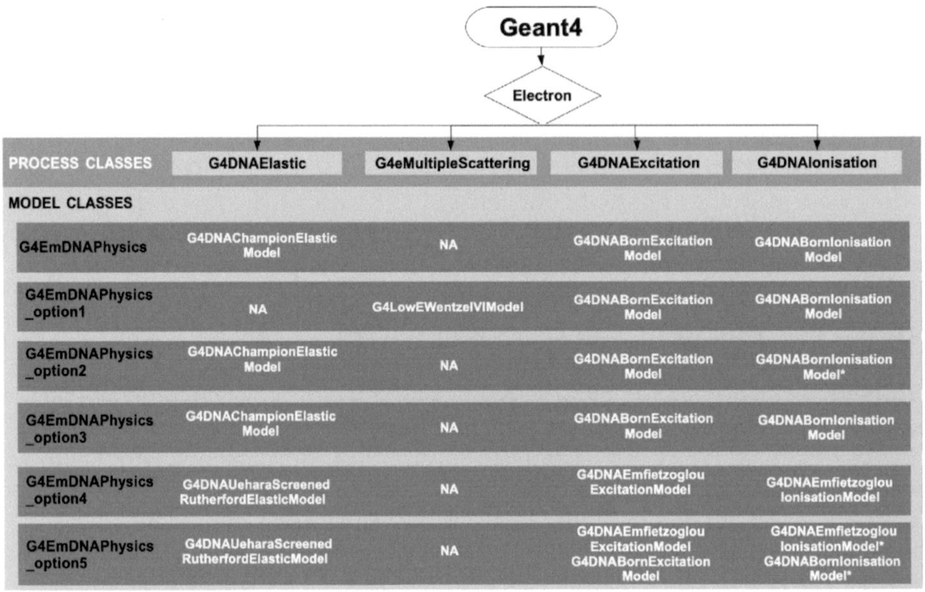

Figure 2.
*Scheme of the processes and models for electron transport and different physics lists. The symbol * indicates that the flag "SelectFasterComputation" was activated. The G4DNAAttachment and G4DNAVibExcitation classes always evoke respectively the G4DNAMeltonAttachmentModel and the G4DNASancheExitationModel classes, so it was nod added to the scheme.*

The total number of histories, the total deposited energy and its statistical fluctuations were recorded at the end of each run. Also, during the simulation, for each interaction, the following information were recorded: pre and post-step position (x, y and z, in nm); deposited energy due to the interaction (in MeV); event, parent particle, track and step identification; process name and particle type. Later, these information were organized/accumulated on bins representing radial deposited energy profiles (henceforth called radial profile) and SP obtained with different physics lists.

To estimate the CDCS, radial profiles based on the position information recorded for 2 MeV incident kinetic energy protons beam and thicknesses from 2 nm to 200 nm were generated. The SP for protons was calculated considering the total energy deposited in each UTL divided by its thickness.

2.3 The reliability evaluation

On this subsection the methodological strategy used and the results for the reliability evaluation are presented.

The CDCS was calculated based on the standard thermally activated model (STAM) [16] taking into account the radial deposited energy profile simulated to generate the probability energy deposition function which was adjusted to estimate the activation energy density value for a specific bond-break in N positions [14]. The simulated-calculated CDCS was compared to experimental ones for 2 MeV H^+ on PMMA and PVC ultra-thin films.

The SP profile as function of the UTL thickness was evaluated by the fitting curve considering all thicknesses for each incident kinetic energy protons beam and each physics list. The fitting TamLog ($y = a + b*\ln(sing*(x-c))$) for all curves presented R-squared coefficient larger than 0.99. The radial profile, mainly formed by secondary electrons, was used to define the simulated electron range which was compared to the CSDA ESTAR electrons range [17]. The simulated SP is a microscopic quantity since the largest water UTL thickness is smaller than the expected electron range which is a limitation in this comparison, but unfortunately it was not possible to find an experimental SP database valid for ultra-thin layers. This comparison was performed using the extrapolation of the fitting TamLog curve considering the macroscopic CSDA ESTAR electrons range as a limit. The NIST CSDA ESTAR electrons range was defined based on the calculated maximum kinetic energy (K_{max}) that can be transferred in a head-on collision with an atomic electron. The Eq. (1) [18] depends on the relativistic velocity parameter of the incident proton ($\beta = \frac{v_{incidentparticle}}{c}$) and the rest-mass energy of the scattered electron (m_0c^2).

$$K_{max} \simeq 2m_0c^2\left(\frac{\beta^2}{1-\beta^2}\right) \tag{1}$$

This equation assumes that the electrons are unbound and is applicable for an incident heavy particle with kinetic energy smaller than its rest-mass energy M_0c^2, which is the condition applied to the study case presented in this chapter. The K_{max} was used as input parameter to estimate the CSDA range from ESTAR database [17], using a log–log interpolation to calculate the data not presented on the database grid. The CSDA ranges estimated from ESTAR were considered maximum limits taking into account the theoretical limitation of the interaction with unbounded electrons. This overestimates electrons range value and limits the comparison between the microscopic simulated range, based on SP, and the macroscopic ESTAR CSDA range. Both conditions underestimate the simulated electrons range and if the simulated value is larger than the ESTAR value, the former is unreliable.

2.3.1 Reliability results based on chemical damage cross section

On this subsection the reliability for the CDCS and the SP considering different Geant4-DNA physics lists are presented and analysed. **Figures 3** and **4** present the

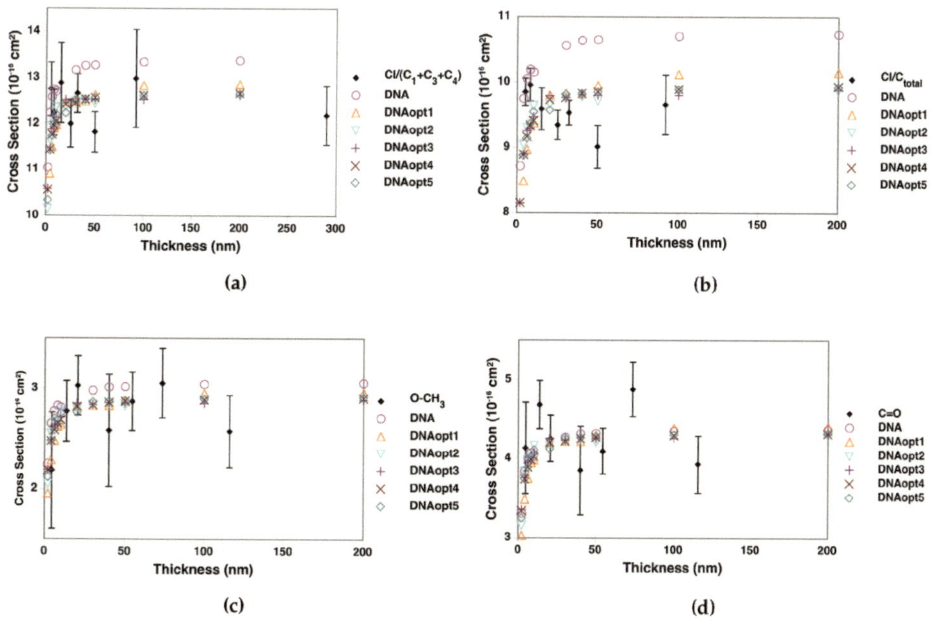

Figure 3.
Chemical damage cross section as function of the UTL thickness considering 2 MeV, bin size of 1 nm and standard water density for bonds Cl/(C_1+C_3+C_4) (a), Cl/C_{total} (b), O-CH_3 (c) and O=C (d) and all physics lists.

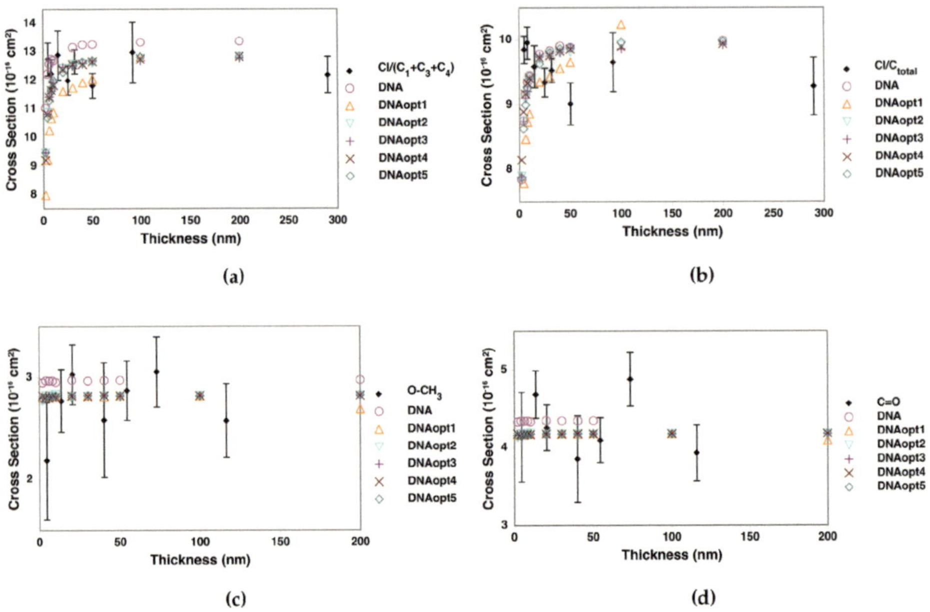

Figure 4.
Chemical damage cross section as function of the UTL thickness considering 2 MeV, bin size of 0.2 nm and standard water density for bonds Cl/(C_1+C_3+C_4) (a), Cl/C_{total} (b), O-CH_3 (c) and O=C (d) and all physics lists.

experimental and simulated data for CDCS and **Table 1** presents the activation energy density considering different bond-breaking for 2 MeV kinetic energy protons beam. Adapting the activation energy published by [10] to the conditions used

	0.2 nm		1.0 nm	
Condition	**Cl/(C_1+C_3+C_4)**	**Cl/C_{total}**	**Cl/(C_1+C_3+C_4)**	**Cl/C_{total}**
water_DNA	2.21	2.60	4.71	5.12
dense_water_DNA	3.14	3.70	6.07	6.65
water_DNAopt1	2.33	2.73	4.69	5.13
water_DNAopt2	2.21	2.62	4.82	5.29
water_DNAopt3	2.21	2.61	4.82	5.29
water_DNAopt4	2.53	2.98	5.10	5.60
water_DNAopt5	2.53	2.98	5.11	5.60
Condition	O-CH_3	C=O	O-CH_3	C=O
water_DNA	26.45	13.01	7.53	6.84
dense_water_DNA	34.26	17.45	9.62	8.66
water_DNAopt1	27.37	13.91	7.43	6.69
water_DNAopt2	28.32	14.41	7.63	6.87
water_DNAopt3	28.34	14.42	7.64	6.87
water_DNAopt4	29.13	14.82	8.09	7.28
water_DNAopt5	29.08	14.80	8.08	7.28

Table 1.
Calculated values of activation energy density (ε_O), in eV/nm^3, for each bond-break situation and condition simulated.

in this study case, the activation energy to get a reliable result must be in the range from 1 eV/nm^3 to 10 eV/nm^3. To analyse the influence of the radial profile step size on CDCS, these profiles were organized in steps of 0.2 nm and 1.0 nm for all studied cases. These two different bins were defined to explore the influence of the extremely strong slope in the first 3 nm of the radial profile curve, where the deposited energy is reduced approximately between 8% and 13% of the total deposited energy, depending on the kinetic energy protons beam.

The CDCSs for water, 1 nm bin size and all transport models for each ultra-thin layer are presented on **Figure 3**. On this figure it is visible that most of the transport models showed the same profile for CDCS as function of the UTL thickness, with the exception of Cl bond-breaks (**Figure 3a** and **b**) and DNAopt1 that showed similar tendency but different amplitude. A similar behaviour for different physics lists can be seen on **Figure 4**, including the DNAopt1 discrepancy observed on CDCSs for 0.2 nm bin size and all physics lists. However, for a complete reliability evaluation it is important to take into account the activation energy used to get the best fitting curve on the estimation of the simulated CDCS (**Table 1**).

Still considering the 0.2 nm bin (**Figure 4**), there is a visible difference on the activation energy values when compared to 1 nm bin. As one can see, the results of activation energy are out of the reliability range presented by [10] for the bonds O-CH_3 and O=C and 0.2 nm bin (**Table 1**). However, it is important to notice that these results are dependent on the accentuated slope of the radial profile discussed on subsection 2.2 which leads to the condition that small changes on the bin size may result in a large change on the activation energy. It is necessary to take this observation into account on further evaluations and to use the most conservative methodology to guarantee the reliability of the results. In this chapter, the total deposited energy calculated for 1 nm bin are reliable because this results presented smaller statistical fluctuations than the ones calculated for 0.2 nm bin, keeping the consistency for the activation energy calculated value.

The dense water when compared to standard water presented, in general, lower CDCSs values, as was expected, due to the increase on this material density.

The activation energies defined to get the best fitting presented on **Table 1** showed values larger than 10 eV/nm^3, out of the reliability range, for bin size 0.2 nm and bond-breaks O-CH3 and C = 0. However, for bin size 1 nm (**Figure 3**) all activation energies evaluated for all bond-breaks are in the reliability range. This significant difference in the activation energy shows the dependency of the CDCS on the bin size defined to generate the radial profile. This happens due to the accentuated slope on the simulated radial profile (**Figure 7b**) where most of the deposited energy is absorbed in the first 5 nm. Because of that, the interpolation method used to integrate the radial deposited energy and its agreement with the simulated data are fundamental.

Another important consideration about CDCS is the shape of the curves for different bin sizes and same bond. In this cases, specially the O=C and O-CH_3 (**Figure 3c**, **d**, **4c** and **d**), the changes on the curve shapes are visible, where 0.2 nm bin presented a flat shape curve which is less reliable based on the experimental data.

To evaluated the effect of material density on CDCS and to observe a condition closer to the experiment setup (PMMA material), the data obtained with dense and standard water were compared to the experimental data. It is visible that the standard water data presented only one case (**Figure 5c**) out of the region defined by the error bars of the experimental CDCS, Cl/C_{total}. Considering the activation energy (**Table 1**) one may see that the values presented by dense water DNA were always larger than the ones presented by standard water DNA. Despite the differences, both descriptions of water presented activation energies in the reliability range, however, taking into account the standard water that presented one case

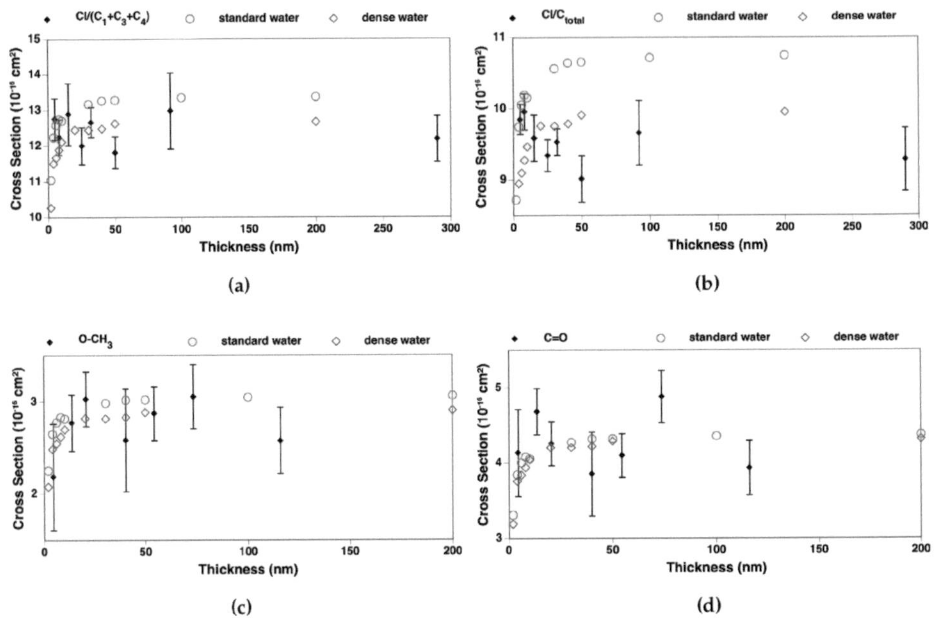

Figure 5.
Chemical damage cross section as function of the UTL thickness considering 2 MeV, different water description, bin size of 1 nm and bonds $Cl/(C_1+C_3+C_4)$ (a), Cl/C_{total} (b), O-CH_3 (c) and O=C (d) simulated with DNA physics list.

(**Figure 5c**) out of the error bars of the experimental CDCS, one may assume that the dense water may be a better option to describe PMMA on simulations.

2.3.2 Reliability based on stopping power

Figure 6 presents the SP values as function of the UTLs for all evaluated energies. All values of SP based on simulated deposited energy were sub-estimated, as expected, since the thickness of the UTLs were inferior to the electron range. According to ESTAR - of National Standards and Technology [17] the observed tendencies for SP at 2 MeV incident protons will achieve the (macroscopic) value for thicknesses around a few hundred nm. For energies of 5 MeV, 10 MeV and 20 MeV differences of 8%, 10% and 18%, respectively, were observed. One can see that the percentage difference increased with the increase on proton incoming energy. The studied cases were in the domain of UTLs, which means that the layers were not thick enough to reach stability on the energy depth profile. This comparison has limitations since the values published by [17] are macroscopic measurements. Since there were no SP data available on literature in the simulated conditions presented on this chapter, the strategy was to compare the data considering ESTAR [17] value as a limit to the tendency curve evaluated as electron range (**Table 2**). Values above this limit were considered inconsistent for the simulation.

As can be seen on **Table 2**, only the electron range presented by DNAopt1 is above the macroscopic limit turning this the unique physics list that can be considered unreliable. Another important observation is that DNAopt2 and DNAopt3, and DNAopt4 and DNAopt5 presented similar electrons ranges due to the similarity on their transport models for electrons in energy range of this simulation (**Figure 2**).

The DNAopt1 simulates more electrons interactions (increasing the running time) than DNAopt4 and DNAopt5 that were the fastest among all physics lists

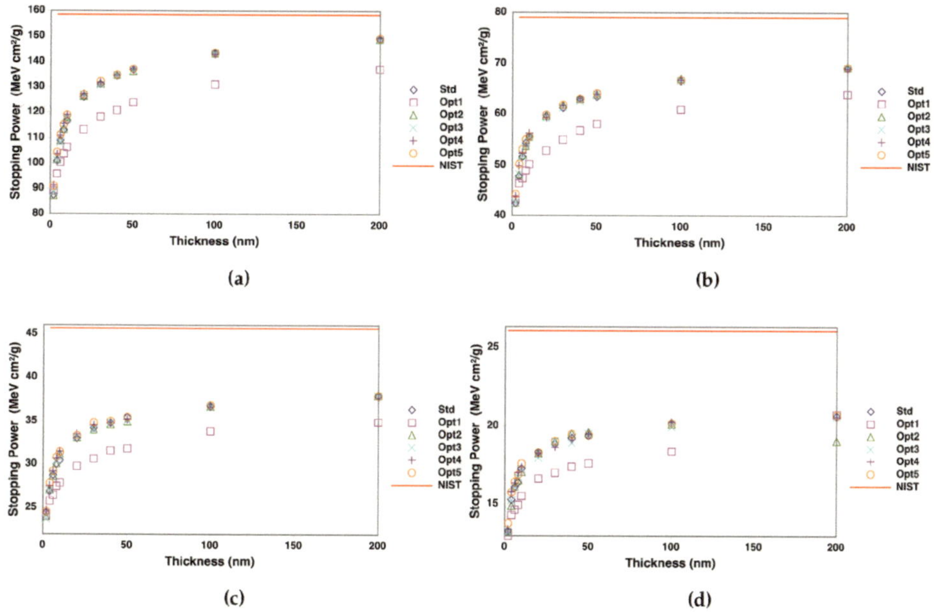

Figure 6.
Stopping power behaviour as function of the UTL thickness for different incident energy: 2 MeV (a), 5 MeV (b), 10 MeV (c) and 20 MeV (d). The red line represents the expected value of stopping power from PSTAR-NIST [19].

K_{proton}	2 MeV	5 MeV	10 MeV	20 MeV
K_{max}	4.34 keV	10.92 keV	21.90 keV	44.03 keV
std	380.41	1336.06	2917.06	10226.08
opt1	1409.78	4556.32	14524.03	7495.98
opt2	393.53	1246.24	3573.87	28166.71
opt3	385.60	1300.47	3911.23	13049.32
opt4	416.72	1513.44	4586.29	19025.65
opt5	415.67	1625.90	4281.31	15316.91
NIST	**581.61**	**2937.84**	**10060.36**	**34562.29**

1. *K_{proton} represents the incident protons kinetic energy that will interact with the atomic electrons.*
2. *K_{max} represents the maximum kinetic energy that can be transferred in a head-on collision with an atomic electron.*

Table 2.
Electrons range estimated with the simulated data and defined using NIST ESTAR [17].

evaluated. It was observed, for the simulated energies, that DNAopt1 was 1.5 to 4.1 more time consuming than DNA.

The tendencies, similarities and differences showed by the results obtained with the different physics lists can be explained reporting to the scheme on **Figure 2**. In this Figure one can see that: * DNAopt1 evokes the multiple scattering class for electrons instead of the elastic scattering (DNA) class for electrons; * DNAopt2 and DNAopt3 evoke the same process and model classes with the only change on the ionization model that had the flag SelectFasterComputer activated on DNAopt2; * DNAopt4 and DNAopt5 evoke similar process and model classes with exception for energy above 10 keV where additional models were evoked for excitation and ionization processes and the flag SelectFasterComputer was activated on DNAopt5.

2.4 The physics lists comparison

On this subsection the methodological strategy used and the comparisons among results of deposited energy profiles, total deposited energy and number of interactions for different Geant4-DNA physics lists is presented and analysed.

All statistical comparisons among different physics lists were performed using two-sample non-parametric statistical tests: Chi-Square test (χ^2) considering the statistical fluctuation of the simulation, independence Anderson-Darling k-Sample test (AD) and Kolmogorov–Smirnov (KS) test to evaluate both profiles distributions. The evaluation of the general conformity of the DNApto physics lists to the reference DNA physics list was performed by Chi-Square contingency Tables (CT) based on the number of cases that passed and failed the statistical χ^2 test. The contingency tables were applied to the total deposited energy and the depth and radial profile evaluations. All the statistical tests were performed for a significance level (SL) of 0.05.

2.4.1 The comparison findings

Figure 7 presents an example of radial and depth profiles according to DNA physics list that exemplifies the behaviour observed in all cases.

The shape of both profiles presented on **Figure 7** is similar to the expected. Usually, depth profiles show a Bragg Peak when the depth is thick enough to stop the incident particle [20]. This consideration cannot be applied in this study case since even the largest thickness evaluated is smaller than the protons and electrons range. Also, for UTLs the influence of the surface properties becomes significant due to the particles that are able to escape from it. In what follows, one may expect a small reduction on deposited energy at entrance surface and then an increase on the deposited energy as function of the depth until it reaches the stability around the range of the particles of interest (in this study case specially electrons). This behaviour is compatible with the example shown on **Figure 7a**. For radial profile one may expect the proportionality $E_{abs} \propto \frac{1}{r^n}$, where r represents the radius which means the distance from the center of the transported protons core and the position of the energy absorption, and $n \approx 2$. However, the n values presented by the simulations are slightly larger than 2 when the data presented on **Figure 6** are fitted to the proportionality equation. The same behaviour was reported by [21] on his validation of radial profiles with Geant4-DNA where this general tendency for the profiles was observed in all cases analysed and n was in the range of 2.1 to 2.38.

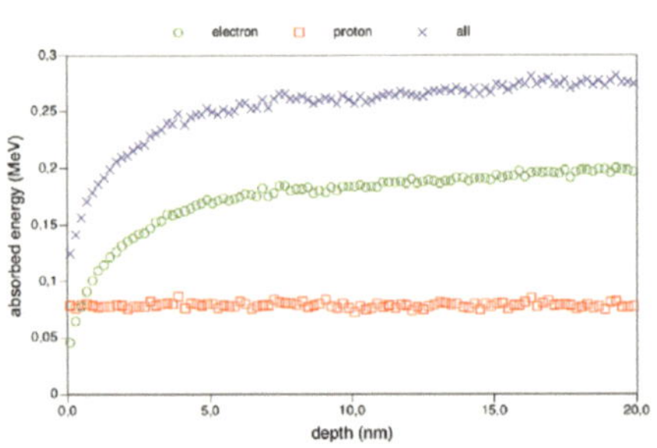

(a)

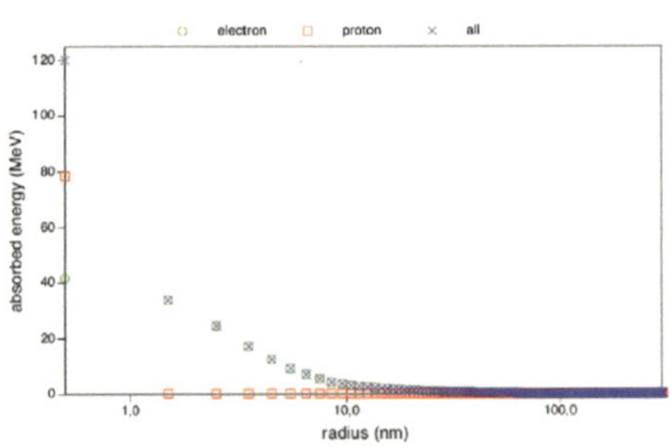

(b)

Figure 7.
Example of depth (a) and radial profile (b) simulated considering protons of 2 MeV passing through 20 nm water UTL by evoking DNA (stable) physics list.

The curves on **Figure 8** show a visual comparison among different transport models evoked. These curves exemplify the behaviour observed for all incident energies and UTL thicknesses. All models presented behaviour similar to the DNA physics list, with the exception of DNAopt1 where the depth profile showed a peak at the end of the water UTL.

Both depth and radial profiles showed lower energy deposition for DNAopt1 when compared to the other physics lists, indicating larger range for the secondary electrons generated by DNAopt1. Also, both profiles presented a localised deposited energy, evident on depth profile (at the end of the exit surface of the water ultra-thin layer) and diluted on the radial profile (near the core). The difference on the particles range (secondary electrons) can be noticed on **Figure 9**, where the DNA presented a smaller electron range when compared to the one showed by DNAopt1. DNAopt2,3,4 and 5 showed a similar behaviour to DNA physics list. Further investigation and statistical analysis are needed to generalize these results and evaluate the significance of these observations.

Tables 3 and **4** present the statistical tests p-values for depth profile by protons and electrons generated with different possible optional physics lists when compared to DNA physics list.

On **Table 3**, for protons, when the DNAopts are compared to DNA physics list, it is observable that χ^2 p-values are always higher than the SL with exception of DNAopt1 considering 2 MeV for thickness 6 nm and 5 MeV for 4 nm, as well as, one case on the limit of SL for 10 MeV and 200 nm. AD and KS statistical tests presented distributions significantly different when 100 nm and 200 nm were evaluated. Since χ^2 test evaluates the fluctuations on average value and the AD and KS tests evaluate the distribution considering only the average data, it reveals that the average data has some differences but they are not significant when the statistical

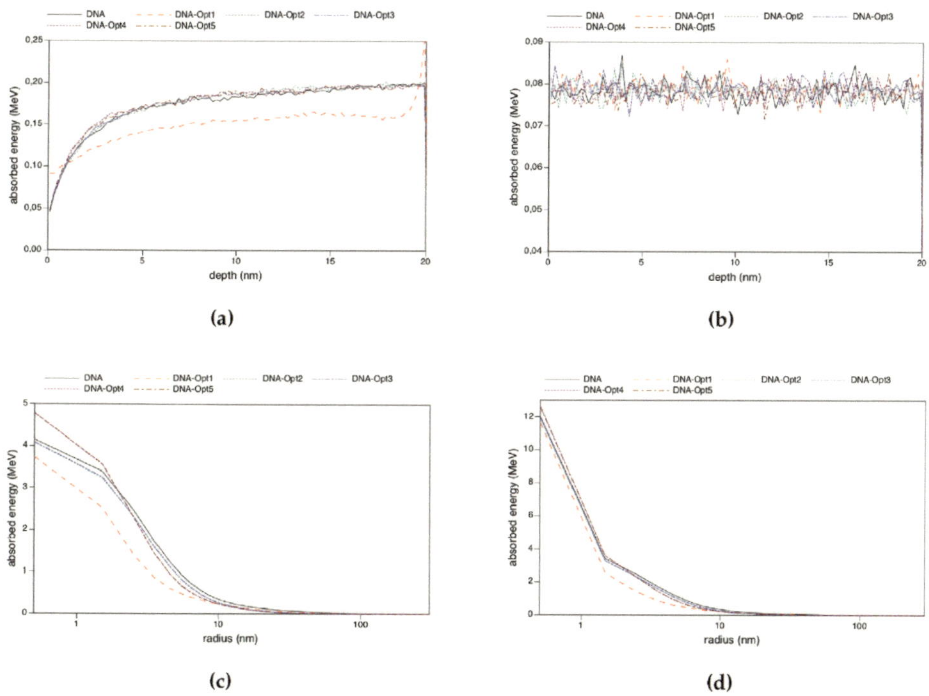

Figure 8.
Example of comparisons among different physics lists evoked on depth (a,b) and radial (c,d) profiles considering the deposited energy by secondary electrons (a,c) and protons (b,d) for incident protons of 2 MeV passing through 20 nm water UTL.

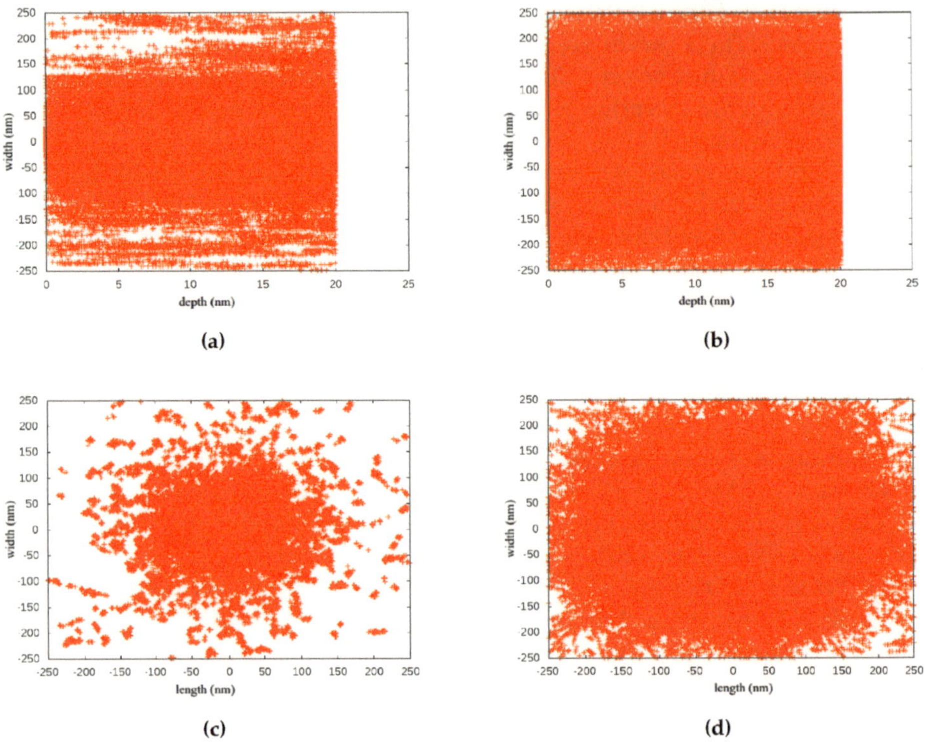

Figure 9.
Typical interaction maps as function of the depth (a,b) and radius (c,d) considering all particles for DNA (a, c), reference physics list and DNAopt1 (b,d) physics list for incident protons of 2 MeV passing through 20 nm water UTL.

Energy	Thickness	DNAopt1		DNAopt2		DNAopt3		DNAopt4		DNAopt5	
(MeV)	(nm)	AD	KS	AD	KS	AD	KS	AD	KS	AD	KS
2	10	0.0191	0.0243	0.0021	0.0008	—	—	—	—	—	—
	100	< 0.001	< 0.001	< 0.001	< 0.001	< 0.001	< 0.001	< 0.001	< 0.001	< 0.001	< 0.001
	200	< 0.001	< 0.001	< 0.001	< 0.001	< 0.001	< 0.001	< 0.001	< 0.001	< 0.001	< 0.001
5	8	—	—	< 0.001	< 0.001	< 0.001	< 0.001	—	—	—	—
10	8	—	—	—	—	< 0.001	< 0.001	—	—	—	—
20	8	< 0.001	< 0.001	—	—	—	—	—	—	—	—

Table 3.
Statistical evaluation of the energy depth profile for protons among the different physics list options when DNA physics lists is the reference, considering all studied cases, but showing only the cases with p-value inferior to 0.02.

fluctuations in each bin are taken into account. The contingency table evaluation shows 0.306 p-value when the physics list DNAopt1 is compared to DNA (higher than the SL). It happens because only few studied cases for DNAopt1 are significantly different from the reference DNA on the comparison among the evaluated physics lists. The physics lists DNAopt2,3,4 and 5 passed 100% of the statistical tests presenting no significant differences when compared to DNA physics list. The contingency table comparing all different optional physics lists presents p-value 0.3961, evidencing no significant difference from the reference physics list.

On **Table 4**, for electrons, when the DNAopts are compared to DNA physics list, it is observable that χ^2 p-values are always higher than the SL. However, for AD and

Energy	Thickness	DNAopt1		DNAopt2		DNAopt3		DNAopt4		DNAopt5	
(MeV)	(nm)	AD	KS	AD	KS	AD	KS	AD	KS	AD	KS
2	2	< 0.001	< 0.001	< 0.001	< 0.001	< 0.001	< 0.001	< 0.001	< 0.001	< 0.001	< 0.001
	6	< 0.001	< 0.001	—	—	—	—	—	—	—	—
	8	< 0.001	< 0.001	< 0.001	< 0.001	< 0.001	< 0.001	< 0.001	< 0.001	< 0.001	< 0.001
	10	< 0.001	< 0.001	—	—	—	—	—	—	—	—
	20	< 0.001	< 0.001	—	—	—	—	—	—	—	—
	40	< 0.001	< 0.001	—	—	—	—	—	—	—	—
	100	0.0019	0.0014	—	—	—	—	—	—	—	—
	200	< 0.001	< 0.001	—	—	—	—	—	—	—	—
5	2	< 0.001	< 0.001	0.0025	0.0010						
	6	< 0.001	< 0.001	—	—	—	—	—	—	—	—
	8	< 0.001	< 0.001	< 0.001	< 0.001	< 0.001	< 0.001	0.0050	0.0128		
	10	< 0.001	< 0.001	—	—	—	—	—	—	—	—
	20	< 0.001	< 0.001	—	—	—	—	—	—	—	—
	40	< 0.001	< 0.001	—	—	—	—	—	—	—	—
	50	< 0.001	< 0.001	—	—	—	—	—	—	—	—
	100	0.0023	< 0.001	—	—	—	—	—	—	—	—
	200	< 0.001	< 0.001	—	—	—	—	—	—	—	—
10	2	< 0.001	< 0.001	—	—	—	—	—	—	—	—
	6	< 0.001	< 0.001	—	—	—	—	—	—	—	—
	8	< 0.001	< 0.001	0.0739	0.0546	< 0.001	< 0.001	—	—	—	—
	10	< 0.001	< 0.001	0.0119	0.0132	0.0519	0.0727	0.0021	0.0016	0.0144	—
	20	< 0.001	< 0.001	—	—	—	—	—	—	0.0055	0.0028
	40	< 0.001	< 0.001	—	—	—	—	—	—	—	—
	50	< 0.001	< 0.001	—	0.0067	—	—	—	—	—	—
	100	0.0023	0.0014	—	—	—	—	—	—	—	—
	200	< 0.001	< 0.001	—	—	—	—	—	—	—	—
20	2	< 0.001	< 0.001	—	—	0.0042	0.0015	—	—	—	—
	6	< 0.001	< 0.001	—	—	< 0.001	< 0.001	—	—	—	—
	10	< 0.001	< 0.001	—	—	—	—	—	—	—	—
	20	< 0.001	< 0.001	—	—	—	—	0.0022	< 0.001	—	—
	30	—	0.0063	—	—	—	—	—	—	—	—
	40	< 0.001	< 0.001	0.0019	< 0.001	< 0.001	0.0016	—	—	0.0128	—
	50	< 0.001	< 0.001	—	—	—	—	—	—	—	—
	100	< 0.001	< 0.001	—	—	—	—	—	—	—	—
	200	< 0.001	< 0.001	—	—	—	—	—	—	—	—

Table 4.
Statistical evaluation of the energy depth profile for electrons among the different physics list options when DNA physics lists is the reference, considering all studied cases, but showing only the cases with p-value inferior to 0.02.

KS statistical tests the physics list DNAopt1 generally shows p-values lower than the SL, presenting significant differences in most cases when compared to DNA physics list. Again, it shows that the average data has some differences but they are not significant when the statistical fluctuations of each bin are taken into account. Considering the χ^2, all the optional physics lists passed 100% of the statistical tests presenting no significant differences when compared to the reference physics list with the statistical fluctuations in the evaluation.

The statistical evaluation shows that despite the visible systematic differences represented on the **Figure 8**, those are not significant. Nevertheless, it is important to consider these systematic differences when one is studying depth profiles on a sensitive case, and it would be better to use any other physics list than DNAopt1, to avoid the influence of the changes in shape and the systematic lower energy deposition on the results.

It is not possible to statistically evaluate the proton radial profile because almost 100% of the energy is deposited on the first bin so, on **Table 5**, only the deposited energy radial profile for electrons is presented. It is observable that χ^2 p-values are always higher than the SL. The physics list DNAopt1 presents significant differences

Energy	Thickness	DNAopt1		DNAopt2		DNAopt3		DNAopt4		DNAopt5	
(MeV)	(nm)	AD	KS	AD	KS	AD	KS	AD	KS	AD	KS
2	4	< 0.001	< 0.001	—	—	—	—	—	—	—	—
	6	0.0032	0.0187	—	—	—	—	—	—	—	0.0033
	10	0.0034	0.0118	—	—	—	—	—	—	—	—
	20	0.0143	—	—	—	—	—	—	—	—	—
	30	< 0.001	< 0.001	—	—	—	—	—	—	—	—
	100	< 0.001	< 0.001	—	—	—	—	—	—	—	—
	200	< 0.001	< 0.001	—	—	—	—	—	—	—	—
5	2	< 0.001	< 0.001	0.0025	0.0010	—	—	—	—	—	—
	6	< 0.001	< 0.001	—		—	—	—	—	—	—
	8	< 0.001	< 0.001	< 0.001	< 0.001	< 0.001	< 0.001	0.0050	0.0128	—	—
	10	< 0.001	< 0.001	—	—	—	—	—	—	—	—
	20	< 0.001	< 0.001	—	—	—	—	—	—	—	—
	40	< 0.001	< 0.001	—	—	—	—	—	—	—	—
	50	< 0.001	< 0.001	—	—	—	—	—	—	—	—
	100	0.0020	< 0.001	—	—	—	—	—	—	—	—
	200	< 0.001	< 0.001	—	—	—	—	—	—	—	—
10	4	< 0.001	< 0.001	—	—	—	—	—	—	—	—
	30	< 0.001	< 0.001	—	—	—	—	—	—	—	—
	100	0.0012	0.0092	—	—	—	—	—	—	—	—
	200	< 0.001	0.0072	—	—	—	0.0092	—	—	—	—
20	4	< 0.001	< 0.001	—	—	—	—	0.0104	—	0.0094	0.0235
	6	< 0.001	0.0033	—	—	—	—	—	—	—	—
	20	< 0.001	0.01176	—	—	—	—	—	—	—	—
	30	< 0.001	< 0.001	—	—	—	—	—	—	—	—
	100	< 0.001	0.0019	—	—	—	—	—	—	—	—
	200	0.0016	0.0118	—	—	—	—	—	—	—	—

Table 5.
Statistical evaluation of the energy radial profile for electrons among the different physics list options when DNA physics lists is the reference, considering all studied cases, but showing only the cases with p-value inferior to 0.02.

in at least 50% of cases when compared to DNA physics list for all evaluated energies on AD and KS statistical tests. Once more, it shows that average data differences are not significant when the statistical fluctuations of each bin are taken into account. The optional physics lists passed 100% of the statistical tests presenting no significant differences when compared to DNA physics list.

The number of interactions of protons and electrons for all evaluated physics lists is presented on **Figure 10**. To analyse the data presented on **Figure 10** it is necessary to consider that on Monte Carlo simulation, as it is performed on Geant4-DNA, the increase on the number of interactions represents consequently an increase on running time.

It is easy to observe that, for the energy range studied, there is no significant change on the number of proton interactions. This can be explained by the processes and models evoked in the energy range of this study case, where incident protons transfer a few eV of their kinetic energy to electrons. Considering the energy range of incident protons, it can be seen on **Figure 1** that the process and model classes evoked by all physics lists evaluated were G4DNAExcitation process with G4DNABornExcitationModel, G4DNAIonisation process with G4DNABornIonisationModel and G4DNAChargeDecrease process with G4DNADingfelderChargeDecreaseMode. The only difference was the activation of the flag "SelectFasterComputation" for ionization transport of DNAopt2.

On the other hand, the number of electrons interactions presents significant changes.The observable differences for different physics lists can be justified by the different process and models evoked for electrons presented on **Figure 2**. DNAopt1 presents larger number of electrons interactions for thicknesses larger than 40 nm with the exception of 20 MeV incident kinetic energy protons beam where the DNAopt2 presents a number of electrons interactions similar to DNAopt1. This behaviour can be explained by the process and model classes evoked by all physics lists evaluated and the maximum kinetic transferred energy to the electrons which was estimated [14] as 4.34 keV for incident protons of 2 MeV, 10.92 keV for incident protons of 5 MeV, 21.90 keV for incident protons of 10 MeV and 44.03 keV for incident protons of 20 MeV. Under these conditions, all transport process classes can be evoked for electrons G4DNAElastic (for DNA and DNAopt2,3,4,and 5) or G4eMultipleScattering (for DNAopt1), G4DNAExcitation, G4DNAIonisation, G4DNAVibExcitation and G4DNAAttachment.

The main difference among the DNAopt1 and the other physics lists is the scattering process and model classes evoked that were a multiple scattering process and model instead of the discrete elastic process class implemented on Geant4-DNA. Taking DNA as reference, one may see that the scattering model was the only one that changed on the DNAopt1 implementation, so the high discrepancies observed on the number of electrons generated, deposited energy and electron

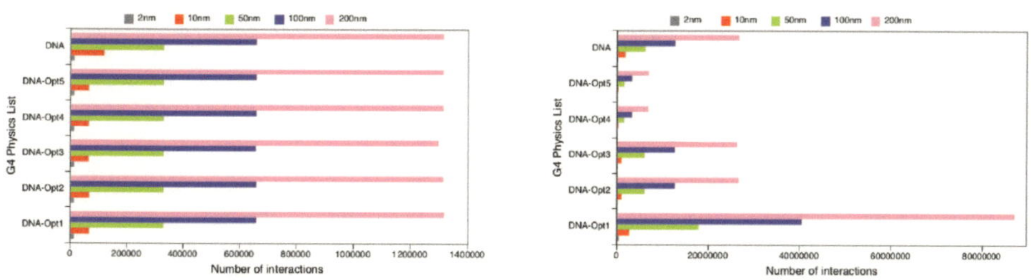

(a) (b)

Figure 10.
Number of interactions for protons (a) and for electrons (b) considering incident protons of 10 MeV and different thicknesses of UTLs.

range are related to the implementation of the G4eMultipleScattering process class and the G4LowEWentzelVIModel model class. The physics lists DNAopt2 and DNAopt3 evoked the same process and model classes as DNA with the exception of DNAopt2 in ionization process where the flag "SelectFasterComputation" was activated. DNAopt4 and DNAopt5 evoked the same model as DNA for the process classes G4DNAVibExcitation and G4DNAAttachment; however, the other process classes G4DNAElastic, G4DNAExcitation and G4DNAIonisation models were different from DNA. Besides that, DNAopt5 process classes G4DNAExcitation and G4DNAIonisation evoked two models to each process, one additional model than the evoked by DNAopt4 for energies above 10 keV.

Taking the DNA physics list as reference, one can see that DNAopt1 presents a larger number of interactions for electrons. Considering the whole dataset simulated, all thicknesses and energies evaluated: DNAopt1 presents 1.5 to 4.1 times interactions; DNAopt2 and DNAopt3 present 0.5 to 1.0 times interactions; and DNAopt4 and DNAopt5 present 0.15 to 0.35 times interactions. The similar behaviour presented by DNAopt2 and DNAopt3 and by DNAopt4 and DNAopt5 was expected due to the similarities on the physics lists evoked to transport the secondary particles (electrons) in the energy range (**Figure 2**).

Figure 11 presents the graphics of the relative differences in the total deposited energies considering the DNA physics list as reference. It is easy to observe that DNAopt1 physics list presents the lowest average total deposited energy, 3rd and 4th quartiles and the largest standard deviation in all cases. This is in agreement to the observed energy depth and radial profiles where the DNAopt1 presents the lowest deposited energy (**Figure 7**).

Table 6 shows the statistical evaluation of the total deposited energy per ultrathin layers presenting χ^2, AD and KS p-values. These values are always higher than the SL for all optional physics list with the exception of DNAopt1 which presents all p-values below the SL. The contingency table for different incident kinetic kinetic energy protons beams presents p-value of 0.0016 (lower than the SL) for DNAopt1 when compared to each of the other optional physics lists. The evaluation of the

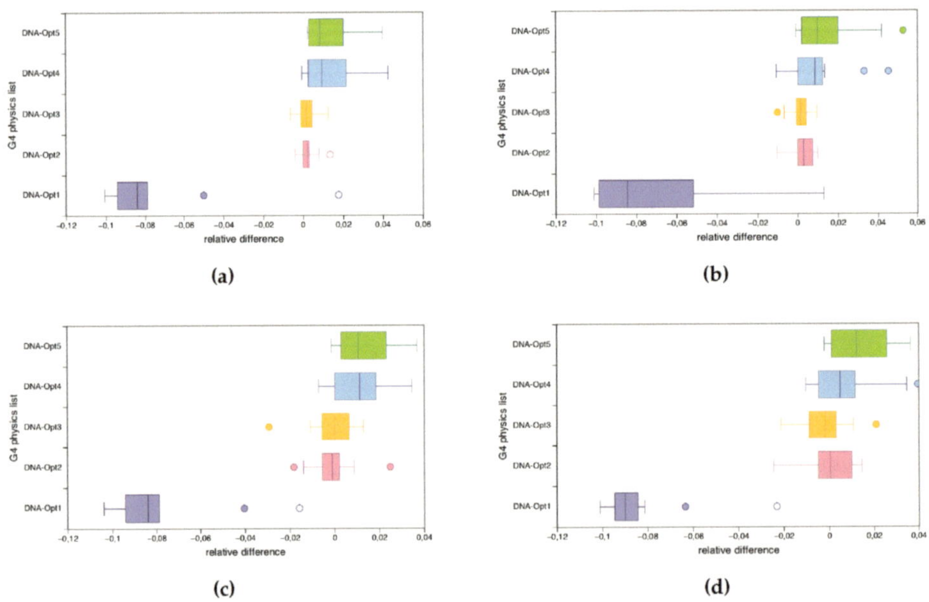

Figure 11.
Box-and-Whisker plots of the relative difference on total deposited energy on UTL considering all cases for incident kinetic energy protons beam of 2 MeV (a), 5 MeV (b), 10 MeV (c) and 20 MeV (d).

	DNAopt1			DNAopt2			DNAopt3		
Energy	χ^2	**AD**	**KS**	χ^2	**AD**	**KS**	χ^2	**AD**	**KS**
All Energies	0.0001	0.0444	0.0449	0.1195	1.0000	1.0000	0.0850	1.0000	1.0000
2 MeV	0.0001	0.0455	0.0449	1.0000	1.0000	1.0000	1.0000	1.0000	1.0000
5 MeV	0.0461	0.0490	0.0449	1.0000	1.0000	1.0000	1.0000	1.0000	1.0000
10 MeV	0.0430	0.0333	0.0449	0.9988	1.0000	1.0000	0.9402	1.0000	1.0000
20 MeV	0.0010	0.0395	0.0449	0.0573	1.0000	1.0000	0.5401	1.0000	1.0000

	DNAopt4			DNAopt5		
Energy	χ^2	**AD**	**KS**	χ^2	**AD**	**KS**
All Energies	1.0000	1.0000	1.0000	1.0000	1.0000	1.0000
2 MeV	1.0000	1.0000	1.0000	1.0000	1.0000	1.0000
5 MeV	1.0000	1.0000	1.0000	1.0000	1.0000	1.0000
10 MeV	1.0000	1.0000	1.0000	1.0000	1.0000	1.0000
20 MeV	0.7128	1.0000	1.0000	0.8225	1.0000	1.0000

Table 6.
Statistical evaluation of the total deposited energy considering all studied cases.

deposited energy considering all optional physics lists and studied conditions presents p-value lower than 0.001 evidencing the significant difference among all models. Taking into account that only DNAopt1 presents p-values below the statistical significance when compared to the DNA physics list one may conclude that the DNAopt1 is the only physics list with significant difference among the optional physics lists.

To get closer to the characteristic of the experimental material used (PMMA) the influence of the water density change on the profiles was analysed. **Figure 12** presents the depth and radial profiles comparing dense to standard water. As expected, the dense water presents higher deposited energies. The statistical evaluation shows that that the significant difference on depth profile is mainly due to secondary electrons interactions. It was not possible to obtain the statistical evaluation of protons radial profile because its deposited energy is almost completely performed on first bin. The logarithmic scale applied on x axis of radial profile makes more evident the differences between the curves, since on linear scale these differences are not visible. The total deposited energy on dense water is always larger than the total deposited energy on standard water, usually 15–20% higher (as was expected).

Table 7 shows p-values for statistical evaluation depth profile for protons and electrons interactions comparing dense to standard water. For protons, the χ^2 p-values are above the SL and AD and KS tests show p-values smaller than the SL, which means that the systematic differences observed in **Figure 12b** are not significant. The differences observed for electrons, in **Figure 12a**, are significant in all studied cases when comparing dense to standard water. For electrons, it is observable that χ^2 p-values are always higher than the SL and AD and KS tests show difference for few cases when comparing dense to standard water. It shows that, again, the average data has some differences but they are not significant when the statistical fluctuations in each bin are taken into account. The contingency table presents significant agreement on depth profile for protons with p-value 0.982 and significant difference on depth profile for electrons with p-value lower than 0.001.

Table 7 shows the statistical evaluation of radial profile for electrons considering standard and dense water indicating no significant statistical differences with exception of AD and KS tests for 2 MeV with 20 nm and 100 nm thicknesses.

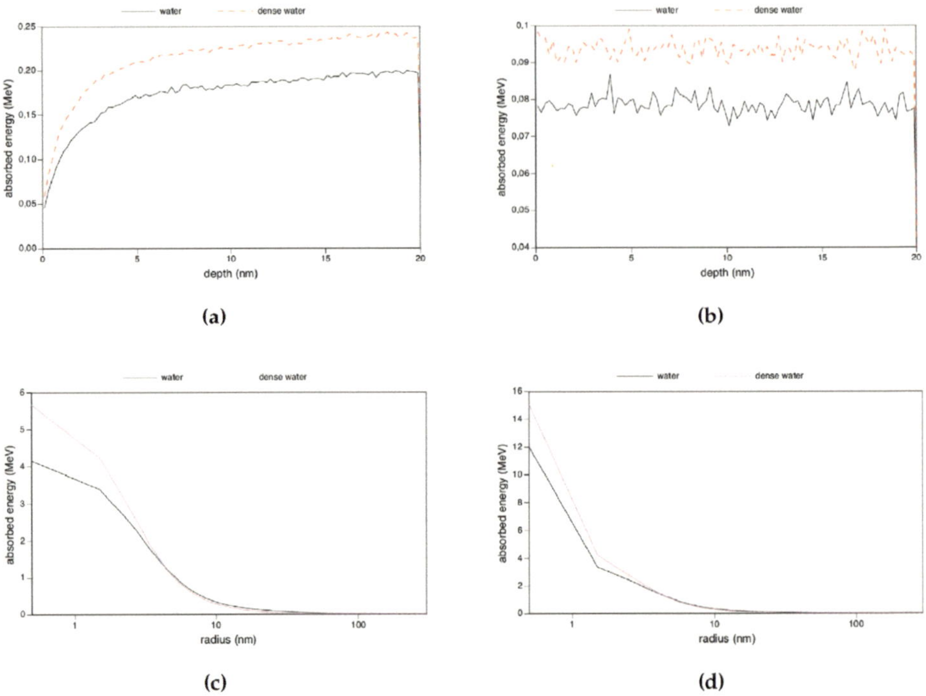

Figure 12.
Comparative depth (a,b) and radial (c,d) profiles of energy deposition due different particles: electrons (a,c), proton (b) and all particles (d).

Thickness	Depth profile for protons			Depth profile for electrons			Radial profile for electrons		
(nm)	χ^2	AD	KS	χ^2	AD	KS	χ^2	AD	KS
2	1.0000	< 0.001	< 0.001	0.0032	0.0881	0.2058	1.0000	0.2224	0.3505
4	1.0000	< 0.001	< 0.001	0.0219	0.0042	0.0063	1.0000	0.0937	0.0666
6	1.0000	< 0.001	< 0.001	0.0032	< 0.001	< 0.001	1.0000	0.2281	0.1163
10	1.0000	< 0.001	< 0.001	0.0010	< 0.001	< 0.001	1.0000	0.8888	0.6350
20	1.0000	< 0.001	< 0.001	0.0024	< 0.001	< 0.001	1.0000	0.0483	0.0362
30	1.0000	< 0.001	< 0.001	0.0210	< 0.001	< 0.001	1.0000	0.4578	0.3505
40	1.0000	< 0.001	< 0.001	0.0330	< 0.001	< 0.001	1.0000	0.2346	0.1163
50	1.0000	< 0.001	< 0.001	0.0043	< 0.001	< 0.001	1.0000	0.9351	0.5727
100	1.0000	< 0.001	< 0.001	0.0012	< 0.001	< 0.001	1.0000	< 0.001	< 0.001
200	1.0000	< 0.001	< 0.001	0.0032	< 0.001	< 0.001	1.0000	0.8697	0.9730

Table 7.
Statistical evaluation comparing the electrons and proton deposited energy depth profile and the electrons deposited energy radial profile for the standard water density to the water with PMMA density, considering all studied cases.

3. Final remarks

The evaluation of the CDCS (based on radial profile) showed that the bin size influences on the CDCS curve shapes. These results presented good agreement between the experimental CDCSs for polymer films and the simulated-calculated values for standard water with 1 nm bin, despite different materials used. When the water density was augmented to the PMMA density value (dense water) the results became even more reliable.

In general, the SP values increased with the increase on the UTLs thickness, as expected, since the water layer thicknesses considered were smaller than the electron range. The simulated SP always presented lower values than NIST, as expected, with DNAopt1 generating the lowest (worst) SP values. Due to this behaviour the DNAopt1 presented unreliable electrons range values. This behaviour is probably due to the evocation of the multiple scattering process (with low energy Wenzel VI model) instead of the DNA elastic process.

Considering the reliability information presented in this chapter, all transport models available in Geant4-DNA presented reliable results for SP and CDCS with exception of DNAopt1. Further investigation is needed to map the differences among the possible physics lists available on Geant4-DNA.

In summary the comparison of energy deposition radial and depth profiles, taking DNA physics list as reference, showed that: (i) DNAopt2 to DNAopt5 presented similar results with percentage differences on simulated values lower than 8%; (ii) DNAopt1 presented the lowest deposited energy in both profiles when compared to the other physics lists, one peak at the end of the depth profile deposited energy, and a significant change on the curve shape on radial profile. In a general analysis the radial deposited energy decreased systematically in ultra-thin layers.

In a general evaluation, no significant differences were observed for the total deposited energy among all models, with exception of DNAopt1 which presented systematic distortions in the profile curves shape with a non-expected behaviour as confirmed by the contingency tables.

DNAopt1 showed itself being more time consuming and generated the lowest total deposited energy in UTLs, which resulted in the worst general agreement to the reference physics list DNA and to the expected data. It is important to emphasize that these conclusions are valid for the evaluated physics lists, energy range and geometrical conditions in this study and just for the Geant4-DNA (version 10.02.p01). Any other generalization requires further evaluation.

Acknowledgements

Research developed with the partial support of the National Supercomputing Center (CESUP), Federal University of Rio Grande do Sul (UFRGS).

Author details

Gabriela Hoff[1*], Raquel S. Thomaz[2], Leandro I. Gutierres[2], Sven Muller[2], Viviana Fanti[3], Elaine E. Streck[4] and Ricardo M. Papaleo[2]

1 National Institute for Nuclear Physics-Sezioni di Genova, Genova, GE, Italy

2 Interdisciplinary Center of Nanoscience and Micro-Nanotechnology, School of Technology, Pontifical Catholic University of Rio Grande do Sul, Porto Alegre, Brazil

3 Università di Cagliari and National Institute for Nuclear Physics-Sezioni di Cagliari, Monserrato, CA, Italy

4 Pontifical Catholic University of Rio Grande do Sul, Porto Alegre, Brazil

*Address all correspondence to: ghoff.gesic@gmail.com

References

[1] S. Agostinelli, J. Allison, K. a. Amako, J. Apostolakis, H. Araujo, P. Arce, M. Asai, D. Axen, S. Banerjee, G. . Barrand *et al.*, "Geant4a simulation toolkit," *Nuclear instruments and methods in physics research section A: Accelerators, Spectrometers, Detectors and Associated Equipment*, vol. 506, no. 3, pp. 250–303, 2003.

[2] J. Allison, K. Amako, J. Apostolakis, H. Araujo, P. A. Dubois, M. Asai, G. Barrand, R. Capra, S. Chauvie, R. Chytracek *et al.*, "Geant4 developments and applications," *IEEE Transactions on nuclear science*, vol. 53, no. 1, pp. 270–278, 2006.

[3] J. Allison, K. Amako, J. Apostolakis, P. Arce, M. Asai, T. Aso, E. Bagli, A. Bagulya, S. Banerjee, G. Barrand *et al.*, "Recent developments in geant4," *Nuclear Instruments and Methods in Physics Research Section A: Accelerators, Spectrometers, Detectors and Associated Equipment*, vol. 835, pp. 186–225, 2016.

[4] S. Incerti, A. Ivanchenko, M. Karamitros, A. Mantero, P. Moretto, H. Tran, B. Mascialino, C. Champion, V. Ivanchenko, M. Bernal *et al.*, "Comparison of geant4 very low energy cross section models with experimental data in water," *Medical physics*, vol. 37, no. 9, pp. 4692–4708, 2010.

[5] M. Bernal, M. Bordage, J. Brown, M. Davídková, E. Delage, Z. El Bitar, S. Enger, Z. Francis, S. Guatelli, V. Ivanchenko *et al.*, "Track structure modeling in liquid water: A review of the geant4-dna very low energy extension of the geant4 monte carlo simulation toolkit," *Physica Medica: European Journal of Medical Physics*, vol. 31, no. 8, pp. 861–874, 2015.

[6] S. Incerti, M. Douglass, S. Penfold, S. Guatelli, and E. Bezak, "Review of geant4-dna applications for micro and nanoscale simulations," *Physica Medica: European Journal of Medical Physics*, vol. 32, no. 10, pp. 1187–1200, 2016.

[7] *Guide For Physics Lists: release 10.4*, Geant4 Collaboration, 2017. [Online]. Available: http://geant4-userdoc.web.cern.ch/geant4-userdoc/UsersGuides/PhysicsListGuide/html/index.html

[8] G.-D. collaboration. The geant4-dna project: Extending the geant4 monte carlo simulation toolkit for radiobiology. [Online]. Available: geant4-dna.in2p3.fr/styled-3/styled-8/index.html

[9] G. Compagnini, R. Reitano, L. Calcagno, G. Marletta, and G. Foti, "Hydrogenated amorphous carbon synthesis by ion beam irradiation," *Applied Surface Science*, vol. 43, no. 1, pp. 228 – 231, 1989, beam Processing and Laser Chemistry. [Online]. Available: http://www.sciencedirect.com/science/article/pii/016943328990216X

[10] R. Barillon, M. Fromm, R. Katz, and A. Chambaudet, "Chemical Bonds Broken in Latent Tracks of Light Ions in Plastic Track Detectors," *Radiation Protection Dosimetry*, vol. 99, no. 1-4, pp. 359–362, 06 2002. [Online]. Available: https://doi.org/10.1093/oxfordjournals.rpd.a006802

[11] A. Licciardello, M. Fragal, G. Compagnini, and O. Puglisi, "Cross section of ion polymer interaction used to individuate single track regime," *Nuclear Instruments and Methods in Physics Research Section B: Beam Interactions with Materials and Atoms*, vol. 122, no. 3, pp. 589 – 593, 1997, nanometric Phenomena Induced by Laser, Ion and Cluster Beams. [Online]. Available: http://www.sciencedirect.com/science/article/pii/S0168583X96007550

[12] R. M. Papaléo, A. Hallén, B. U. R. Sundqvist, L. Farenzena, R. P. Livi, M.

A. de Araujo, and R. E. Johnson, "Chemical damage in poly(phenylene sulphide) from fast ions: Dependence on the primary-ion stopping power," *Phys. Rev. B*, vol. 53, pp. 2303–2313, Feb 1996. [Online]. Available: https://link.aps.org/doi/10.1103/PhysRevB.53.2303

[13] R. Papalo, L. Farenzena, M. de Arajo, R. Livi, M. Alurralde, and G. Bermudez, "Cratering in pmma induced by gold ions: dependence on the projectile velocity," *Nuclear Instruments and Methods in Physics Research Section B: Beam Interactions with Materials and Atoms*, vol. 148, no. 1, pp. 126 – 131, 1999. [Online]. Available: http://www.sciencedirect.com/science/ article/pii/S0168583X98008775

[14] R. Thomaz, P. Louette, G. Hoff, S. Müller, J. J. Pireaux, C. Trautmann, and R. M. Papaléo, "Bond-breaking efficiency of high-energy ions in ultrathin polymer films," *Phys. Rev. Lett.*, vol. 121, p. 066101, Aug 2018. [Online]. Available: https://link.aps.org/doi/10.1103/ PhysRevLett.121.066101

[15] G.-D. Collaboration. (2018) Geant4 cross reference. [Online]. Available: http://www-geant4.kek.jp/lxr/source/

[16] U. Hossain, V. Lima, O. Baake, D. Severin, M. Bender, and W. Ensinger, "On-line and post irradiation analysis of swift heavy ion induced modification of pmma (polymethyl-methacrylate," *Nuclear Instruments and Methods in Physics Research Section B: Beam Interactions with Materials and Atoms*", vol. 326, pp. 135 – 139, 2014, 17th International Conference on Radiation Effects in Insulators (REI). [Online]. Available: http://www.sciencedirect.com/ science/article/pii/S0168583X14000950

[17] N.-N. I. of Standards and Technology. (2009) Estar: stopping power and range tables for electrons. [Online]. Available: https://physics.nist.gov/ PhysRefData/Star/Text/ESTAR.html

[18] F. H. Attix, *Introduction to radiological physics and radiation dosimetry*. John Wiley & Sons, 1986.

[19] N.-N. I. of Standards and Technology. (1998) Pstar: stopping power and range tables for protons. [Online]. Available: https://physics.nist.gov/PhysRefData/Star/Text/PSTAR.html

[20] R. Garcia-Molina, I. Abril, P. de Vera, I. Kyriakou, and D. Emfietzoglou, "A study of the energy deposition profile of proton beams in materials of hadron therapeutic interest," *Applied Radiation and Isotopes*, vol. 83, pp. 109–114, 2014.

[21] H. Wang and O. N. Vassiliev, "Radial dose distributions from protons of therapeutic energies calculated with geant4-dna," *Physics in Medicine & Biology*, vol. 59, no. 14, p. 3657, 2014.

Chapter 7

Applications of Simulation Codes Based on Monte Carlo Method for Radiotherapy

Iury Mergen Knoll, Ana Quevedo and Mirko Salomón Alva Sánchez

Abstract

Monte Carlo simulations have been applied to determine and study different parameters that are challenged in experimental measurements, due to its capability in simulating the radiation transport with a probability distribution to interact with electrosferic electrons and some cases with the nucleus from an arbitrary material, which such particle track or history can carry out physical quantities providing data from a studied or investigating quantities. For this reason, simulation codes, based on Monte Carlo, have been proposed. The codes currently available are MNCP, EGSnrc, Geant, FLUKA, PENELOPE, as well as GAMOS and TOPAS. These simulation codes have become a tool for dose and dose distributions, essentially, but also for other applications such as design clinical, tool for commissioning of an accelerator linear, shielding, radiation protection, some radiobiologic aspect, treatment planning systems, prediction of data from results of simulation scenarios. In this chapter will be present some applications for radiotherapy procedures with use, specifically, megavoltage x-rays and electrons beams, in scenarios with homogeneous and anatomical phantoms for determining dose, dose distribution, as well dosimetric parameters through the PENELOPE and TOPAS code.

Keywords: Monte Carlo, codes, radiotherapy

1. Introduction

The constant development of medical applications using ionizing radiation requires an understanding of the transport of particles through materials, such as tissues, organs, patients, imaging devices. For this reason, computational simulations, using the Monte Carlo Method, have been extensively used in several areas, specifically, in Radiological Physics, where this tool is applied for modeling a treatment or medical examinations, for example, that for some regions of interest are difficulties and complexities to making experimentally.

Several computational codes, based on Monte Carlo simulation, have been used in radiological protection, radiotherapy source dosimetry, planning systems, and other applications PENELOPE [1–6], MCNP [7–11], EGSnrc [12–15], FLUKA [16–20], TOPAS [21–25], GAMOS [26–29], Geant [30, 31].

In this work, some applications will be presented, in different scenarios to determine dose, relative dose, dose distribution, as well as to determine dosimetric parameters used in radiotherapy, using computational codes through Monte Carlo simulation.

2. Applications of computational codes in radiotherapy

In this section, some applications of computational codes in radiotherapy will be presented.

2.1 Determination of dosimetric parameters of clinical sources of high dose rate brachytherapy using the PENELOPE package

The PENELOPE (PENetration and Energy LOss of Positrons and Electrons) computational code includes several computational codes written in FORTRAN 77 [32]. The package simulates the transport of electrons, photons, and positrons in arbitrary materials and energy values from 100 eV to 1 GeV, in geometries and materials defined by the user [33]. PENELOPE has a database of cross-sections for materials involving elements with atomic numbers from 1 to 92, and 180 other compounds and mixtures of interest in Radiological Physics.

This computational code has been used in several applications in Radiotherapy, such as determining parameters of brachytherapy sources [4, 34–36].

According to the TG-43 protocol, later revised and entitled TG-43 U1 [37, 38], the calculation of the dosimetric parameter Anisotropy Function is performed through Eq. (1):

$$F(r,\theta)=\left[\dot{D}(r,\theta)/\dot{D}(r,\theta_0)\right]*\left[G_L(r,\theta_0)/G_L(r,\theta)\right] \qquad (1)$$

where D is the dose rate; r is the distance from the center of the source to the point of interest (in cm), θ is the polar angle that specifies the point of interest, θ_0 is the reference angle (90°); G_L is the geometric factor, determined analytically.

Quevedo et al. [4] determined a dosimetric parameter, using Monte Carlo simulation with the PENELOPE package. The high dose-rate 192Ir brachytherapy source commonly used in gynecological brachytherapy was modeled, and the Anisotropy Function in regions close to the source was determined. **Figure 1** shows (a) the source geometry, modeled in the PENELOPE package, and (b) the Anisotropy Function in regions close to the irradiation source.

To validate the results obtained, using Monte Carlo simulation with the PENELOPE package, dose profiles in the longitudinal direction of the source were compared with data from the BrachyVision planning system. **Figure 2** shows the comparisons of relative doses, as a function of distances, between the PENELOPE package and the BrachyVision treatment planning system for distances (a) 0.4 cm from the center of the source towards the source cable, (b) center of the source and (c) 0.4 cm from the center of the source towards the top of the source encapsulation, adapted from Quevedo et al. [4].

From the comparisons in the three plans, it is possible to verify that the data obtained in the simulations with the PENELOPE package shows agreement greater than 98% of the points, with the data obtained from the treatment planning system, indicating that the PENELOPE package has great potential in the determination of dosimetric parameters of high dose rate brachytherapy sources.

2.2 Applications of the TOPAS code in ocular brachytherapy

The TOPAS (Tool for Particle Simulation) computational code is based on physical Geant4 models and low energy electromagnetic models of the PENELOPE code [39] and has shown great potential for simulations in medical and quality control applications [22, 40–43].

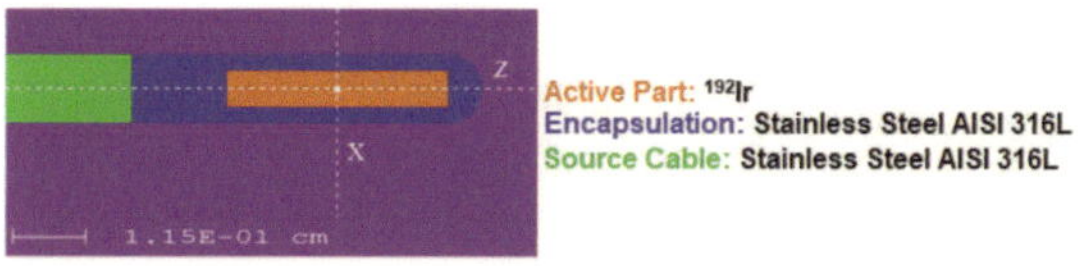

(a)

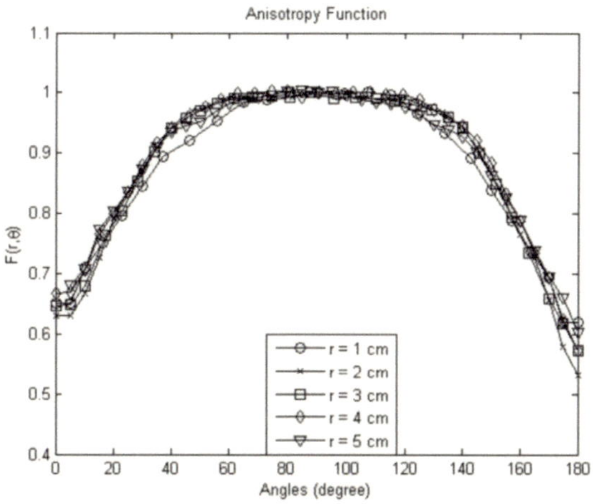

(b)

Figure 1.
(a) Source geometry, modeled in the PENELOPE package, adapted from Quevedo et al. [4] and (b) anisotropy function in regions close to the irradiation source.

TOPAS is based on an innovative parameter control system, which allows the modeling of therapy and imaging devices without the need for programming knowledge, in order to reduce user errors. The code has a vast library of predefined modules for geometry, physical models, and detection. Despite this, it is possible for more experienced users to write their own modules in C++. You can also create patient geometries based on computed tomography (CT) images. Therefore, TOPAS presents itself as a highly extensible and accessible tool for medical physicists for modeling therapy and imaging devices.

Another innovative feature of this code is that it handles time-dependent amounts, that is, it has 4D resources through its Time Feature System. Time-varying amounts are essential for modeling advanced radiotherapy treatment techniques, the Image-Guided Radiotherapy (IGRT) technique is an example of this, where the positioning of the patient or target organ varies. Therefore, TOPAS proposes an approach to 4D simulation to deal with various time-dependent quantities in a single simulation, such as volume change, rotational movement, current variation, magnetic field, etc. [44].

Knoll et al., determined dose deposition from a source of brachytherapy used in ophthalmic treatments [23]. In this work, the applicator used in the treatment, which uses the $^{90}Sr/^{90}Y$ source, was modeled in the TOPAS code, including the active part, encapsulation, and simulator object filled with water. A relative dose profile was obtained in a high dose gradient region, normalized at the reference point to 1 mm and, for validation purposes, the data obtained with TOPAS were compared with data from the ICRU (International Commission on Radiation & Measurements) with the same conditions. **Figure 3** shows (a) geometry of the $^{90}Sr/^{90}Y$ source with applicator and simulator object and (b) comparison of relative dose, as a function of distance, between TOPAS and ICRU data (Adapted from [23]).

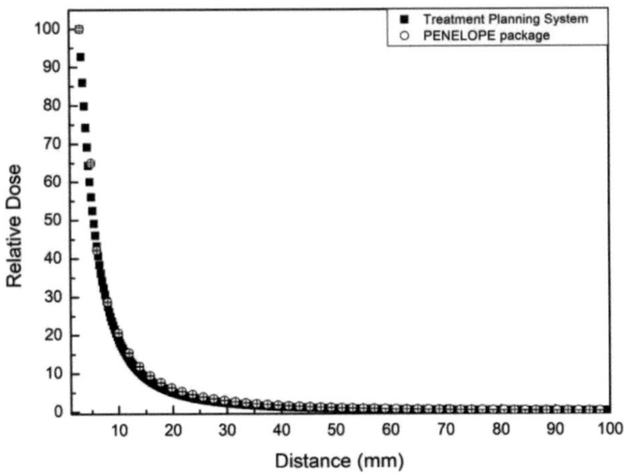

(a)

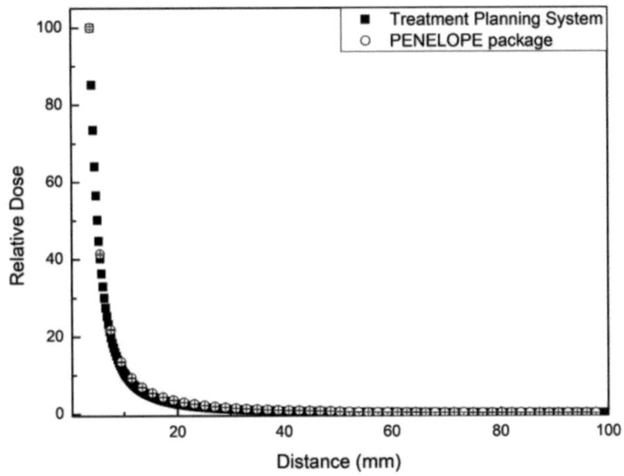

(b)

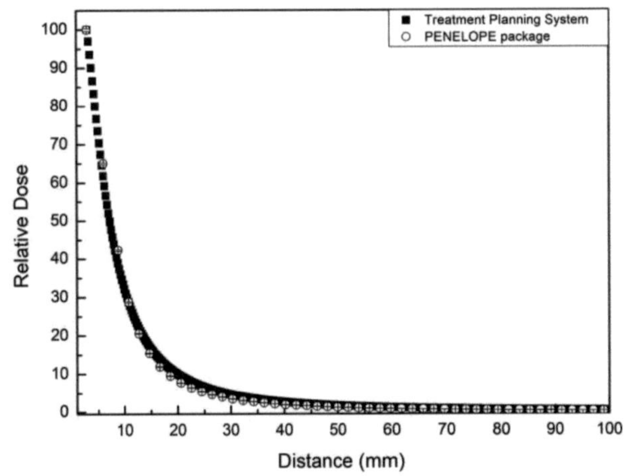

(c)

Figure 2.
Relative dose comparisons, as a function of distances, between the PENELOPE package and the BrachyVision treatment planning system for distances (a) 0.4 cm from the center of the source towards the source cable, (b) center of the source, and (c) 0.4 cm from the center of the source towards the top of the source encapsulation, adapted from Quevedo et al. [4].

The greatest uncertainty obtained in computer simulations with the TOPAS code was approximately 0.01% at the point where it was normalized. When compared to the ICRU, the greatest difference found was approximately 4% at 1.6 mm depth. Thus, the TOPAS code has been shown to be a promising tool for dosimetry in brachytherapy and radiological applications.

2.3 Applications of the TOPAS-nBio code

Although the TOPAS code provides a wide range of tools for use in radiotherapy at patient scale or clinically applicable geometries, the fundamental unit of biological response to the effects of ionizing radiation is at the cellular or subcellular scale. For this reason, TOPAS has an extension dedicated to the study of the biological effects of radiation in micrometer and nanometer scale, designed based on Geant4-DNA, the extension of Geant4, the basis of TOPAS, so that very low energy are included in the interactions.

When interacting with matter, excitations and ionizations can be caused due to ionizing radiation. In the context of radiotherapy, incident particles cause radiolysis of water and subsequent chemical interactions, inducing molecular damage to the cell, more specifically to DNA, which is the critical target for most biological effects

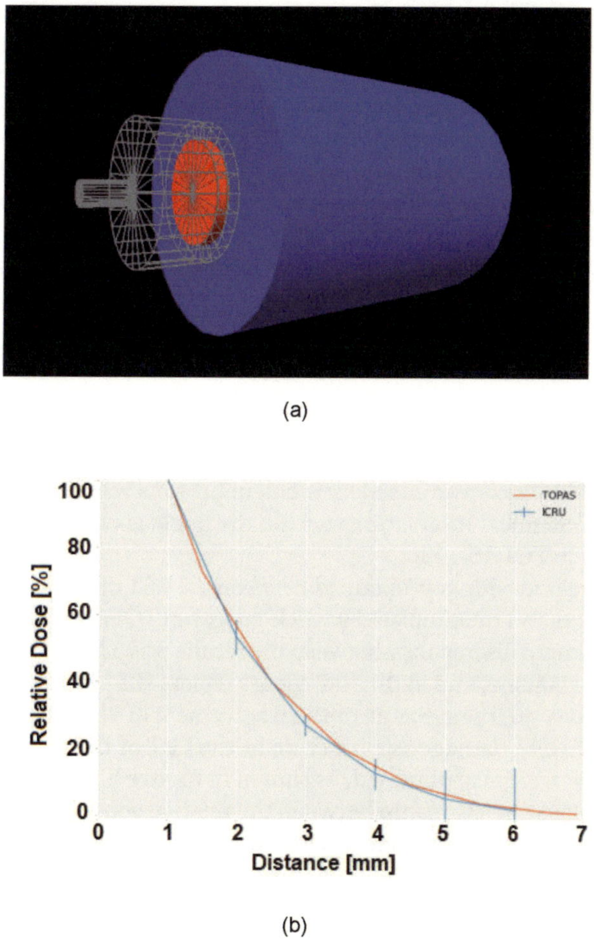

Figure 3.
(a) 90Sr/90Y source geometry with applicator and simulator object and (b) comparison of relative dose, as a function of distance, between TOPAS and ICRU data (adapted from [23]).

of radiation [45]. For this reason, TOPAS-nBio has several models of cellular and subcellular geometries (such as blood cells and single and double-stranded DNA models, for example) specialized in a pre-defined way [46]. Furthermore, with regard to biological modeling, the code inherits the chemical parameters provided by the Geant4-DNA toolkit and also includes mechanistic DNA repair models to perform water radiolysis simulations. With this, it is possible to develop complete modeling from the initial physical events to the final observed biological result [47].

According to Semenenko et al. [48] the combination of runway structure simulations with core geometry models is considered the gold standard for predicting the spectrum of DNA damage induced by ionizing radiation. Therefore, Hongyu et al. [43] determined the cellular response after proton irradiation using the TOPAS-nBio code for damage induction and repair modeling with MEDRAS, which is a model capable of predicting the main final biological damage in a variety of cell types, including repair kinetics, chromosomal aberrations, and cell survival.

To determine the DNA damage yield, the results were scored in the SSD format and quantified by strand break (SBs), single-strand break (SSBs), and double-strand break (DSBs) yields, compared to published and experimentally measured data.

The initial DNA damage after proton irradiation (0.5–500 MeV, corresponding to the LET region of 60–0.2 keV/μm) was simulated with the code. The core model used was placed in the center of a cubic world with a side length of 14 μm, containing the core, filled with water. Primary protons were randomly initiated on the surface of the nucleus and propagated within the nucleus in a random direction. Induced DNA damage caused by direct and indirect interactions in the physical and chemical stages was quantified as SBs, SSBs, or DSBs and sent in standard DNA damage data format (SDD). To get enough statistics, 100 stories were performed for each energy point. Each simulation had a fixed number of primary particles and deposited a dose of 1 Gy inside the nucleus. Statistical uncertainty associated with DSB dose and yield was less than 2%.

The average LET was recorded as a radiation quality index and calculated by the equation:

$$\mathrm{LET} = \varepsilon / d \tag{2}$$

where d is the average length of the proton path inside the nucleus and ε is the energy deposition of primary and secondary particles inside the nucleus.

The initial DNA damage induced by incident protons was simulated by modeling the physical and chemical interactions within the nucleus with standard process models available in TOPAS-nBio.

As a result, a relationship was obtained between the LET of the proton according to literature references and the simulated particle energy in TOPAS-nBio. In low energy regions, the maximum discrepancy between the results was 32.5%, probably due to the size of the scoring volume, and in this low energy region, the protons do not cross the entire nucleus. However, there was an optimal agreement of 96%, as shown in **Figure 4**.

The results of DNA damage as a function of the LET of the proton simulated with TOPAS-nBio were also obtained, as shown in **Figure 5**.

The figure shows a relationship between the relative contribution of direct and hybrid damage as a fraction of each type of SB, SSB, and DSB break. Thus, it was shown that most SBs and SSBs would be caused by indirect damage and the indirect contribution rate would increase from approximately 60% to approximately 75% at 4.5 keV/μm (10 MeV proton energy) and, then decrease to higher LET values where radiolysis is denser, causing a greater number of chemical interactions. Furthermore, it was shown that most DSB damage was classified as a hybrid type, caused by the combination of direct and indirect damage. Simulations using

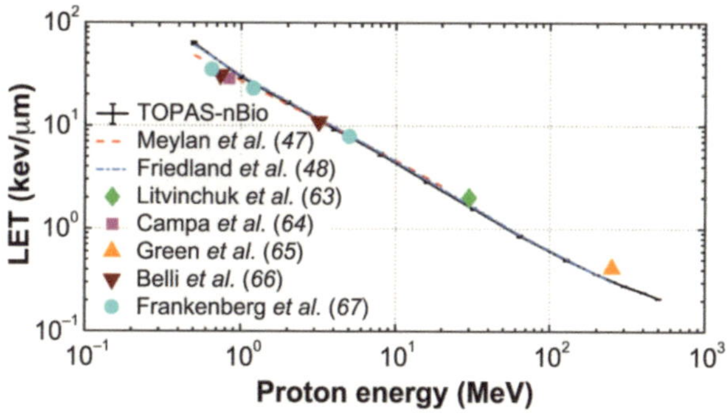

Figure 4.
Proton LET as a function of proton energy compared to experimental data [48].

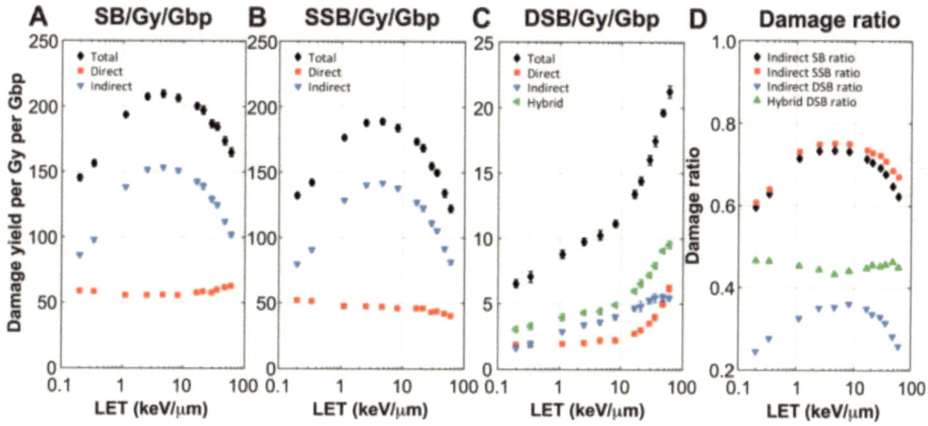

Figure 5.
DNA damage obtained with TOPAS-nBio. Eml A: Total, direct, and indirect SB yield per Gy per Gbp of DNA. In B: Total, direct and indirect SSB yield per Gy per Gbp of DNA. In C: Total, direct, indirect, and hybrid DSB yield per Gy per Gbp of DNA. In D: Contribution of indirect or hybrid damage to SB, SSB, and DSB [43].

TOPAS-nBio showed that Monte Carlo tools can predict DNA damage and can be used to interpret experimental data and design new theories.

3. Conclusion

Monte Carlo simulations have been applied to determine and study different parameters that are challenged in experimental measurements, due to its capability in simulating the radiation transport. In this chapter were presented applications for radiotherapy procedures, in scenarios with homogeneous and anatomical phantoms determining dose values, dose distribution, and dosimetric parameters through the PENELOPE and TOPAS code, showing itself as a useful tool for radiotherapy.

Author details

Iury Mergen Knoll[1], Ana Quevedo[2] and Mirko Salomón Alva Sánchez[3]*

1 Undergraduate in Medical Physics/Federal, University of Health Sciences of Porto Alegre, Porto Alegre, Brazil

2 Department Health Science, Oeste Paulista University, Jau, Brazil

3 Department of Exact Science and Applied Social, Federal University of Health Sciences of Porto Alegre, Porto Alegre, Brazil

*Address all correspondence to: mirko@ufcspa.edu.br

References

[1] Mariotti V, Gayol A, Pianoschi T, Mattea F, Vedelago J, Pérez P, et al. Radiotherapy dosimetry parameters intercomparison among eight gel dosimeters by Monte Carlo simulation. Radiation Physics and Chemistry. 2022;**190**:109782. DOI: 10.1016/j.radphyschem.2021.109782

[2] Alva-Sánchez MS, Quevedo A, Bonatto A, Pianoschi T. Preliminary Monte Carlo simulation of non-laser light sources for photodynamic therapy. Journal of Physics Conference Series. 2021;**1826**:012052. DOI: 10.1088/1742-6596/1826/1/012052

[3] Alva-Sánchez MS, Pianoschi T. Study of the distribution of doses in tumors with hypoxia through the PENELOPE code. Radiation Physics and Chemistry. 2020;**167**:108428. DOI: 10.1016/j.radphyschem.2019.108428

[4] Quevedo A, Borges LF, Nicolucci P. Evaluation of dosimetric parameters for brachytherapy source in regions close to the source. Scientia Plena. 2018;**14**(4):1-12. DOI: 10.14808/sci.plena.2018.046001

[5] Verbeek N, Wulff J, Baumer C, Smyczek S, Timmermann B, Brualla L. Single pencil beam benchmark of a module for Monte Carlo simulation of proton transport in the PENELOPE code. Medical Physics. 2021;**48**(1):456-476. DOI: 10.1002/mp.14598

[6] Bosman DF, Balcasa VG, Delgado C, Principi S, Duch MA, Ginjaume M. Validation of the MC-GPU Monte Carlo code against the PENELOPE/penEasy code system and benchmarking against experimental conditions for typical radiation qualities and setups in interventional radiology and cardiology. Physica Medica. 2021;**82**:64-71. DOI: 10.1016/j.ejmp.2021.01.075

[7] Forster RA, Cox LJ, Barrett RF, Booth TE, Briesmeister JF, Brown FB, et al. MCNP version 5. Nuclear Instruments and Methods B. 2004;**213**:82-86. DOI: 10.1016/S0168-583X(03)01538-6

[8] Vahabi SM, Zafarghandi MS. Applications of MCNP simulation in treatment planning: A comparative study. Radiation and Environmental Biophysics. 2020;**59**(2):307-319. DOI: 10.1007/s00411-020-00841-2

[9] Leal-Acevedo B, Gamboa-deBuen I. Dose distribution calculation with MCNP code in a research irradiator. Radiation Physics and Chemistry. 2020;**167**:108320. DOI: 10.1016/jradphyschem.2019.05.010

[10] Kolacio MS, Brkic H, Faj D, Radojcic DS, Rajlic D, Obajdin N, et al. Validation of two calculation options built in Elekta Monaco Monte Carlo based algorithm using MCNP code. Radiation Physics and Chemistry. 2021;**179**:109237. DOI: 10.1016/j.radphyschem.2020.109237

[11] Kim MJ, Sung SH, Hr K. Spectral resolution evaluation by MCNP simulation for airborne alpha detection system with a collimator. Nuclear Engineering and Technology. 2021;**53**(4):1311-1317. DOI: 10.1016/j.net.2020.09.009

[12] Yani S, Rizkia I, Kamirul RMF, Haekal M, Haryanto F. EGSnrc application for IMRT planning. Reports of Practical Oncology and Radiotherapy. 2020;**25**(2):217226. DOI: 10.1016/j.rpor.2020.01.004

[13] Jayamani J, Osman ND, Tajuddin AA, Noor NM, Aziz MZA. Dosimetric comparison between Monaco TPS and EGSnrc Monte Carlo simulation on titanium rod in 12bit and 16bit image format. Journal of Radiation Research and Applied Science. 2020;**13**(1):496-506. DOI: 10.1080/16878507.2020

[14] Tessier F, Ross CK. Technical Note: Implications of using EGSnrc instead of EGS4 for extracting electron stopping powers from measured energy spectra. Medical Physics. 2021;**48**(4):1996-2003. DOI: 10.1002/mp.14567

[15] Aamri H, Fielding A, Aamry A, Sulieman A, Tamam N, Alkhorayef M, et al. Comparison between PRIMO and EGSnrc Monte Carlo models of the Varian True Beam linear accelerator. Radiation Physics and Chemistry. 2021;**178**:109013. DOI: 10.1016/j.radphyschem.2020.109013

[16] Embriaco A, Attili A, Bellinzona EV, Dong Y, Grzanka L, Mattei I, et al. FLUKA simulation of target fragmentation in proton therapy. Physica Medica. 2020;**80**:342346. DOI: 10.1016/j.ejmp.2020.09.018

[17] Chattaraj A, Selvam TP. Applicability of pure Propane gas for microdosimetry at brachytherapy energies: A Fluka study. Radiation Protection Dosimetry. 2020;**189**(3):286-293. DOI: 10.1093/rpd/ncaa041

[18] Soltani-Nabipour J, Khorshidi A, Shojai F, Khorami K. Evaluation of dose distribution from C-12 ion in radiation therapy by FLUKA code. Nuclear Engineering and Technology. 2020;**52**(10):2410-2424. DOI: 10.1016/j.net.2020.03.010

[19] Sharma A, Singh B, Sandhu BS. A compton scattering technique for wood characteristics using FLUKA Monte Carlo code. Radiation Physics and Chemistry. 2021;**185**:109364. DOI: 10.1016/j.radphyschem.2021.109364

[20] Chattaraj A, Selvam TP. Microdosimetry-based relative biological effectiveness calculations for radiotherapeutic electron beams: A FLUKA-based study. Radiological Physics and Technology. 2021;**14**(3):297-308. DOI: 10.1007/s12194-021-00627-1

[21] Souza LS, Alva-Sánchezb MS, Bonatto A. Computational simulation of low energy x-ray source for photodynamic therapy: A preliminary study. Brazilian Journal of Radiation Science. 2021;**9**(1):1-15. DOI: 10.15392/bjrs.v9i1.1639

[22] Berumen F, Ma YZ, Ramos-Mendez J, Perl J, Beaulieu L. Validation of the TOPAS Monte Carlo toolkit for HDR brachytherapy simulations. Brachytherapy. 2021;**20**(4):911-921. DOI: 10.1016/j.brachy.2020.12.007

[23] Knoll I, de Souza L, Ramon P, Quevedo A, Alva-Sanchez MS. Determination of dose deposition from an Ocular Brachytherapy source: Simulation data with TOPAS. Radiotherapy and Oncology. 2021;**158**(1):S182-S183

[24] Hahn MB, Villate JMZ. Combined cell and nanoparticle models for TOPAS to study radiation dose enhancement in cell organelles. Scientific Reports. 2021;**11**(1):6721. DOI: 10.1038/s41598-021-85964-2

[25] Wu JA, Xie YQ, Ding Z, Li FP, Wang LH. Monte Carlo study of TG-43 dosimetry parameters of GammaMed Plus high dose rate IR-192 brachytherapy source using TOPAS. Journal of Applied Clinical Medical Physics. 2021;**22**(6):146-153. DOI: 10.1002/acm2.13252

[26] Ozbay T, Yourt A, Ozsoykal I. Simulation of water equivalency of polymer gel dosimeters with GAMOS. Journal of Basic Clinical Health Sciences. 2020;**4**(1):51-58. DOI: 10.30621/jbachs.2020.899

[27] Pistone D, Auditore L, Italiano A, Mandaglio G, Minutoli F, Baldari S, et al. Monte Carlo based dose-rate assessment in F-18-Choline Pet Examination: A comparison between

Gate and Gamos Codes. Atti Accademia Peloritana Dei Pericolanti-Classe Di Scienze Fisiche Matematiche e Naturali. 2020;**98**(1):A5. DOI: 10.1478/AAPP.981A5

[28] Dubois PA, Thao NTP, Trung NT, Azcona JD, Aguilar-Redondo PB. A tool for precise calculation of organ doses in voxelised geometris using GAMOS/Geant4 with a graphical user interface. Polish Journal of Medical Physics and Engineering. 2021;**27**(1):31-40. DOI: 10.2478/pjmpe-2021-0005

[29] Al-Tuweity J, Sadiq Y, Mouktafi A, Arce P, Fathi I, Mohammed M, et al. GAMOS/GEANT4 simulation and comparison study of X-ray narrowspectrum series at the national Secondary Standard Dosimetry Laboratory of Morocco. Applied Radiation and Isotopes. 2021;**175**:109789. DOI: 10.1016/j.apradiso.2021.109789

[30] Chrobak A, Konefal A, Wronska A, Magiera A, Rusiecka K, et al. Comparison of various models of Monte Carlo Geant 4 code in simulations of prompt gamma production. Acta Physica Polonica, B. 2017;**48**(3):675-678. DOI: 10.5506/APhysPolB.48.675

[31] Baumann KS, Kaupla S, Bach C, Engenhar-Cabillic R, Zink K. Monte Carlo calculation of perturbation correction factors for air-filled ionization chambers in clinical proton beams using TOPAS/GEANT. Zeitschrift für Medizinische Physik. 2021;**31**(2):175-191. DOI: 10.1016/j.zemedi.2020.08.004

[32] Salvat F, Fernández-Varea J, Sempau J, Llovet X. Monte Carlo simulation of bremsstrahlung emission by electrons. Radiation Physics and Chemistry. 2006;**75**:1201-1219. DOI: 10.1016/j.radphyschem.2005.05.008

[33] Sempau J, Fernández-Varea JM, Acosta E, Salvat F. Experimental benchmarks of the Monte Carlo Code PENELOPE. Nuclear Instruments and Methods in Physics B. 2003;**207**:107-123. DOI: 10.1016/S0168-583X(03)00453-1

[34] Rodriguez EAV, Alcon EPQ, Rodriguez ML, Gutt F, de Almeida E. Dosimetric parameters estimation using PENELOPE Monte-Carlo simulation code: Model 6711 I-125 brachytherapy seed. Applied Radiation and Isotopes. 2005;**63**(1):41-48. DOI: 10.1016/j.apradiso.2005.02.004

[35] Casado FJ, Garcia-Pareja S, Cenizo E, Mateo B, Bodineau C, Galan P. Dosimetric characterization of an Ir-192 brachytherapy source with the Monte Carlo code PENELOPE. Physica Medica. 2010;**26**(3):132-139. DOI: 10.10.16/j.ejmp.2009.11.001

[36] Almansa JF, Guerrero R, Torres J, Lallena AM. Monte Carlo dosimetric characterization of the Flexisource Co-60 high-dose-rate brachytherapy source using PENELOPE. Brachytherapy. 2017;**16**(5):1073-1080. DOI: 10.1016/j.brachy.2017.04.245

[37] Nath R, Anderson LL, Luxton G, Weaver KA, Williamson JF, Meigooni AS. Dosimetry of interstitial brachytherapy sources: Recommendations of the AAPM radiation therapy committee Task Group No. 43. Medical Physics. 1995;**22**(2):209234. DOI: 10.1118/1.597458

[38] Rivard MJ, Ballester F, Butler WM, DeWerd LA, Ibbott GS, Meigooni AS, et al. Supplement 2 for the 2004 Update of the AAPM Task Group No. 43: Report: Joint Recommendations by the AAPM and GEC-ESTRO. Medical Physics. 2017;**44**(9):e297-e338. DOI: 10.1002/mp.12430

[39] Perl J, Shin J, Schümann J, Faddegon B, Paganetti H. TOPAS: An innovative proton Monte Carlo Platform for research and clinical applications.

Medical Physics. 2012;**39**(11):6818-6837. DOI: 10.1118/1.4758060

[40] Hall D, Perl J, Schuemann J, Faddegon B, Paganetti H. Meeting the challenges of quality control in the TOPAS Monte Carlo Simulation Toolkit for proton therapy. Medical Physics. 2016;**43**(6):3493-3494. DOI: 10.1118/1.4956275

[41] Liu HD, Zhang L, Chen Z, Liu XG, Dai ZY, Li Q, et al. A preliminary Monte Carlo study of the treatment head of a carbon-ion radiotherapy facility using TOPAS. EPJ Web of Conferences. 2017;**153**:04018. DOI: 10.1051/epjconf/201715304018

[42] Baumann KS, Kaupa S, Bach C, Engenhart-Cabillic R, Zink K. Monte Carlo calculation of beam quality correction factors in proton beams using TOPAS/GEANT4. Physics in Medicine and Biology. 2020;**65**(5):055015. DOI: 10.1088/1361-6560/ab6e53

[43] Zhu H, McNamara AL, McMahon SJ, Ramos-Mendez J, Henthorn NT, Faddegon B, et al. Cellular response to proton irradiation: A simulation study with TOPAS-nBio. Radiation Research. 2020;**194**(1):9-21. DOI: 10.1667/RR15531.1

[44] Shin WG, Testa M, Kim HS, Jeong JH, Lee SB, Kim YJ, et al. Independent dose verification system with Monte Carlo simulations using TOPAS for passive scattering proton therapy at the National Cancer Center in Korea. Physics in Medicine and Biology. 2017;**62**(19):7598-7616. DOI: 10.1088/1361-6560/aa8663

[45] Hall EJ, Giaccia AJ. Radiobiology for the Radiologist. 8th ed. Philadelphia, PA: Lippincott Williams & Wilkins; 2018

[46] Schuemann J, McNamara AL, Ramos-Méndez J, Perl J, Held KD, Paganetti H, et al. TOPAS-nBio: An extension to the TOPAS simulation toolkit for cellular and sub-cellular radiobiology. Radiation Research. 2019;**191**(2):125-138. DOI: 10.1667/RR15226.1

[47] McNamara A, Geng C, Turner R, Mendez JR, Perl J, Held K, et al. Validation of the radiobiology toolkit TOPAS-nBio in simple DNA geometries. Physica Medica. 2017;**33**:207-215. DOI: 10.1016/j.ejmp.2016.12.010

[48] Semenenko V, Stewart R. A fast Monte Carlo algorithm to simulate the spectrum of DNA damages formed by ionizing radiation. Radiation Research. 2004

Chapter 8

Physical Only Modes Identification Using the Stochastic Modal Appropriation Algorithm

Maher Abdelghani

Abstract

Many operational modal analysis (OMA) algorithms such as SSI, FDD, IV, ... are conceptually based on the separation of the signal subspace and the noise subspace of a certain data matrix. Although this is a trivial problem in theory, in the practice of OMA, this is a troublesome problem. Errors, such as truncation errors, measurement noise, modeling errors, estimation errors make the separation difficult if not impossible. This leads to the appearance of nonphysical modes, and their separation from physical modes is difficult. An engineering solution to this problem is based on the so-called stability diagram which shows alignments for physical modes. This still does not solve the problem since it is rare to find modes stable in the same order. Moreover, nonphysical modes may also stabilize. Recently, the stochastic modal appropriation (SMA) algorithm was introduced as a valid competitor for existing OMA algorithms. This algorithm is based on isolating the modes mode by mode with the advantage that the modal parameters are identified simultaneously in a single step for a given mode. This is conceptually similar to ground vibration testing (GVT). SMA is based on the data correlation sequence which enjoys a special physical structure making the identification of nonphysical modes impossible under the isolating conditions. After elaborating the theory behind SMA, we illustrate these advantages on a simulated system as well as on an experimental case.

Keywords: in-operation modal analysis, modal appropriation, spurious modes, SMA

1. Introduction

Operational modal analysis (OMA) is a good complement to classical modal analysis where the structure is installed in a laboratory and excited under well-controlled conditions. For structures under their operating conditions, the excitation cannot be measured, random, complex in nature, and can be nonstationary. Examples are offshore structures under swell, aircraft under turbulence, etc.

Several algorithms exist to extract the modal parameters from the output, only measurements. Most of these algorithms are stochastic realization algorithms, such as SSI, BR, CVA, FDD. These algorithms are based on the separation of two orthogonal subspaces, namely the signal subspace and the noise subspace. Although in theory, this is a trivial problem, in the practice of using them, strictly speaking, it is impossible to separate them. Errors, such as finite sample length errors, estimation errors, modeling errors, noise, ... make the separation impossible leading to the problem of model order estimation. In order to solve this problem, the stability diagrams are used where the modal model is estimated at increased orders leading to alignments for physical modes. However, numerical modes, noise modes, spurious modes, harmonics, etc. appear and the challenge is how to reject these modes especially that the modal model has to be identified in unique model order.

The stochastic modal appropriation algorithm (SMA) is based on rotating and stretching the outputs correlation sequence which was derived based on physical background. We show that based on this idea, SMA rejects automatically nonphysical modes as well as harmonics. Harmonics are assumed to be modes with zero damping and we show that such a mode can never be appropriated because the phase angle between the input and output is always different from zero. On the other hand, the physical structure of the correlation sequence is respected if and only if the mode is physical. We illustrate this on a simulated example as well as experimentally.

2. The stochastic modal appropriation algorithm (SMA)

We describe here quickly the basics of the SMA algorithm. The considered system is a quarter car model excited with the unmeasured white noise of a certain variance. The impulse response of the system may be written as [1]:

$$h(t) = C_h e^{-\xi\omega_n t} \sin(\omega_d t) \tag{1}$$

where ξ is the system damping ratio, ω_n is the system natural frequency, and ω_d is the damped natural frequency.

Computing the correlation sequence of the system based on the above impulse response leads to the following expression [1]:

$$R(t) = C_r e^{-\xi\omega_n t} \sin(\omega_d t - \phi(\xi)) \tag{2}$$

where $\phi(\xi)$ is a known parametric function that depends on the system damping ratio. The impulse response, as well as the correlation sequence, may be considered as two rotating vectors in the complex plane but with decaying amplitudes (spirals).

In the INOPMA algorithm [2], it is assumed that the outputs correlation sequence is the system impulse response. As a consequence, it has been shown that the mode is appropriated at a frequency $\omega^* = \omega_n\sqrt{1-4\xi^2}$ and not the natural frequency ω_n. This is considered a limitation of INOPMA, especially that the natural frequency has to be estimated in two steps. In this work, we propose a different approach that allows us to overcome this limitation and we show that it is still possible to appropriate the mode at its natural frequency using a dynamic transformation on the correlation sequence.

Let $\overline{R}(t,\alpha)$ be the image of $R(t)$ by a linear anti-symmetric function that depends on a certain design parameter α and consider the following sequence:

$$H(t,\alpha) = R(t) + \overline{R}(t,\alpha) \tag{3}$$

$H(t,\alpha)$ may be interpreted as a combination of two transformations on the correlation sequence namely a rotation and stretching. By varying only α, one rotates and stretches the correlation sequence and hence it is possible to modify the phase shift as well as the amplitude of the correlation sequence and consequently modify the damping ratio leading to a pure sinusoid. At this stage, the mode is appropriated and the modified correlation sequence is the system impulse response up to an unknown factor.

In this work, we propose to use the following anti-symmetric transformation:

$$F(R(t), \alpha) = j\alpha R(t) \tag{4}$$

The key idea is similar to the INOPMA algorithm in the sense that one takes the convolution of a driving harmonic force with the modified correlation sequence; notice, however, that with SMA one varies two parameters namely the driving frequency and the parameter alpha.

In the frequency domain, this means that the system transfer function (the Laplace transform of the correlation sequence) is multiplied by a complex factor $(1 + j\alpha)$. It can easily be shown that the transfer function phase angle is zero exactly at the following condition:

$$\begin{cases} \omega = \omega_n \\ \alpha = 2\xi \end{cases} \tag{5}$$

Geometrically, one interpretation is that when the mode is appropriated the correlation sequence vector describes a circle in the complex plane meaning that the conservative part of the system is isolated. The nonconservative part follows immediately. Consequently, the system modal parameters are identified simultaneously at the same step. This is one advantage of the SMA algorithm.

3. Harmonics rejection

Harmonics are assumed here to be modes with zero damping. We show that the algorithm SMA automatically rejects these modes. This avoids hand-based removal of these harmonics as done in practice.

Let us consider the correlation sequence of an SDOF system excited with unmeasured white noise. We propose to show in the sequel that if the damping ratio is zero then the mode cannot be appropriated (the phase angle is never zero) hence rejected.

The SMA algorithm starts by considering the following modified parametric correlation sequence:

$$H(t, \alpha) = (1 + j\alpha)R(t)$$

The Laplace transform of this function can be shown to write as:

$$G(s) = (1 + j\alpha)\frac{s + 2\xi\omega_n}{s^2 + 2\xi\omega_n s + \omega_n^2}$$

The imaginary part of the frequency response is:

$$I = (2\xi\alpha\omega_n + \omega)\left(\omega_n^2 - \omega^2\right) - (2\xi\omega_n - \alpha\omega)(2\xi\omega\omega_n)$$

While the real part is:

$$\text{Re} = \left(\omega_n^2 - \omega^2\right)(2\xi\omega_n - \alpha\omega) + 2\omega^2\xi\omega_n$$

When the damping ratio is zero the tangent of the phase angle of the frequency response reduces to:

$$tg = -\frac{\omega\left(\omega_n^2 - \omega^2\right)}{\alpha\omega\left(\omega_n^2 - \omega^2\right)} = -1/\alpha$$

Which is always different from zero. Consequently, for a harmonic, the angle between the input and the output is never zero meaning that the harmonic is never identified (no zero-crossing).

4. Spurious modes rejection

Spurious/numerical modes appear in an OMA procedure due to many reasons, such as finite sample length effects, truncation orders, measurement noise, ... These modes appear because they are fitted to the system characteristic equation and rejecting them is a challenge. This leads to a spurious frequency and damping that we still denote in the sequel as wn and zeta. The correlation sequence of the system output is given by [3, 4]:

$$R(t) = e^{-\xi\omega_n t}\left[\cos\left(\omega_d t\right) + \frac{\xi}{\sqrt{1-\xi^2}}\sin\left(\omega_d t\right)\right] \tag{6}$$

The phase shift in this correlation sequence is given by:

$$tg(\theta) = \frac{\xi}{\sqrt{1-\xi^2}} \tag{7}$$

This particular expression of the phase shift is valid for physical modes only [3]. We propose to show in the sequel that if the phase shift of a correlation sequence enjoys this particular expression, then the mode is necessarily physical.

Consider the following correlation sequence:

$$R_x(t) = e^{-\xi\omega_n t}\left[\cos\left(\omega_d t\right) + x\sin\left(\omega_d t\right)\right]$$

The Laplace transform of $(1 + j\alpha)R_x(t)$ is:

$$G_x(s) = (1 + j\alpha)\frac{s + \omega_n\left(\xi + x\sqrt{1-\xi^2}\right)}{s^2 + 2s\xi\omega_n + {\omega_n}^2}$$

The numerator writes as:

$$(1 + j\alpha)\left(j\omega + \omega_n\left(\xi + x\sqrt{1-\xi^2}\right)\right)\left(-\omega^2 + 2j\omega\xi\omega_n + {\omega_n}^2\right)$$

And the imaginary part writes as:

$$\mathrm{Im} = -\omega^3 + \omega\omega_n^2 - \alpha\omega^2\omega_n\left(\xi + x\sqrt{1-\xi^2}\right) + \alpha\omega_n^3\left(\xi + x\sqrt{1-\xi^2}\right)$$
$$+2\xi\omega_n^2\omega\left(\xi + x\sqrt{1-\xi^2}\right) - 2\xi\omega^2\omega_n\alpha$$

Under the appropriation conditions

$$\begin{cases} \omega = \omega_n \\ \alpha = 2\xi \end{cases}$$

$$\omega_n^3\left(2\xi\left(\xi + x\sqrt{1-\xi^2}\right) - 4\xi^2\right) = 0$$

Leading to:

$$x = \frac{\xi}{\sqrt{1-\xi^2}}$$

This proves that under the SMA isolating conditions, the mode is necessarily physical.

5. Simulation validation

We propose in this section to study the performance of the SMA algorithm on a simple simulated example. A SDOF system is taken as an example. The considered system parameters are taken as m = 2 kg, k = 10,000 N/m, and c = 8 Ns/m. The excitation is a white noise with unit variance. This leads to the following modal parameters; ω_n = 11.254 Hz and ξ = 2.83%. The output is then simulated using a sampling frequency of Fs = 64 Hz and 2% measurement noise is added to the output. **Figure 1** shows the identification results of this data set.

5.1 Harmonics rejection

We propose to study in the section the ability of SMA to reject harmonics. Let us consider an SDOF system excited with unmeasured white noise. We add a harmonic component with frequency 5 Hz and amplitude 0.1 N. **Figure 2** shows the phase angle corresponding to the identification results and we notice that the harmonic component is rejected and only the system frequency is identified.

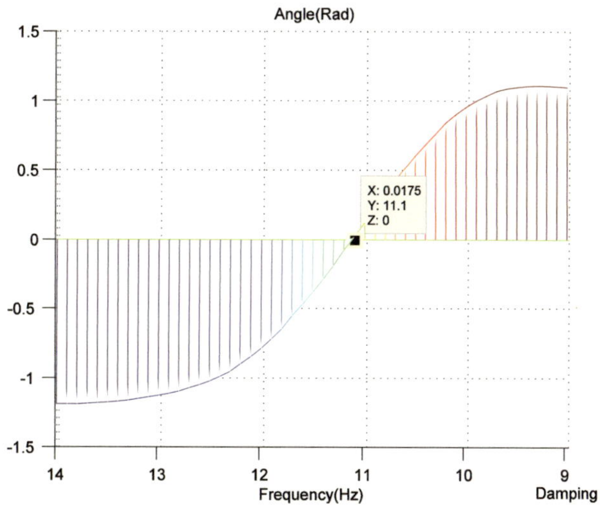

Figure 1.
Phase angle as a function of frequency and alpha.

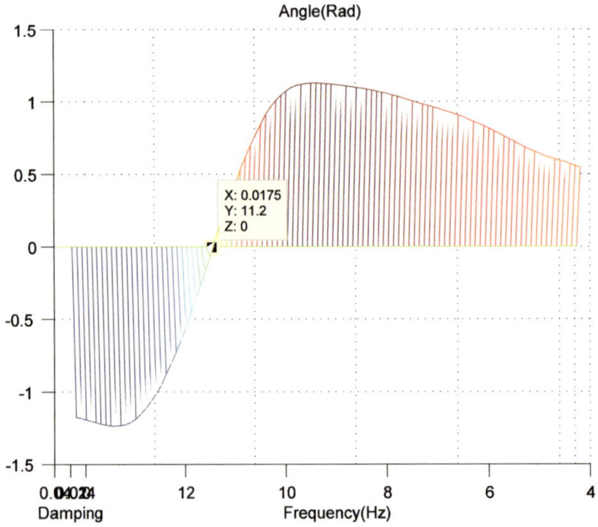

Figure 2.
Phase angle for harmonics rejection.

5.2 Spurious modes rejection

Spurious modes may arise from different sources such as noise, measurements, errors, ... In order to simulate spurious modes, we consider adding noise to the system as well as introducing colored noise. We drive a unit of white noise through an AR [5] process whose output serves as the excitation to the system. **Figure 3** shows that SMA is robust against spurious modes and only the physical mode is identified.

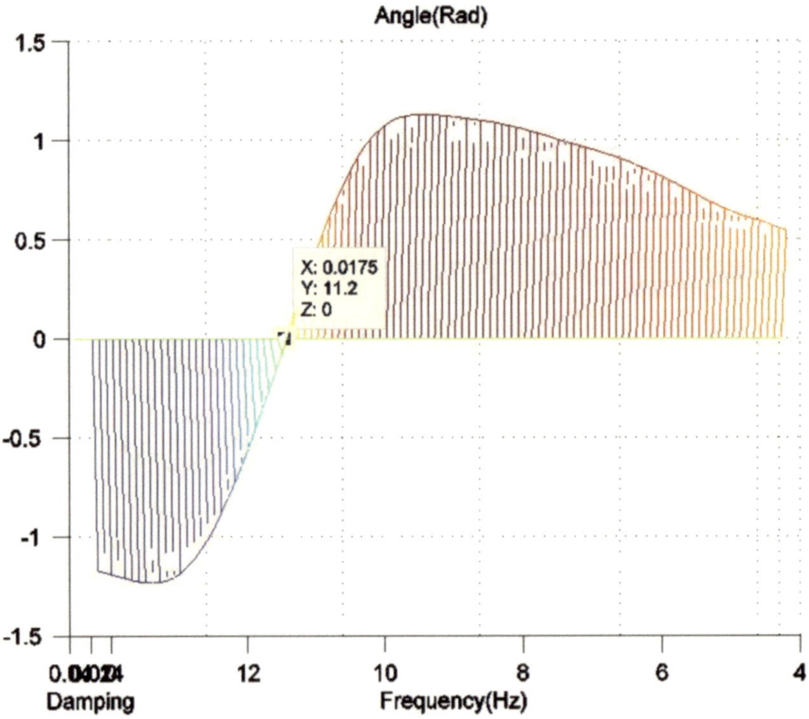

Figure 3.
Spurious modes rejection.

6. Experimental validation

The considered test object is a standard B&K demo plate (WA0846), which is a rectangular aluminum plate with dimensions 290 × 250 × 8 mm^3; for the test, the plate was placed on soft foam. A B&K demo motor WB 1471 with an unbalanced rotor was attached to the plate; the motor was set to operate at 374 rps (**Figure 4**). For the experiment, the plate was excited by tapping its surface by the tip of a plastic pen. About 16 monoaxial accelerometers B&K Type 4507 were mounted equidistantly on the plate on the grid points of a 4 × 4 grid, oriented to measure in the direction perpendicular to the plate surface. The data acquisition was performed by B&K LAN-Xi DAQ, the sampling frequency was set to 4096 Hz, and 60 seconds of the acceleration data were recorded, which was used as an input to both SMA and SSI algorithms.

Figure 4.
The test setup.

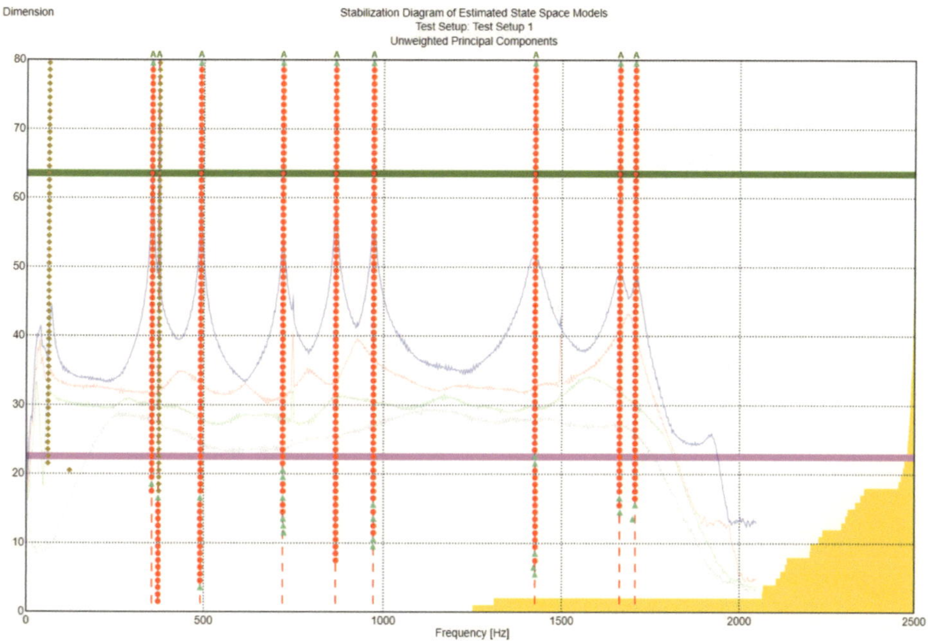

Figure 5.
The stability diagram.

Frequency [Hz]	Damping [%]	Comment
353.4	0.57	1st torsional mode (along Y-axis)
371.9	0.05	Harmonic, automatically identified as a noise mode
491.2	0.62	1st bending mode (along Y-axis)
720.3	0.98	1st bending mode (along X-axis)
866.7	0.47	2nd torsional mode (along Y-axis)
971.7	0.60	2nd torsional mode (along X-axis)
1424	1.1	2nd bending mode (along Y-axis)
1663	0.69	2nd membrane mode
1706	0.67	3rd torsional mode (along Y-axis)

Table 1.
Identification results.

Frequency [Hz]	Damping [%]	Comment
354	0.57	1st torsional mode (along Y-axis)
372.5	0.05	Harmonic, automatically identified as a noise mode
492	0.62	1st bending mode (along Y-axis)
721.4	0.98	1st bending mode (along X-axis)
871	0.47	2nd torsional mode (along Y-axis)
972	0.60	2nd torsional mode (along X-axis)
1422	1.1	2nd bending mode (along Y-axis)
1661	0.69	2nd membrane mode
1705	0.67	

Table 2.
Identification results.

To validate the results of the SMA algorithm, we used the commercial OMA software package "PULSE Operational Modal Analysis 5.1.0.4—x64"; the software was used in automatic identification mode, that is, all default settings were applied; OMA-SSI-UPC method was employed. The stabilization diagram is shown in **Figure 5**, and the modal identification results are presented in **Table 1**.

The SMA algorithm was used with 256 correlation lags. Sensors 1 and 5 were used for the identification. The modes were identified as the angle crossings with zero. The results are reported in **Table 2**.

Notice that the harmonics as well as spurious/numerical modes were not identified and were automatically rejected.

7. Conclusion

Nonphysical modes, as well as harmonics, present a challenge in OMA. Although stability diagrams help in solving this problem, rejecting these modes is not trivial. Although stability diagrams help to solve this problem, the results will remain user-dependent.

The SMA algorithm seems to present an advantage. Not only the correlation sequence has a physical meaning, but also the simultaneity in the identification of the modal parameters makes a constraint on the modes to be exclusively physical.

This was illustrated on a simulated example as well as experimentally.

Author details

Maher Abdelghani[1,2]

1 University of Sousse, Sousse, Tunisia

2 LASMAP, EPT, La Marsa, Tunisia

*Address all correspondence to: maher.abdelghani@gmail.com

References

[1] Balmès E, Chapelier C, Lubrina P, Fargette P. An evaluation of modal testing results based on the force appropriation method. In: International Modal Analysis Conference; Orlando; 1996

[2] Abdelghani M, Inman DJ. Modal appropriation for use with in-operation modal analysis. Journal of Shock and Vibration. 2015;**2015**:537030. DOI: 10.1155/2015/537030

[3] Meirovitch L. Elements of Vibration Analysis. McGraw-Hill; 1986

[4] Gerradin M, Rixen D. Theory of vibrations. Masson; 1993

[5] Abdelghani M, Friswell MI. Stochastic modal appropriation (SMA). In: IMAC'2018, USA; 2018